The Heirs of Archimedes

DISCOVERY TECHNOLOGY HISTORY SCIENCE

DIBNER
INSTITUTE
FOR THE HISTORY
OF SCIENCE AND
TECHNOLOGY

Dibner Institute Studies in the History of Science and Technology
Jed Z. Buchwald, general editor

Isaac Newton's Natural Philosophy
Jed Z. Buchwald and I. Bernard Cohen, editors

Histories of the Electron: The Birth of Microphysics
Jed Z. Buchwald and Andrew Warwick, editors

Science Serialized: Representations of the Sciences in Nineteenth-Century Periodicals
Geoffrey Cantor and Sally Shuttleworth, editors

Natural Particulars: Nature and the Disciplines in Renaissance Europe
Anthony Grafton and Nancy Siraisi, editors

The Enterprise of Science in Islam: New Perspectives
J. P. Hogendijk and A. I. Sabra, editors

Instruments and Experimentation in the History of Chemistry
Frederic L. Holmes and Trevor H. Levere, editors

Systems, Experts, and Computers: The Systems Approach in Management and Engineering, World War II and After
Agatha C. Hughes and Thomas P. Hughes, editors

The Heirs of Archimedes: Science and the Art of War through the Age of the Enlightenment
Brett D. Steele and Tamera Dorland, editors

Ancient Astronomy and Celestial Divination
N. L. Swerdlow, editor

The Heirs of Archimedes

Science and the Art of War through the Age of Enlightenment

Brett D. Steele and Tamera Dorland, editors

The MIT Press
Cambridge, Massachusetts
London, England

MIT Press books may be purchased at special quantity discounts for business or sales promotional use. For information, please email special_sales@mitpress.mit.edu or write to Special Sales Department, The MIT Press, 5 Cambridge Center, Cambridge, MA 02142.

Set in Bembo by The MIT Press. Printed and bound in the United States of America.

Library of Congress Cataloging-in-Publication Data

The heirs of Archimedes : science and the art of war through the Age of Enlightenment / Brett D. Steele and Tamera Dorland, editors.
p. cm. — (Dibner Institute studies in the history of science and technology)
Includes bibliographical references and index.
ISBN 0-262-19516-X (alk. paper)
1. Military art and science—Europe—History. 2. Science—History. 3. Technology—History. I. Steele, Brett D. II. Dorland, Tamera. III. Series.
U43.E95H45 2005
623'.094—dc22 2004055164

10 9 8 7 6 5 4 3 2 1

Contents

Acknowledgements

We wish to thank Geoffrey Symcox, Peter Reill, UCLA's Center for Seventeenth- and Eighteenth-Century Studies, and the Clark Library for organizing the conference Science and Warfare in the Old Regime (1998), and Bert Hall and the Dibner Institute for the History of Science and Technology for organizing the conference Colonels and Quartermasters (1999). The contributors to this volume first presented their work at these conferences. We also wish to express appreciation to Jeremy Black, Mary Henninger-Voss, Thomas Arnold, and John Guilmartin for contributing to the success of the conferences.

This volume depended on the Dibner Institute's postdoctoral fellowship and on a research grant from the National Science Foundation's Program of Societal Dimensions of Engineering, Science, and Technology. Appreciation also goes to RAND project managers Jefferson Marquis and Thomas Szayna for their stimulating questions, and to the GAO and ANSER's Homeland Security Institute for their flexibility during the last stage of this project. Finally, this volume owes much to Edwin T. Layton Jr. of the University of Minnesota for his mentorship, friendship, and encouragement to take risks.

This volume is dedicated to the memory of Gail Elizabeth Steele (1934–1970), an inspiring teacher and a loving mother.

The Heirs of Archimedes

Introduction

Brett D. Steele and Tamera Dorland

> "What right have you," he asked the algebraists and the arithmeticians, "to occupy the post in the vanguard of science? . . . All Europe is cutting its throat; what are you doing to stop this butchery? Nothing. What am I saying? It is you who perform the means of destruction; you who direct their use in all the armies."—Comte de Saint-Simon, 1813[1]

On August 18, 1776, Voltaire composed a personal letter to Jacques Antoine Baratier de Saint Auban, a veteran artillery general who was famous for his polemics against Gribeauval's Austrian-inspired artillery system. Perturbed by the sober reality of scientific knowledge, Voltaire confessed:

> You allow me only to moan at my age of 82 years at seeing that the techniques of destroying men and towns results from all the mathematical sciences, and that it is necessary to be a great physicist, as you are Monsieur, in order to be a fine murderer, but it was so at the time of Archimedes—like him, you defend your country and the happy peace from which we enjoy ourselves. . . .[2]

By characterizing military combat as "fine" murder, Voltaire was hardly trying to offend Saint Auban with his dark humor. As much as the great spokesman of Newtonian science and the quintessential Enlightenment philosophe despaired over the seemingly pointless addiction of Christian monarchs to warfare, he could simultaneously salivate over the prospect of Catherine the Great exterminating the Ottoman Turks. Voltaire's ostensibly incongruous coupling of theoretical physics and strategic warfare appears disturbingly modern, but he was actually making a historical statement. Writing a month after the Declaration of Independence, the French philosophe maintained that this connection pertained not only to the era of the American Revolution but also to that of the Second Punic War, as embodied by Archimedes (287–212 B.C.), the most renowned mathematician of the Hellenistic world.

Archimedes' transformation of mechanics into a rigorous geometrical science emerged as the paragon of scientific rigor for early modern natural philosophers such as Galileo and Descartes. His intellectual triumphs include the formulation of the center of gravity and its use in deriving the lever law; the determination of the quadrature of a parabola; the discovery of specific gravity, or density; the articulation of fundamental principles of hydrostatics; and, finally, the invention of the hydraulic screw pump.[3] Yet, from a late medieval or early Renaissance perspective, Archimedes' fame derived from his status as the most celebrated military engineer of antiquity. This occurred in spite of Plutarch's snobbish effort to downplay such a practical orientation and instead to portray Archimedes as an indifferent Platonic idealist.[4] The mathematician's military reputation was sealed by his design and coordination of Syracuse's defensive siegecraft during the Roman siege of the Second Punic War, when that city-state sided with the invading Carthaginian forces under Hannibal. The defensive campaign nonetheless concluded catastrophically because the Romans unexpectedly attacked Syracuse during a religious festival. In the subsequent sack of the city, a Roman soldier murdered Archimedes. Nevertheless, Archimedes' defensive system shattered Rome's initial frontal assaults from both land and sea, and transformed their siege into a frustrating blockade.

According to the great classical historian Polybius (c. 200–118 B.C.), a near contemporary of Archimedes, the Romans under Marcellus' command initially launched a massive naval assault that deployed ships armed with heavy catapults, as well as bows and slings.[5] Once the Romans had succeeded in clearing away the active defenses, Marcellus planned to storm the city walls after transporting the heavy assault troops by way of floating siege towers, or sambucae. Yet, as Marcellus' naval siegecraft approached the walls, Archimedes released a withering bombardment of projectiles, followed by lighter stones and heavy cranehooks. The intensity of the defensive fire was so devastating that a shaken Marcellus soon ordered all survivors to regroup. To evade Archimedes' deadly machinery, he then decided to attack more subtly under the cover of darkness. As Marcellus' forces crept up to the walls, however, Archimedes subjected them to the intense "small-arms" fire of archers and dart launchers, who, with utter impunity, discharged their instruments through concealed loopholes. A similar fate awaited the land-based attack under Pulcher's command. Both his siege engines and his assault troops were decimated by "combined-arms" defenses that Archimedes had prepared.

The tactical military success of Archimedes at the Siege of Syracuse was not based on innovative new machinery. Virtually all the cranes, catapults, and fortification elements he deployed were well-established military instruments.[6]

His prowess relied instead on mastering three crucial dimensions of siege warfare: correctly predicting all the ways the Romans would attack; optimizing the defensive siegecraft designs to counter each specific threat; and integrating all the machinery and troops into a synergistic "defense in depth," which subjected their Roman opponents to three distinct layers of attack. Not surprisingly, when Plutarch wrote his famous account of Archimedes, he focused more on his military engineering capability than on his mathematical theorems and theoretical mechanics. Likewise, when the Archimedean mathematical revival commenced in the Italian city-states during the Habsburg-Valois Wars, it was spearheaded by mathematicians oriented towards military engineering, such as Maurolico, who sought to gain conceptual insights into the mind of antiquity's greatest military engineer.[7] Archimedes emerged, of course, as a Renaissance embodiment of scientific rigor that would initially inspire Commandino, Guidobaldo, and Stevin in the sixteenth century with respect to statics, and ultimately Galileo, Huygens, and Newton with respect to kinematics and kinetics. Yet Archimedes also embodied for them the empowerment that stems from infusing military practice with rigorous mechanistic reasoning.

This esteem raises a question: Was Archimedes' integration of scientific knowledge and military power only institutionalized in the twentieth century (as demonstrated by organic chemists in World War I and nuclear physicists in World War II), or was this integration already occurring in the early modern Christian states?[8] Did Renaissance and Enlightenment mathematicians, engineers, and soldiers invoke Archimedes only for rhetorical inspiration, or did they realize his ideal in practice, as Voltaire seemed to indicate? In short, when did Archimedes have real intellectual heirs who re-created for themselves his personal union of science and the art of war? To answer these questions, we must first determine the fundamental dimensions of this union.

The essential military accomplishment of Archimedes at the Siege of Syracuse was the development of a highly effective defensive system based on optimizing, constructing, and coordinating the siegecraft of his day. This accomplishment demanded consideration of the "acquisitional," operational, tactical, and political domains of the art of war. "Acquisitional" here refers to the technological transformation of civilian resources into military assets, including weapons, ammunition, and armor, in addition to suitable vehicles, food, and fodder. In Archimedes' situation, such civilian resources could have included the use of animal tendons or human hair for the elastic elements of torsion catapults, as well as the use of raw stones for projectiles and of wood, metal, and hemp for cranes and catapults.[9] By applying the laws of the lever and the center of gravity, not to mention the empirical catapult rule (a cube-root operation), Archimedes presumably minimized the material demands of the machines by formulating the most efficient designs.[10] Our first Archimedean relationship between science and warfare therefore entails the acquisitional use of scientific theory to optimize the conversion of raw materials into products that can be directly deployed in military actions.

The "operational" domain of the art of war refers to the distribution and coordination of military assets in conducting a campaign. A concept that was formally developed in the Soviet Union before World War II, it focuses on setting up an army to effect the systemic disruption or entropy of the opposing enemy force.[11] This domain thus involves concentrating, maneuvering, and synchronizing military units against select enemy targets primarily before the actual combat begins. From Archimedes' perspective, operational planning first entailed selecting and situating siegecraft and troops, along with determining the systemic vulnerabilities or "Achilles' heels" of the Roman assault. For Syracuse, this meant targeting the Romans as they approached the city walls, before they could unleash the overwhelming power of their disciplined heavy infantry. Integral to operational planning is the deception or deliberate confusion of enemy forces (also known as "stratagem") to maximize their psychological stress. The operational realm thus demands an assessment of the strengths and vulnerabilities of the enemy force in both a physical and a mental sense.

Archimedes' ability to shock the Romans with his unexpectedly powerful defenses, his correct anticipation of the initial Roman assault plans, and his ultimate success in deterring additional assaults for over two years represents operational thinking of the highest order. Did a direct relationship exist between his mechanical prowess and his ability to discourage the Romans so effectively? Was it simply a coincidence that antiquity's greatest mathematical mind was so proficient in not only the acquisitional but also the operational realm? Did his ingenious analytical brain—capable of designing a mechanical system that allowed his own body to counter the enormous static force of a beached ship—also aid him in devising a defensive system with modest acquisitional resources that overturned a dynamic Roman military force? From another perspective: Was there a connection between the ability of Archimedes to calculate the center of gravity of complex geometrical objects and the organizational "center of gravity" of the Roman army under Marcellus?[12] Or are these merely figurative comparisons? The primary sources required to address such questions have long been lost. Nevertheless, this concurrence of mathematical genius and operational prowess highlights the second Archimedean relationship we will invoke between science and the art of war: the use of theory in operational planning and execution.

In this context, "tactical" denotes the process of controlling, including destroying, enemy military and civilian resources—the realm that so obsessed

Clausewitz in his classic *On War.*[13] It involves maintaining discipline and command to maximize the control or destruction of enemy forces and resources as well as to minimize the friendly (or neutral) casualties suffered and equipment destroyed. There is little evidence, of course, that the elderly Archimedes played an active tactical role in the Siege of Syracuse. This, presumably, was the responsibility of the professional military commanders. Nevertheless, he supposedly designed the siegecraft to fulfill specific tactical missions and supervised the training of soldiers in their efficient and disciplined operation. In this realm, we see a central scientific function in the tactical realm: the use of theory to optimize the operating performance of the weapon systems, subject to the quantity and quality of available troops. Archimedes unmistakably selected the ranges of his siegecraft to maximize the destructive effect of each shot and the odds of hitting each target. He also restricted the use of light arrows, darts, and cranes to locations along the walls, where they would always strike their target with virtual impunity; in turn, he restricted the use of the medium-weight and heavyweight projectiles to the middle and distant ranges, where their launchings were most convenient and the Romans were most vulnerable. What emerged is the use of the classic tactic of a combined-arms action to achieve an active defense in depth, while making only moderate demands on the Syracusian troops for discipline and self-sacrifice.

The fourth aspect of Archimedes' integration of science and the art of war lies in the political domain. This is where civilian resources—financial, human, and material—are secured for military use. Archimedes was involved in the political domain of warfare on two levels: inventing machinery to help boost tax revenues and developing a process for preventing the theft of such financial resources. Archimedes is reputed to have invented the hydraulic screw pump, which soon promoted agricultural productivity, especially in Egypt (and possibly in Sicily as well), and boosted tax revenues accordingly. Archimedes' most famous hydraulic study occurred within the second level of the political domain. In an early campaign to combat "waste, fraud and abuse" among governmental contractors and to conserve fiscal resources, Herion, the military dictator or Tyrant of Syracuse, formally requested that Archimedes calculate the actual percentage of silver and gold in a recently commissioned ornamental wreath. As a by-product of his successful solution to Herion's problem, Archimedes stumbled upon the scientific concept of specific weight (or density) while taking his famous bath. The significance here is that the interaction of scientific knowledge and military power may lead to fundamental new discoveries or inventions that transcend immediate political, acquisitional, operational, or tactical significance. A similar situation arose when Archimedes extended the implications of the lever law by proclaiming "Give me a place to stand and I will move the world." Herion then challenged the self-assured engineering theorist to invent a machine that would allow a man to launch a naval vessel single-handedly. The fruit of this challenge was the famous polyspaston, whose success conclusively demonstrated the universality of Archimedes' statics, at least on earth.

For Archimedes, the science of mechanics may have also facilitated the optimum coordination of the tactical, operational, acquisitional, and political domains. The diagram in figure I.1 models their complex interactions. One may reasonably conjecture that Archimedes initially conducted a preliminary operational study that established where and with what strength the Romans would attack. Then, with those intelligence assessments in mind, he could have ascertained the greatest vulnerabilities in the Roman attack and calculated the size and the range of projectiles or other destructive weapons required to maximize their impact. Archimedes could then have decided how best to utilize the available political resources to determine both the quantity and the performance of the siege engines to be built, as well as their distribution along the city walls. This is an iterative process that would have demanded basic mathematical analysis and mechanical considerations to furnish reasonably accurate cost and performance predictions. Such a "systems analysis" would also have to consider basic tactical constraints of the Syracuse military force, including the number of troops available and their level of fitness and training.[14]

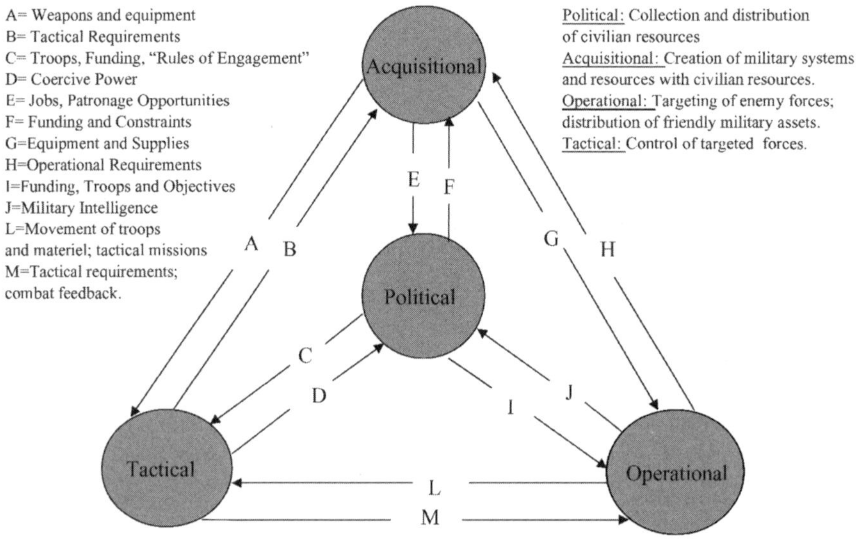

Figure I.1
The military organization as a complex system.

The operating requirements of the siegecraft would be informed by such human considerations. With basic design and production requirements emerging from such an operational study, Archimedes could then have moved to the acquisitional domain and applied such mathematical theories as the catapult rule, the lever law, and the concept of the center of gravity to meet the established performance and operating specifications while minimizing the construction costs. The rigorous geometrical science of mechanics, in short, may have been more than just an efficient design tool for Archimedes. It also may have provided an integral means of establishing optimum design and production requirements for his siegecraft.

There is nothing new about addressing the relationships between scientific knowledge and military power in the pre-modern world. Nevertheless, a basic motivation for this volume lies in the neglect and dismissals this topic has endured from both military historians and historians of science during the latter half of the twentieth century. From the perspective of the history of science, the military context of early modern science attracted serious attention before the advent of the Cold War—beginning with Boris Hessen. A prominent Soviet physicist, he actively promoted Einstein's theory of relativity during the early Stalinist era—a position that ultimately resulted in his execution by the late 1930s. In an initial attempt to assuage the Stalinist authorities, he presented a pioneering historical account of early modern mechanics in 1931.[15] There he argued that the military needs of early modern Europe, specifically in navigation, gunnery, and military engineering, determined the development of modern science in general and Newton's *Principia* in particular.[16] Hessen's influence is apparent in Robert Merton's landmark study in the sociology of science, *Science, Technology and Society in Seventeenth Century England* (1938). Famous for his arguments about the Puritan influence on natural philosophy in Restoration England, Merton nevertheless devoted significant attention to military issues. Not only were Hooke, Halley, and Petty concerned with navigation and artillery in their research, Newton himself was conscious of the practical implications of the *Principia*.[17] According to Merton, Newton reveals a practical orientation in his analyses of tides, lunar motion, hydrostatics, hydrodynamics, and projectile motion in a resisting medium (to which we could add his membership in the Longitude Board and his position as director of the Mint).

Whereas Hessen and Merton focused on England, Henry Guerlac, at the onset of World War II, expanded the scope of this debate to include Italy and France. Before he became a leading historian of chemistry through his work on Lavoisier, Guerlac devoted his lengthy Ph.D. dissertation to describing how military engineering and artillery developments in the Renaissance and

Enlightenment eras helped shape the research agenda of many prominent mathematicians and natural philosophers.[18] Leonardo da Vinci, Tartaglia, Galileo, Petit, Desargues, Blondel, and Huygens were among those whose work reflected the militant nature of their states. Guerlac went further than Hessen and Merton, however, by describing in great detail the utility of high-level science and mathematics for the military engineering and artillery academies of Enlightenment France. In other words, Guerlac focused not only on what war did for scientific research but on what scientific knowledge did for military institutions.

Guerlac's dissertation is an astonishingly sophisticated study of the cultural and political history of science. He succeeded in integrating the history of science into European history as few works have done since. Nevertheless, Guerlac never published the work. Other than an early article on the mathematical and empirical nature of Vauban's military practice, he published virtually nothing on the topic of early modern science and warfare.[19] Whether or not this reticence to publish reflected the harsh scientific realities he observed during World War II, his unwillingness to be associated with the low academic prestige of military history or the controversial Marxist arguments of Hessen will probably never be resolved.[20] But his silence implicitly encouraged the Cold War generation of historians of science, following Alexandre Koyré's lead, to ignore—if not deny outright—the significance of both science for war and war for science during the era of the Scientific Revolution.[21]

A. Rupert Hall's *Ballistics in the Seventeenth Century* (1952) remains a highly influential introduction to the early development of the science of ballistics and early modern gunnery.[22] Few works have had greater influence in both the history of science and military history. Hall's interest in this subject reflected more than his active military service during World War II. It was in ballistics that Hessen, Merton, and Guerlac appeared to make their most convincing arguments about the fruitful interactions between scientific knowledge and military power. It was here that Alexandre Koyré also made perhaps his weakest arguments for the sufficiency of the purely philosophical context of early modern science. This extremely influential historian of science declared, for example, that the science of Galileo and Descartes was "made not by engineers or craftsmen, but by men who seldom built or made anything more real than a theory. The new ballistics was made not by artificers and gunners, but against them. And Galileo did not learn his business from people who toiled in the arsenals and shipyards of Venice. Quite the contrary: he taught them theirs."[23] To decouple ballistic theory from gunnery practice, Hall first argued that the ballistics theory of Galileo was incapable of reasonably modeling artillery trajectories due to its neglect of air resistance. Secondly, Hall maintained that fundamental hardware problems existed with smooth-bore artillery, which made it impossible to satisfy the underlying assumptions of a mechanical theory. This includes controllable initial conditions, especially that of muzzle velocity.

Hall also argued for the scientific ignorance of early modern engineers in general. Not only were such practitioners largely oblivious to the utility of mathematics, "there was little enough for the practical man to compute: but what he did he did badly, clinging to antiquated rules-of-thumb."[24] Although Hall did admit that military engineers relied on Euclidean geometry in designing and constructing fortifications, he asserted that their mathematics was scarcely more complex than that used by medieval masons.[25] At best, Hall declared, early modern soldiers employed scientific instruments and theories only as status-enhancing symbols. In spite of his considerable contribution to the history of science through the 1990s, Hall has done little to modify in print his initial position on early modern science and warfare.[26] To his credit, however, he now concedes that real changes were occurring during the eighteenth century that promoted highly constructive relationships between ballistics theory and military practice.

Hall's thesis concerning the impracticality of scientific knowledge for early modern warfare has received much support. It has encouraged historians of science and technology to ignore the scientific implications of early modern artillery that Hessen, Merton, and Guerlac had maintained.[27] It certainly legitimized Richard Westfall's dismissal of Jerome Ravetz's Marxist position, which suggested that early modern science was a "means of oppression" by virtue of its connection with military technology. Westfall retorted, "I confine myself to saying that there is no way in which I am willing to describe such empirically derived knowledge and practices, which had no connection with a theoretical framework, as part of science." Westfall continued to assert that, since Ravetz admitted that science was largely unsuccessful in influencing practical life, "his comments on science as oppression rest, not on sand, but on no foundation whatsoever."[28]

This scholarly unwillingness to stomach the idea of a fruitful Archimedean synthesis of scientific theory and military practice during the early modern era has not lessened since the Cold War concluded. Steven Shapin addressed the issue of scientific utility in *The Scientific Revolution*.[29] Although he acknowledged the pervasive influence of technological motives among seventeenth-century natural philosophers, he denied such inspiration was successfully translated into practice:

> The question of the real historical relation between the growth of scientific knowledge and the extension of technological control has been endlessly debated by historians and economists. On the one hand it now appears

> unlikely that the "high" theory of the Scientific Revolution had any substantial direct effect on economically useful technology in either the seventeenth century or the eighteenth. . . . On the other hand, we have already noted intimate links between the "mixed" (or "impure") mathematical sciences and military and productive technology going back to antiquity, and there is no reason to think that such links were not strengthened through the early modern period. Moreover, there can be little doubt that the vast expansion of natural historical and geographical knowledge that attended the voyages of exploration and conquest contributed significantly to the making of empires and fortunes. It is the link between "theory" as a cause and technical change as an effect that remains at issue.[30]

While Shapin conceded that undeniable interactions existed between early modern institutions of scientific research and military power, he maintained that the triumphant theoretical breakthroughs of the Scientific Revolution were largely irrelevant for technological change. He further argued that "historians have had great difficulty in establishing that any of these spheres of technologically or economically inspired science bore substantial fruit. Baconian rhetoric, that is to say, translated poorly into practical reality, and the military-industrial-scientific complex is more properly regarded as a creation of the nineteenth and twentieth centuries."[31] In other words, the Archimedean mathematical revival of the early modern era was relevant only for theory, and not for practice; hence, it was not a full-fledged renaissance. Shapin is not alone with his negative assessment.

James McClellan and Harold Dorn deserve much credit for having synthesized the history of science and the history of technology in the early modern era.[32] They commented on the monumental Renaissance innovations in cartography, gunpowder weaponry, full-rigged sailing vessels, drilled-infantry formations, and military engineering, as well as the global military impact of such innovations. Nevertheless, they refused to believe that such military power had much to do with scientific knowledge:

> What role did scientific thought play in this immensely consequential European (military) development? The answer is essentially none. Some of the basic inventions (such as gunpowder and the compass), as we have seen, originated in China, where they were developed independently of any theoretical concerns. In Europe itself, with its established tradition of Aristotle, Euclid, and Ptolemy, none of contemporary natural philosophy was applicable to the development of the new ordnance or any of its ancillary techniques. In retrospect, theoretical ballistics might have been useful, but a science of ballistics had not yet been deduced; it awaited Galileo's law of falling bodies, and even then the applicability of theory to practice in the

> seventeenth century can be questioned. . . . Scientific cartography probably did play a supporting role in early European overseas expansion, but navigation remained a craft, not a science. The gunners, foundry men, smiths, shipbuilders, engineers, and navigators all did their work and made their inventions and improvements with the aid of nothing more (and nothing less) than experience, skill, intuition, rules of thumb, and daring.[33]

These sweeping declarations are by no means unanimous. Given David Water's classic studies of quantitative navigation practices in the sixteenth and seventeenth centuries, and continuing with Mario Biagioli's and Mary Henninger-Voss' investigations of the interactions between mathematical and military communities in sixteenth-century Italy, it is hard to ignore the existence of a flourishing relationship between early modern scientific and military practitioners.[34] Henninger-Voss, in particular, revealed the contemporaneous military utility of such fundamental theoretical constructs as Tartaglia's geometrical ballistic trajectories and Guidobaldo's mechanics of machinery, while Serafina Cuomo described the vast influence of Tartaglia's theoretical artillery treatise *Nova Scientia* (1637) during the sixteenth century.[35] Nevertheless, there remains much academic skepticism toward the notion that full-fledged Archimedean relationships thrived before the onset of industrial modernity.

While historians of science are ambivalent about tackling early modern military issues, military historians conversely choose to ignore early modern science. The debate concerning the historical significance of military change from the Renaissance through the Enlightenment is virtually devoid of scientific considerations. The classic examples include Carlo Cipolla's *Guns, Sails and Empires* and William McNeill's *The Pursuit of Power*.[36] This reluctance to examine scientific contexts is especially striking among those scholars grappling with the Military Revolution thesis, which assesses the revolutionary versus evolutionary nature of early modern warfare.[37] This debate was formally instigated in 1955 by Michael Roberts. To his credit, he acknowledged the application of theoretical science to early modern warfare; this is especially evident with cartography and the mathematical education of officers. Roberts even asserted that science had a revolutionary effect on the celebrated military reforms of Maurice of Nassau and Gustavus Adolphus.[38] Unfortunately, the hard evidence he presented as support for such bold claims was marginal: he pointed to the invention of the telescope, but little more. Hence, the subsequent proponents and critics of the Military Revolution thesis have had little to say about its scientific context.[39] Nonetheless, David Eltis has described the use of mathematics in infantry command during the sixteenth century, and Black has noted the military efforts of such eighteenth-century scientific figures as Denis Papin,

Thomas Desaguliers, William Congreve Jr., and Leonhard Euler, but these references are tangential to their central arguments.[40] A prominent exception to this trend lies with Alex Roland's classic account of the early development of submarines.[41] Although this monograph analyzed the role of hydraulic theory in stimulating this invention from the sixteenth through the eighteenth century, it has attracted little scholarly attention, presumably due to the tactical failure of submarines until the American Civil War. A more recent exception is Erik Lund's *War for the Every Day*. Armed with sources from Vienna's Kriegsarchiv, he argued extensively for the operational relevance of mathematical science for senior military commanders in the Habsburg service during the late seventeenth and early eighteenth centuries.[42] In spite of such admirable work, the history of science and the military history communities remain largely oblivious of each other to the detriment of their respective historiographies. The central objective of this volume is thus to encourage the synthesis of these ostensibly independent branches of early modern history.

The contributions to this volume are divided into four sections. The first addresses the innovation of gunpowder weaponry in the late Medieval and Renaissance worlds of both Christian and Islamic domains. The second addresses the interaction of mathematical science and naval power. The third investigates the role of chemistry in the manufacturing of gunpowder. Finally, the fourth section confronts the utility of mathematics and mechanics for military engineering and artillery institutions. These essays were originally presented at two academic conferences. The first was hosted by UCLA's Center for Seventeenth- and Eighteenth-Century Studies and presented at the Clark Library in October 1998. Organized by Geoffrey Symcox of UCLA, it included the papers of Cormack, Alexander, Mahoney, Mauskopf, Langins, and Steele, which all grappled with the relationships between the "Scientific" and "Military" Revolutions of early modern Europe. The second conference, hosted by and presented at MIT's Dibner Institute for the History of Science and Technology in May 1999, was organized by Bert Hall of the University of Toronto. Featuring the papers of DeVries, Baumgartner, Hildred, Hacker, Ágoston, Buchanan, and Kaiserfeld, it addressed the issue of medieval and early modern military innovation from a global perspective.

The essays in part I support the assertion that formal scientific knowledge played a nominal role in the transformation of gunpowder weapons into essential military instruments. A great deal of cultural conflict was generated during this innovative process, both in the West and the East, yet it was largely resolved during the sixteenth century. A lack of scientific structure or inspiration did not prevent the expression of considerable technical creativity on both instrumental and systemic levels, however. The essays in part I thus suggest that formal science was not a necessary condition for the invention, innovation, and diffusion of the fundamental weapons systems that dominated the late medieval and Renaissance eras. These systems included fortresses designed to accommodate gunpowder weaponry, infantry forces armed with firearms, and sailing ships designed to fight with artillery. An implicit argument about the relationship between science and warfare, however, emerges in part I: as long as scientific theory and methodologies were not employed in the acquisitional, operational, and tactical domains of warfare, the Islamic states remained militarily competitive, if not superior, to their Christian counterparts.

Kelly DeVries commences part I by discussing the early development of gunpowder fortifications in France and England during the fourteenth and fifteenth centuries. Fortifications were designed to resist siege artillery actively by integrating smaller firearms that were both shoulder-mounted and wall-mounted. DeVries directly challenges the commonly held perception that the famous *trace italienne* of the sixteenth-century Italian city-states represented a revolutionary response to the threat of offensive siege artillery. Instead, he argues that this innovation reflected the culmination of an intense evolutionary development in fortification design that began with the gunports that Edward II added to his fortresses in the fourteenth century. Although this architectural precocity declined in England during the fifteenth century, such system integration was actively pursued by the French in response to the rise of the Duchy of Burgundy. Powerful fortification elements such as artillery towers (tall masonry structures riddled with gunports) and boulevards (low, earthen cannon-resistant gun platforms) were innovated in response to the devastating Burgundian siege artillery. Equally important, DeVries emphasizes, are the French distribution and coordination of these new design elements to furnish flanking fire, eliminate dead zones, and cover vulnerable gates or positions. In short, the late medieval practices in military architecture demonstrate a shrewd operational awareness that the best way to respond to offensive gunpowder weaponry is to deploy firearms actively. What also emerges from DeVries' account is evidence of operational awareness in the positioning and coordination of these design elements to achieve a synergistic increase in defensive resistance. Such an achievement did not rely on a sophisticated scientific or mathematical awareness, however. Only in the sixteenth century did Italian military engineers begin to employ formal geometrical methodologies on a routine basis in order to maximize both active and passive defensive power for the least cost.[43] However much this development appears to be a renaissance in Archimedean military engineering, it nonetheless signifies an evolutionary medieval process.

Whereas DeVries' essay discloses the astonishing integration of firearms with fortifications in late medieval France, Baumgartner calls attention to France's cultural resistance to incorporating firearms in field warfare. Baumgartner highlights the extraordinary unwillingness of the French military leadership to raise and deploy native French arquebusiers during the era of the Habsburg-Valois Wars in the late fifteenth and early sixteenth centuries. He substantiates this neglect by contrasting the thousands of arquebuses housed in the arsenals of the Holy Roman Emperor Maximilian I with the hundreds scattered among the urban militias of France. Even when subjected to the devastating firepower of entrenched Spanish arquebusiers, both the French cavalry and the allied Swiss pikemen still insisted on reckless medieval-style frontal charges. Such refusal to adapt to new technological conditions, Baumgartner argues, ultimately forced the French to relinquish their control of Italy to the victorious Spaniards, who displayed tactical creativity in deploying *arquebusiers* on the battlefield. Even during the latter phase of the conflict with Spain, when Francis I acknowledged the grievous vulnerability of his infantry firepower, the French could rely only on foreign mercenaries to redress that tactical shortcoming. They continued to fall short in their ability to transform native recruits into competitive arquebusiers. As a consequence, the French were forced to pay ruinous prices to secure foreign mercenaries.

By examining a variety of literary sources, Baumgartner assigns the cause of such neglect to a deep medieval ambivalence within the French aristocracy rather than to a technical deficiency among French artisans. It was extremely difficult to accept the new reality that individual infantrymen had acquired highly potent engines with which to strike down armored "men of arms." Equally perturbing was the fact that individualistic warrior skill with a sword, lance, or halbard amounted to little in the face of entrenched arquebusier fire. French aristocratic resistance to firearms in field warfare was mitigated during the Religious Wars, when mounted forces enthusiastically deployed the wheel-lock pistol. Yet, as Baumgartner notes, frustrations with its unreliability led to its rejection in the seventeenth century in favor of aggressive action with cold steel.

Baumgartner's analysis, like DeVries', shows that the original process of incorporating firearms into field warfare was hardly an Archimedean process of formal applied science. The Spaniards accomplished this process by soberly reflecting on the costly tactical mistakes they experienced during the early phases of the Habsburg-Valois Wars in Italy. And yet the operational process of forming offensive arquebusiers and defensive pikemen into a *tercio*— a central core of pikemen that supported detached units of arquebusiers—did involve an Archimedean element. As Thomas Arnold of Yale University argued, the *tercio* was inspired by formal geometrical principles used in the engineering of *trace italienne* fortresses, including the equitable distribution of flanking fire along fortress walls.[44] By the late sixteenth century, the command of a *tercio* also required the ability to calculate its dimensions by taking the square root of the ratio of the total number of troops and the quotient of its length and width—one of the many mathematical operations Galileo could solve with his "military compass."[45] As Baumgartner reminds us, however, the basic invention of the arquebus and the cultural process of deploying it in a field battle did not rely directly on such intellectual sophistication. Instead, it hinged upon a cultural process of shedding medieval warrior mindsets and adopting more bureaucratic soldier norms, as exemplified by the Roman centurion.

In their attempt to adopt infantry firearms, the French were not alone in encountering severe cultural impediments. Barton Hacker describes the challenge Islamic military forces similarly encountered when seeking to incorporate gunpowder weaponry into their cavalry-dominated armies. The military, gender, and even religious identities of Moslem warriors were intimately associated with their highly skilled use of the sword and the compound bow while riding a horse, as modeled by the Prophet Mohammed himself. To convince such warriors to negate such talent and take up firearms and marching was inconceivable. It was the Ottoman Turks who discovered how to overcome such cultural resistance. Hacker outlines how they recruited Balkan Christian youths during the fifteenth century to serve directly under the Sultan as arquebusiers in the newly institutionalized Janissary Corps. Although the bulk of the Ottoman army remained committed to the traditional cavalry forces, the Janissary Corps gave the Sultan a heavy concentration of tactical firepower that helped ensure their rapid expansion at the expense of the Mamluks, Hungarians, and Safavids during the early sixteenth century. Even more significant were the new opportunities that gunpowder weaponry made for centralized political authority, as opposed to the tribal structure that traditional cavalry skills encouraged. As Hacker argues, the political centralization associated with firearms in the Islamic world, as symbolized by the Janissary Corps, forever disempowered the nomadic steppe horsemen that had devastated the Asian world for thousands of years.

The example of the Ottoman Turks in adopting infantry firepower was quickly imitated by the Moghuls in Northern India and the Safavids in Persia, in contrast to the Mamluks who never recovered from their resistance to such innovation. The Safavids responded to the crushing defeats from the Ottomans by seeking firearms from both Islamic and Christian sources; but, as Hacker explains, they never integrated their arquebusiers as deeply into their military culture as the Turks did. The Moghuls, in contrast, successfully introduced the arquebus as well as heavy artillery into their armies. Although much of the empire's power rested on the administrative and political competence of Akbar, its monopoly on heavy siege artillery and its development of an elite standing army of arquebusiers proved decisive. Hacker observes how such a diffusion of Christian military hardware into the Islamic world was also accompanied by tactical innovations, including the Hussite *Wagenburg*—the heavy yet mobile wagons used as portable field fortifications. Yet the more conceptual Western innovations of the sixteenth century, such as the mechanization of infantry tactics through the geometrical discipline of the *tercio*, and the thinner, more linear Dutch formations were absolutely rejected—as Marsiglia concluded in his famous study of Islamic military institutions in the late seventeenth century.[46]

While Hacker analyzes the success of Islamic military empires in adopting firearms from a tactical and operational perspective, Gábor Ágoston focuses on the acquisitional dimension of this issue within an exclusively Ottoman context. Ágoston argues that a necessary condition for the Ottoman military success with firearms was its domestic development of acquisitional capabilities. These included the construction of largely self-sufficient gunpowder production facilities throughout the empire, the employment of Christian technical expertise, and the manufacturing of largely adequate supplies of competitive arms and military equipment. Ágoston also confronts a number of Eurocentric views of Ottoman military stagnation, especially the assertion that the growing military weakness of the Turks was based on their dependence on Christians for basic acquisitional resources such as gunpowder and firearms. The political overextension of the Ottoman Empire, as well as the ability of such Christian enemies as the Habsburgs and Romanovs to coordinate their military actions, played a far greater role in the Turks' reversals than any supposed shortfall in military hardware.

What Ágoston does acknowledge, however, was the complete unwillingness of the Ottoman Turks to adopt the new mathematical and mechanistic military "software" of the Christians. This is evinced by the persistent difficulty of the Ottomans to secure capable artillerists during the seventeenth century. Their gunnery competence was based exclusively on combat experience. The Christians, by contrast, incrementally augmented their gunnery training with geometrical techniques and mechanical theories.[47] In short, the Ottoman Turks had few cultural inhibitions about adopting Christian hardware, but they utterly rejected the Archimedean military mindset that was flourishing in Christian Europe during the sixteenth century. It was only with the traumatic loss of the Crimea to Russia in the late eighteenth century that the Ottomans under Sultan Selim III seriously confronted this shortcoming. In response to this debacle, Selim III authorized the formation of a new military force to be trained according to Christian standards in both scientific theory and military practice. But such an Archimedean cultural transfer proved too much for the now reactionary janissaries to accept, thus resulting in their destruction of this "New Model Army," followed by their assassination of Selim in 1807.[48] The Turkish adoption of alien military hardware proved to be far less problematic than the adoption of a scientific military culture that presumed a secular if not blasphemous natural philosophy.

The essays in part II address the relationships between early modern science and naval power. Here we see the intersection of two of the most revolutionary developments of early modern Europe from a global perspective: mathematical science and oceanic navigation. Beginning with Prince Henry the Navigator in the early fifteenth century and culminating with the Greenwich Observatory in the late seventeenth century, the linkage of mathematical and astronomical knowledge to navigational practice was unmistakable. This marriage was integral to Renaissance Europe's transformation from a set of backward, if not barbaric, tribes—at least from an Asian perspective—into a global military and commercial civilization. What then was the actual nature of this relationship between mathematical knowledge and naval power? Three of the papers presented here emphasize the intense political demand for greater oceanic navigational accuracy, given the lucrative commercial profits and tax revenues such a capability fostered. The achievements described in these papers, however, rest more on what the naval powers did to promote fundamental scientific knowledge, rather than on what the mathematicians did for naval power. The two operational "holy grails" that inspired much of the relevant patronage, the Northwest Passage and the determination of longitude at sea, proved to be nonexistent and unsolvable, respectively, during the early modern era. Not until the middle of the eighteenth century did Harrison's horology, not to mention Euler's and Meyer's celestial mechanics, overcome the longitude problem. Likewise, only in the early twentieth century did the Panama Canal finally satisfy the demand for a shortcut to Asia. Nevertheless, the knowledge generated from attempting to solve these formidable operational problems directly fueled the scientific revolution in astronomy and mechanics during the seventeenth century.

If such a lucrative relationship between science and naval warfare existed on both political and operational levels, what then can be said about its relevance for the acquisitional and tactical domains? Galileo's career as a scientific advisor for the Republic of Venice, where he maintained close ties to its famous arsenal, appears to have directly inspired his pioneering research in solid mechanics, the first of his "two new sciences."[49] Such an acquisitional relationship in naval production was more the exception than the rule, however. As David McGee has argued, attempts to apply formal mechanical principles to naval architecture, such as Euler's exhaustive analyses, proved insufficient well into the nineteenth century.[50] Even less interaction between scientific theory and naval practice appears in the tactical domain—that is, at least until Robins' proposal for a low-velocity, high-caliber naval gun effected the invention of the carronade in the 1770s.[51] The limited relationship between scientific theory and naval tactics is demonstrated in Alexandra Hildred's archeological description of the *Mary Rose*, a highly innovative English warship that sank in 1545. This vessel represented among the most revolutionary of innovations in Christian Europe's naval development: the integration of gunpowder artillery with the full-rigged sailing vessel.

Hildred's essay discusses those naval ordnance innovations of early sixteenth-century England that proceeded with astonishing intensity. The physical remains of the *Mary Rose*, whose excavation Hildred has directly participated in, reveal this unmistakably. By comparing the guns recovered with those described in the ship's logistical documents, Hildred not only identifies a number of unrecognizable guns, but she also details the immense variation in the ordnance deployed, as well as the coexistence of seemingly obsolete medieval weapons and contemporaneous guns. What emerges is a naval system that deployed a wide variety of ordnance, ranging from light hail-shot deck sweepers to heavy cast-iron hull smashers. The gunners operated these arms through the newly invented gunports, thereby signifying a combined-arms system that delivered an active defense in depth. With the *Mary Rose*, an integrated, Archimedean-style fortification system emerges, albeit along evolutionary rather than formal scientific lines. This mix of ordnance, according to Hildred, reflected the tactical consideration of arming a ship with the strongest possible defenses, although subject to strict weight restrictions. Yet Hildred also argues that the *Mary Rose* represented the broader operational criterion of the English naval establishment under Henry VIII: how best to distribute the limited supply of heavy ordnance among its mobile naval and static shore-based defenses in the face of potential French invasions. As already noted with respect to land-based military power, a renaissance in Archimedean engineering was not needed to persuade the Atlantic navies to substitute artillery-fire assaults for infantry-boarding attacks. Such a renaissance, on the other hand, would become a prerequisite to the English navy's transition from a "brown-water" coastal defensive force into a "blue-water" offensive power, as the Portuguese and later the Spaniards had pioneered.[52]

Lesley Cormack describes the rise of those naval-oriented mathematical practitioners in Elizabethan England who succeeded in surpassing their

Iberian counterparts. Recipients of British patronage, this new breed of mathematicians, which included Edward Wright and Thomas Harriot, became integral participants in the Empire's early naval expansion. Although educated at Cambridge and Oxford, respectively, Wright and Harriot turned their backs on medieval academic careers and pursued the new patronage of wealthy naval-obsessed aristocrats who sought both intellectual prestige and operational utility. The two mathematicians more than satisfied the expectations of such patrons as Walter Ralegh and Crown Prince Henry. Wright derived the first mathematical formulation of Mercator's projection for an easy calculation of straight rhumb lines in Northern climes, in addition to contributing many practical insights into Gilbert's famous work on magnetism. Although he published little, Harriot worked incessantly on solving the longitude problem through magnetic variation (a technique the Dutch East India Company finally made practical in the eighteenth century) and a host of other navigational and mechanical problems. He also actively participated in the early exploration and colonization of Virginia. Cormack contends that such mathematical practitioners failed to resolve the fundamental operational problems they confronted; they nevertheless succeeded in demonstrating to their English political establishment that their practical yet rigorous mathematical research furnished sufficient navigational solutions to risk oceanic naval enterprises. These included the Muscovy Company and Drake's circumnavigation of the globe. Drake's dramatic expedition underscored the relevance of such science, given his reliance on captured Portuguese navigational data and expertise.[53] Incidentally, such demonstrable research utility for naval purposes was institutionalized at Gresham College in 1598 and through the funding of the Sevilian Chair in Geometry at Oxford University in 1619.[54] Cormack not only portrays how this new breed of mathematicians embraced both the intellectual rigor and the naval utility of advanced mathematics but also demonstrates how they established the institutional foundations for the Scientific Revolution in England during the seventeenth century.

Amir Alexander also discusses the relevance of mathematical practitioners in Elizabethan England. Unlike Cormack, however, he compares Harriot to a more senior contemporary: John Dee. A leading Neoplatonist whose legendary exploits in Hermetic magic included discourses with angels and deceased spirits, Dee was also a consummate practitioner and advocate of "natural practical magic," or mathematically grounded engineering and technology, as demonstrated in his famous introduction to the first English translation of Euclid's *Elements*.[55] As a close friend of the prominent Portuguese mathematics and navigation professor, Pedro Nuñez, Dee was also instrumental in transferring the scientific navigation tradition of the Portuguese to a mathematically backward England during the middle of the sixteenth century. It is even maintained that Dee served as the inspiration for Prospero in *The Tempest*. Unlike Cormack's comparison of Harriot and Wright, which revealed many similarities, Alexander's juxtaposition of the consciously Archimedean Harriot with the Neoplatonic Dee reveals vast differences. These include their outlooks on how the "British Empire" (a term coined by Dee) should expand, as well as their foundational or even metaphysical orientations toward mathematics. By exploring such differences, Alexander displays how these seemingly unrelated perspectives toward power and knowledge actually flourished from the same intellectual framework. Alexander, like Cormack, thus sheds new light on the construction of the necessary conditions for Newton's writing of the *Principia*. Instead of focusing on the relationship between imperial naval power and mathematics with respect to patronage, however, he demonstrates how such a synthesis helped shape the mathematical foundations on which Newton's invention of the calculus would depend.

Michael Mahoney relocates part II's investigations of navigational practice and mathematical theory to the Dutch Republic and France. He furnishes a succinct analysis of Huygens' research effort to resolve the longitude problem. Why it signified such a central naval operational problem is his first concern. Navigating a ship along straight latitudinal lines, using the Portuguese solar declination technique, is a relatively straightforward process, suitable for exploration and routine trade. Mahoney emphasizes, however, that such an approach was extremely hazardous in wartime conditions because it allowed enemy forces to know exactly where to lie in wait for their targets. Determining longitude at sea thus opened the possibility for fleets to engage in precise operational maneuvers. Such clarification helps explain why Huygens' father, Constantine (the famous political advisor to the House of Orange, friend of Galileo, and educator of William III), affectionately referred to his son as "my Archimedes."[56] Mahoney describes how Christiaan Huygens developed fundamental principles of rigid-body dynamics, simple harmonic motion, and feedback regulation in his innovation of the pendulum clock in order to determine longitude at sea. Mahoney also stresses how much practical engineering effort Huygens expended in attempting to make his precision instrument capable of functioning at sea. This longitudinal work had especially profound implications for natural philosophy: Huygens' measurement of global variations in the gravitational acceleration rate directly refuted Newton's theory of universal gravity.

Mahoney also suggests another scientific effect of Huygens' effort to enhance naval operational power: it helped convince Colbert to set up the Paris Observatory and the Academy of Science (the premier scientific academy of the Enlightenment). Huygens' advice to Colbert gave the Academy an initial research focus as well. Thus, Huygens manifested the Archimedean ideal in his formidable research not only in mathematics and mechanics but also in navigational practice. Huygens in turn helped instigate the idea of making use of the power of the scientific institution to furnish new acquisitional resources. He proved far more successful in securing state-sponsored scientific research than in improving the operational power of the French navy, especially given the ultimate failure of his longitude program. Nevertheless, his seemingly useless mechanics theories proved to be critical to Benjamin Robins' breakthrough gunnery experiments with the ballistic pendulum in the middle of the eighteenth century.[57]

The essays in part III address the relationship between chemical knowledge and gunpowder production. The acquisitional process of converting raw materials into gunpowder was of enormous relevance for early modern military organizations. For both land and sea forces, gunpowder fueled the "internal combustion" engines of firearms and artillery. The process of acquiring gunpowder was directly related to broader military transformations, of course. The growing mechanistic military culture of early modern Christian Europe, inspired by both a classical military heritage and a mechanical worldview, placed high value on the deterministic performance of individual soldiers and tactical formations, as well as on their interchangeability. This orientation prevailed with the Spanish Army of Flanders under the Duke of Parma and later with the opposing Dutch forces under Maurice of Nassau and Simon Stevin.[58] This cultural mindset would culminate decisively with the Swedish army under Gustavus Adolphus during the Thirty Years' War. These military reformers demanded gunpowder that not only furnished sufficient killing power but also performed consistently enough to reinforce the deterministic nature of their tactical formations. Such mechanistic aspirations were partially fulfilled by the use of uniform clothing (i.e., "uniforms"), the use of common drills and tactical maneuvers, the inculcation of mathematical reasoning, and the deployment of standardized weaponry. Nevertheless, the procurement of consistent gunpowder charges proved to be frustratingly difficult. This is evidenced by the growing demand for scientifically competent master gunners who could analyze the strength of a particular supply of powder and calibrate their artillery pieces accordingly.[59]

The military pressure to increase gunpowder's strength and production rate, while also decreasing its production cost, placed enormous burdens on the early modern state. Such problems encouraged military bureaucracies to seek solutions from the early practitioners of chemistry, who were already demonstrating their utilitarian orientation in medicine and mining, in addition to alchemy. The actual relationship between chemical knowledge and munitions production is the focus of the following three essays. Here the authors essentially argue that manufacturers gradually made use of chemical methodologies to enhance the physical quality and production rate of gunpowder. By the late eighteenth century, not only were powdermakers employing formal chemical theory to improve their processes but the leading chemical theorists were receiving lucrative patronage and publicity because of the state's insistence on controlling the gunpowder production process. Early modern chemists and gunpowder producers, in short, increasingly embodied the Archimedean ideal of employing science to enhance military acquisitions while simultaneously inspiring new theoretical innovations.

Brenda Buchanan starts off this part with an analysis of English gunpowder production from the sixteenth century through the middle of the eighteenth. In an account that is richly informed with economic history, she argues that an Archimedean incorporation of chemical knowledge into gunpowder production did not abruptly begin with Lavoisier and Congreve in the late eighteenth century. The process of replacing the "art and mystery," or traditional craft orientation, in gunpowder manufacturing with formal scientific considerations was a gradual evolution over the course of two centuries. Buchanan shows that the early modern development of English munitions production comprised four phases. The first phase began in the sixteenth century with the import of crucial craft skills from Continental Europe. The second phase encompasses the adoption of domestic and Continental manufacturing innovations in the middle of the seventeenth century—the era of the English Civil War and the later Dutch Wars. The third phase, beginning in the late seventeenth century, witnessed the growth of flexible provincial partnerships that responded to the growing gunpowder demands of mining and trade—especially the slave trade and privateering. This free-market development coincided with the Royal Society's early research investigations into gunpowder explosion, but such work did little to resolve English gunpowder procurement problems. By contrast, the East India Company began to secure access to vast nitrate deposits in India during the Wars of Louis XIV. Their ability to import nitrates, or saltpeter, efficiently to England led to the demise of domestic nitrate collection and processing by the early eighteenth century; England thus gained a significant acquisitional advantage over France. Nevertheless, serious problems with the processing of gunpowder persisted, as revealed in the fourth phase, during which Charles Frederick directed the procurement of supplies by the Ordnance Board. Through a careful analysis of primary sources, Buchanan reveals his efforts to enforce high standards in a wide range of production facilities. Frederick demanded detailed accounting of the quality and quantity of ingredients, as well as that of the incorporation process. Yet Buchanan also discloses that he encouraged these manufacturers to conduct scientific experiments in order to explore opportunities for cost reduction and performance enhancement. Frederick's dedicated, yet relatively amateur, scientific orientation would prove insufficient for resolving the considerable quality-control problems the Ordnance Board faced during the American Revolutionary War. His replacement, Captain Congreve of the Royal Artillery Corps, would not be hindered by such limitations, as Seymour Mauskopf argues later in this volume.

The Swedes did not share England's good acquisitional fortune of controlling the vast nitrate deposits of the East India Company. Thomas Kaiserfeld articulates the technical challenges the Swedes experienced during the seventeenth and eighteenth centuries when they sought to extract sufficient saltpeter from their agricultural communities in order to meet their vast military needs. These challenges incited deep political tensions between the military and the domestic agricultural and mining interests, and exacerbated relations between the monarchy and the parliamentary estates. The Swedish acquisition and processing of nitrates, Kaiserfeld argues, signified a national effort to maximize societal utility by formally considering both economic prosperity and national security, as well as employing both traditional medieval and mechanistic scientific means. The first part of Kaiserfeld's essay explores how the Swedes developed a system of collecting manure directly from the farms and processing it locally into coarse saltpeter. This process reflected the medieval tradition of collecting material resources for taxes. It slowly evolved through the early eighteenth century while meeting the heavy acquisitional demands of Swedish imperialism. By the middle of the eighteenth century, however, a new system of saltpeter extraction attracted Swedish attention, according to Kaiserfeld. To ensure higher production rates and lower production costs, the Swedish monarchy began offering subsidies for saltpeter barns or facilities where manure could be stored under carefully controlled conditions to maximize their saltpeter yield. It was through this innovation that formal chemical discourse entered into the domain of Swedish munitions production. The theories ranged from alchemical transformation frameworks to mineral growth theories, and helped fuel the debates over gunpowder procurement policy. By the end of the eighteenth century, the Swedes were invoking both phlogiston theories of chemical reactions and Lavoisier's theories of combustion and conservation of caloric subtle fluids. Such theories helped promote the initial investment and management of the saltpeter barns; in addition, they increased the sophistication of public policy debates and strengthened the macroeconomic discussions of what proportion of Sweden's manure should be devoted to agricultural versus military production. We thus see the Swedes pursuing the Enlightenment's utilitarian ideal of using scientific theory to orient public policy towards addressing national rather than individualistic interests.

The philosophes helped popularize the image of the enlightened public servant who enlists science to reform inefficient, if not corrupt, public institutions. This ideal emerges unmistakably in the English and French ordnance departments. Seymour Mauskopf describes the astonishing success of Antoine Lavoisier, the great French chemist, and William Congreve Sr., the innovative English artillery officer, who simultaneously improved the performance of their nations' gunpowder production facilities by the eve of the French Revolutionary Wars.[60] Lavoisier headed the new gunpowder administration in response to Turgot's reorganization efforts, while Congreve served as the Comptroller of the Royal Laboratory at Woolwich. Mauskopf offers a comparative analysis of their achievements with quality control, organizational reform, and scientific research. Although the basic medieval formula for gunpowder remained unaltered throughout this era, Lavoisier and Congreve nevertheless introduced a series of new processes that dramatically boosted its explosive strength. These included the generation of charcoal through iron-cylinder distillation and the experimental determination of optimum wood types. Other changes included the employment of screw presses in incorporating ingredients and the use of different gunpowder grain sizes in artillery and firearms.

Mauskopf observes how Lavoisier and Congreve relied on a variety of instruments, starting with the simpler mortar *éprouvette* and continuing with the sophisticated ballistic and gun pendulums. Although well versed in mathematics and mechanics, Congreve relied less on advanced chemistry. Instead he utilized the parameter-variation technique to guide his optimization efforts—a technique popularized by John Smeaton to double the efficiency of the Newcomen engine during the middle of the eighteenth century.[61] Lavoisier, on the other hand, achieved key breakthroughs by leveraging his formidable theoretical prowess. He dramatically increased the rate of French gunpowder production by substituting pure potash for the traditional wood ash in the processing of saltpeter. As Mauskopf argues, Lavoisier's experiment involving nitric acid in 1776 furnished the insights for this innovation. Subsidized, if not inspired, by Lavoisier's work in the gunpowder administration, this experiment also represented a crucial step in his discovery of combustion. Mauskopf thus illustrates the profound symbiosis of two aspects of Lavoisier: the theoretical chemist, who revolutionized chemistry with his discovery of combustion, and the powerful civil servant, who dramatically strengthened France's acquisitional power by ending its dependence on imported nitrates. Yet Mauskopf ultimately denies that the resulting powder represented a revolutionary breakthrough. What was revolutionary, he argues, was the formal research and development that Lavoisier and Congreve institutionalized within their acquisitional institutions. In short, both administrators were prime representatives of the Archimedean ideal. Lavoisier carried this especially far; his achievements in chemistry certainly matched those of Archimedes in geometry and mechanics, while his scientific leverage of the acquisitional power of France (not to mention that of the dependent American revolutionaries) possibly matched Archimedes' augmentation of Syracuse's military strength.

The essays in part IV address the military engineering and artillery forces of the Enlightenment. These constituted the most prominent technical military branches during the eighteenth century and were associated with some of the most sublime mathematical minds of that era, especially in France. The essays by Janis Langins and Brett Steele illustrate how engineering and artillery officers used advanced scientific knowledge when confronting the tactical issues of gunnery practice and fortification defense. This scientific mindset also guided their acquisitional and operational planning. These officers thus appear to be operating within the Archimedean tradition, although their scientific constructs and applications proved to be contentious. Returning to the sub-theme of the practical implementation of Enlightenment ideals, Langins and Steele show how the Renaissance synthesis of science and the art of war was intensified and expanded to vast proportions during the eighteenth century.

Langins first argues that the relationship between scientific knowledge and military engineering practice was less direct than historians have suggested. In spite of the vast theoretical output of such mathematical engineering luminaries as Bélidor, Bossut, Bézout, Coulomb, Monge, and Meusneir, the direct utility of scientific theory in fortification engineering diminished during the actual design process. Like a classic systems architect, Langins maintains that the simultaneous consideration of the tactical demands of a fortress's defense, the operational needs of an army, and the physical constraints of geography could not be reduced to a deterministic algorithm. He documents how Vauban railed against a supposedly deductive "science" of fortifications modeled after Euclidean geometry. The frustrating limitations of scientific analysis in fortification design are most evident in the raging polemics between Montalembert and the French military engineering corps in the late eighteenth century. Montalembert appears to have been inspired by the Swedish fortification system of casemates and multiple levels of artillery fire, as designed by Erik Dahlberg.[62] The engineering corps, by comparison, remained committed to the legacy of Vauban (an exact contemporary of Dahlberg), who emphasized the low-lying bastion design incorporating long-range crossfire and short-range flanking fire. Langins describes how Montalembert attacked the engineers for their abstract scientific rhetoric. Yet Montalembert was not averse to promoting his own scientific authority as a member of the Paris Academy of Science. Langins likewise exposes the irrationality of certain senior engineering officers who attempted to quantify the total-system performance of a fortress in order to discredit scientifically Montalembert's proposals.

The relationship between knowledge and power within the French engineering corps went beyond a rhetorical defense of bureaucratic authority and privilege. By examining Duportail's arguments for military reform, Langins shows how a senior engineering officer maintained that the talents of military engineers should not be restricted to the mundane operational issue of distributing cannons within a fortress. Instead, engineers should apply their scientific expertise to the higher operational concerns of coordinating the placement of fortresses and the maneuvering of armies. In effect, Duportail was arguing for a scientifically proficient corps of technical officers who could directly guide senior commanders. As with the fate of Montalembert's proposals, Duportail's advice was quickly dismissed by the French yet eventually embraced by the Prussians. It was hardly a coincidence that a science-minded artillery professor in Hanover, who fought in vain against the French Revolutionary armies, founded the Prussian General Staff—the first organization to institutionalize fully Duportail's recommendations.[63] Langins thus reveals how the roots of modern Western-command structures emerged from the early modern Archimedean tradition of unifying scientific theory and military practice.

Steele explicitly addresses the convergence of Enlightenment ideals of progress and military practice in the eighteenth century. He contends that this development was exemplified by the utilization of ballistics theory in artillery organizations throughout Enlightenment Europe. Yet the convergence of philosophical ideals and military practice was also evident in the global transfer of Western military education and the expansion of secular nationalism during the nineteenth century. In the middle of the eighteenth century, Benjamin Robins and Leonhard Euler transformed ballistics from a kinematic Galilean discipline, useful only for mortars, into a kinetic Newtonian science, useful for all gunpowder weapons. Steele suggests that their work had direct implications for the tactical domain. With their advanced experimental techniques and their approximate solutions to nonlinear differential equations, Robins and Euler dramatically enhanced the ability of officers to calculate artillery trajectories. Such scientific prowess helped satisfy the basic tactical objective of neutralizing enemy targets with minimal exposure to defensive responses. The great certainty such Newtonian ballistics offered also helped commanders resist the psychological pressure to squander ammunition at sub-optimum ranges or rely exclusively on canister fire. Even more crucial was the minimization of deadly technical errors when managing powerful heavy machinery under terrifying and exhausting conditions of combat. A rigorous education in Newtonian science therefore went far in transforming combat commanders into disciplined if not deterministic mechanisms highly suitable for hierarchical control. Yet, as Steele reminds us, their knowledge of Robins' and Euler's advanced ballistics also had considerable operational implications. Theories that could accurately model the interior, exterior, and terminal domains of a smooth-bore projectile's trajectory could also reduce the political costs of negotiating between the tactical demand for firepower and the operational demand for maneuverability. This issue lay at the center of the Piedmontese and Austrian artillery reform efforts during the War of the Austrian Succession, and reached ferocious polemical heights in France with the Gribeauval reform controversy following the Seven Years' War.[64]

Steele specifically describes the tactical relationship between Robins' and Euler's "ballistics revolution" and the teaching of calculus or fluxions in artillery academies. By detailing the gunnery tables computed by Lombard at the regimental artillery school of Auxonne, he demonstrates how such mathematical "software" could improve the tactical performance of artillery units. Such software also helped conserve shot and gunpowder, thereby lessening military operational and ultimately acquisitional pressures. Nevertheless, some may argue that master gunners during the sixteenth and seventeenth centuries did not need this "software" to achieve astonishing feats with their guns.[65] The key difference is that, with the institutionalization of calculus and Newtonian mechanics in the artillery academies, the Christian European states were able to "mass-produce" competent gunnery commanders with less practical training and combat experience. This was manifested by Napoleon, who maintained the performance of his artillery units to the bitter end by offsetting his heavy loss of veteran artillery officers with new recruits straight out of the combined programs of the École polytechnique and the École du Metz. Although the cadets were to be rushed through their studies, Napoleon nevertheless insisted that they be well versed in Robins' and Lombard's theories.[66]

The Ottoman Turks, by contrast, continuously suffered from the relatively lengthy apprenticeship required to generate master gunners—that is, until they too established scientific artillery schools based on the Christian model during the eighteenth century.[67] This ability to enlist scientific knowledge to "mass-produce" competent officers represents a powerful tactical advantage, especially in the wars of attrition that were so common in the eighteenth century.

Finally, Steele sheds light on a key relationship between scientific and technological innovations within the acquisitional domain. Such mathematicians as Bernoulli, Robins, Euler, and Lombard were not the only Archimedean figures associated with artillery in the eighteenth century. Steele reveals how Shrapnel and Villentroy, who were active-duty artillery officers on opposing sides during the Napoleonic Wars, furnished basic artillery hardware innovations within the scientific context of Newtonian ballistics.

It is tempting to conclude that the early modern Archimedean relationships between knowledge and power culminated in the midst of the French Revolutionary Wars with a mathematician like Lombard or an officer like Shrapnel. We must consider, however, one of Lombard's most scientifically capable artillery students at Auxonne, to whom he entrusted a series of full-scale ballistics experiments. That young Corsican officer would emerge from the French Revolution as a triumphant general and a scientific member of the Institute of France, before seizing dictatorial power as First Consul and overturning Europe as Emperor Napoleon I. Yet he also confessed that, had he not pursued a military career, he would have gladly settled for an academic appointment in theoretical mathematics.

Notes

1. John U. Nef, *War and Human Progress* (Harvard University Press, 1950), p. 325, as taken from a statement made by Saint-Simon in 1813 towards the end of the Napoleonic Wars. See Henri, Comte de Saint-Simon, *Oeuvres de Saint-Simon et d'Enfantin*, volume I (E. Dentu, 1865), pp. 54–55.

2. "Copie d'une lettre de M. Voltaire à M. de Saint Auban du 18 août 1776," Materie Militari, Int. gen. Artiglieri, Mezzo 2, No. 21 d'addizione. Archivio di Stato di Torino.

3. For a technical description of Archimedes' scientific work, see E. J. Dijksterhuis, *Archimedes* (Princeton University Press, 1987). For a broad overview, see G. E. R. Lloyd, *Greek Science after Aristotle* (Norton, 1973), 40–50.

4. Plutarch, *Makers of Rome: Marcellus*, volume 14 (Penguin, 1965), p. 102.

5. Polybius, *The Rise of the Roman Empire* (Penguin, 1979), pp. 365–367.

6. D. L. Simms, "Archimedes the Engineer," *History of Technology* 17 (1995), p. 68.

7. W. R. Laird, "Archimedes among the Humanists," *Isis* 82 (1991): 634–635. Also see Paul Lawrence Rose, *The Italian Renaissance of Mathematics: Studies on Humanists and Mathematicians from Petrarch to Galileo* (Droz, 1975).

8. For an argument that denies the existence of such an early modern Archimedean integration, see Alex Keller, "Technological Aspirations and Motivations of Natural Philosophy in Seventeenth-Century England," *History of Technology* 15 (1993): 76–92.

9. For a general discussion of classical catapults, see Barton C. Hacker, "Greek Catapults and Catapult Technology: Science, Technology, and War in the Ancient World," in *Technology and the West*, ed. T. Reynolds and S. Cutcliffe (University of Chicago Press, 1997).

10. For a detailed discussion of the analysis associated with ancient catapult design, see J. G. Landels, *Engineering in the Ancient World* (University of California Press, 1978), pp. 99–123.

11. For a thorough analysis of the operational domain, see Shimon Naveh, *In Pursuit of Military Excellence: The Evolution of Operational Theory* (Frank Cass, 1997).

12. The operational concept of "center of gravity," also known in modern military circles as *Schwerpunkt*, originally comes in a fairly unintelligible form from Clausewitz's *On War*, meaning an army's geographical center of strength or concentration of power. In the hands of Tukhachevsky and other Soviet developers of Deep Operation Theory before World War II, it signified the links in a military system that induces maximum systemic entropy or chaos when disrupted. Robert Leonhard defines it as an army's critical vulnerability on pp. 20–24 of *The Art of Maneuver* (Presidio, 1991).

13. Carl von Clausewitz, *On War* (Penguin Books, 1968). Curiously, this celebrated treatise of military strategy has remarkably little to say about the "acquisitional" domain of warfare, as far as resource transformation is concerned.

14. For an excellent discussion of systems analysis, see E. S. Quade, "Principles and Procedures of Systems Analysis," in *Systems Analysis and Policy Planning: Applications in Defense* (American Elsevier, 1968). Though systems analysis was a Cold War analytical approach, developed primarily at the RAND Corporation, Quade (p. 2) asserts that its practice has ancient roots.

15. Loren R. Graham, "The Socio-Political Roots of Boris Hessen: Soviet Marxism and the History of Science," *Social Studies of Science* 15 (1985): 705–722.

16. Boris Hessen, *The Social and Economic Roots of Newton's Principia* (H. Fertig, 1971). See, in particular, pp. 89–99 of *Science, Technology and Freedom*, ed. W. Truitt and T. Graham Solomons (Houghton Mifflin, 1974).

17. Robert K. Merton, *Science, Technology and Society in Seventeenth-Century England* (Harper & Row, 1970), pp. 191–198.

18. Henry Guerlac, Science and War in the Old Regime, Ph.D. dissertation, Harvard University, 1941.

19. Henry Guerlac, "Vauban: The Impact of Science on War," in *Makers of Modern Strategy from Machiavelli to the Nuclear Age*, ed. P. Paret (Princeton University Press, 1986).

20. For evidence of Guerlac's sensitivity to any Marxist affiliation, see Roger Hahn, "Changing patterns for the support of scientists from Louis XIV to Napoleon," *History and Technology* 4 (1987): 401–411. It has recently been argued that, because of his employment as a historian for the MIT Radiation Laboratory during World War II, Guerlac became disillusioned about the mundane, if not cynical, reality of scientific research, thus prompting him to focus more on idealistic aspects of the history of science. I am indebted to Michael Dennis of Cornell University for these insights.

21. Guerlac's unwillingness to develop his thesis on war and science is reflected in H. Floris Cohen's outstanding study of the historiography of early modern science, *The Scientific Revolution: A Historiographical Inquiry* (University of Chicago Press, 1994). Although Cohen summarizes Hessen's and Merton's positions on the role of military conflict in scientific research, as well as Leonardo Olschki's and Edgar Zilsel's broader discussions of the interaction between technological and theoretical developments during the Renaissance, there is no mention of Guerlac's early work.

22. A. Rupert Hall, *Ballistics in the Seventeenth Century* (Cambridge University Press, 1952).

23. Alexandre Koyré, "Galileo and Plato," in *Metaphysics and Measurement* (Harvard University Press, 1968) (original publication: *Journal of the History of Ideas* 4, 1943: 400–428), p. 17. Koyré conveniently forgot, however, that Descartes was a very practical student of military engineering at the camp of Maurice of Nassau in Breda before serving in the early campaigns of the Thirty Years' War. Only after failure to secure a senior military command did he dedicate himself to a formal career in natural philosophy. Galileo, likewise, received his first formal instruction in the mathematics of Euclid and the mechanics of Archimedes from Ostilio Ricci, a senior military engineer and court mathematician for the Medicis in the Grand Duchy of Tuscany.

24. A. Rupert Hall, "Engineering and the Scientific Revolution," *Technology and Culture* 2 (1961), no. 4, p. 335.

25. A. Rupert Hall, "Science, Technology, and Warfare, 1400–1700," in *Science, Technology, and Warfare*, ed. M. Wright and L. Paszek (Government Printing Office, 1971).

26. A. Rupert Hall, "Gunnery, Science and the Royal Society," in *The Uses of Science in the Age of Newton*, ed. J. Burke (University of California Press, 1983).

27. For endorsements of Hall's thesis, see J. D. Bernal, *Science in History* (Watts, 1954), p. 346; commentaries by J. R. Hale and J. B. Wolf in *Science, Technology, and Warfare*; Peter Mathius, "Who Unbound Prometheus? Science and Technological Change," in *Science, Technology and Economic Growth in the Eighteenth Century*, ed. A. Musson (Bungay, 1972), pp. 86–87.

28. Jerome Ravetz and Richard S. Westfall, "Marxism and the History of Science," *Isis* 72 (1981): 393–405.

29. Steven Shapin, *The Scientific Revolution* (University of Chicago Press, 1996).

30. Ibid., pp. 140–141.

31. Ibid., p. 141.

32. James E. McClellan III and Harold Dorn, *Science and Technology in World History* (Johns Hopkins University Press, 1999).

33. Ibid., p. 200.

34. David Waters, *The Art and Science of Navigation in Elizabethan and Early Stuart Times* (Hollis and Carter, 1958); Mario Biagioli, "Social Status of Italian Mathematicians," *History of Science* 27 (1989): 41–95; Mary Henninger-Voss, Between the Cannon and the Book, Ph.D. dissertation, Johns Hopkins University, 1995. Also see Henninger-Voss, "Working Machines and

Noble Mechanics: Guidobaldo del Monte and the Translation of Knowledge," *Isis* 91 (2000), June: 233–259.

35. Serafina Cuomo, "Shooting by the Book: Notes on Niccolò Tartaglia's Nova Scientia," *History of Science* 35 (1997): 155–188.

36. William H. McNeill, *The Pursuit of Power: Technology, Armed Force and Society since A.D. 1000* (University of Chicago Press, 1982); Carlo Chipola, *Guns, Sails and Empires: Technical Innovation and the Early Phases of European Expansion, 1400–1700* (Sunflower University Press, 1985).

37. *The Military Revolution Debate: Reading on the Military Transformation of Early Modern Europe*, ed. C. Rogers (Westview, 1995).

38. Michael Roberts, "The Military Revolution, 1560–1660," in *Essays in Swedish History* (Weidenfeld & Nicolson, 1967), pp. 211–212.

39. Examples include Geoffrey Parker, *The Military Revolution: Military innovation and the rise of the West, 1500–1800* (Cambridge University Press, 1988); *The Military Revolution Debate: Readings on the Military Transformation of Early Modern Europe*, ed. C. Rodgers (Westview, 1995); David Eltis, *The Military Revolution in Sixteenth Century Europe* (Barnes & Noble, 1995); Jeremy Black, *European Warfare, 1660–1815* (Yale University Press, 1994); Bert Hall, *Weapons and Warfare in Renaissance Europe: Gunpowder, Technology, and Tactics* (Johns Hopkins University Press, 1997).

40. Eltis, pp. 54–61; Black, pp. 44–45, 54–55.

41. Alex Roland, *Underwater Warfare in the Age of Sail* (Indiana University Press, 1978).

42. Erik Lund, *War for the Every Day: Generals, Knowledge and Warfare in Early Modern Europe, 1680–1740* (Greenwood, 1999).

43. This issue has received little attention from historians of mathematics. Nevertheless, it has been addressed by such architectural historians as Horst de la Croix ("Military Architecture and the Radial City Plan in Sixteenth Century Italy," *Art Bulletin* 62.4, 1960: 263–290), Catherine Wilkinson ("Renaissance Treatises on Military Architecture and the Science of Mechanics," in *Les traités d'architecture de la renaissance*, ed. J. Guillaume, Picard, 1988), and Kim Veltman ("Military Surveying and Topography: The Practical Dimension of Renaissance Linear Perspective," *Revista da Universidade de Coimbra* 27, 1979: 329–368).

44. Thomas Arnold, "Fascinating Angles: The Mental Place of Flanking Fire in Early Modern Warfare," presented at Clark Library Conference on War and Science in the Old Regime, 1998.

45. Galileo Galilei, *Operations of the Geometric and Military Compass*, ed. S. Drake (Smithsonian Institution Press, 1978); Eltis, pp. 54–57.

46. Luigi Ferdinando Marsili, *Stato militare dell'imperio Ottomanno* (Akadem. Druck-u. Verlagsanst, 1972). Originally published in The Hague and Amsterdam in 1732.

47. For a broad overview of the development and application of mathematical theory in artillery and military engineering through the Wars of Louis XIV, see Brett Steele, The Ballistics Revolution, Ph.D. thesis, University of Minnesota, 1994, pp. 27–63.

48. Stanford Shaw, *Between Old and New: The Ottoman Empire under Selim III, 1780–1807* (Harvard University Press, 1971).

49. Galileo Galilei, *Two New Sciences* (1638), tr. S. Drake (Wall and Emerson, 2000).

50. David McGee, "From Craftmanship to Draftsmanship: Naval Architecture and the Three Traditions of Early Modern Design," *Technology and Culture* 40.2 (1999): 209–236.

51. Steele, The Ballistics Revolution, pp. 113–115, 212–213.

52. Daniel Banes, "The Portuguese Voyages of Discovery and the Emergence of Modern Science," *Journal of the Washington Academy of Sciences* 78.1 (1988): 47–58; David Waters, "Columbus's Portuguese Inheritance," *The Mariner's Mirror* 78 (1992), no. 4: 385–405; David Goodman, "Philip II's Patronage of Science and Engineering," *British Journal for the History of Science* 16.52 (1983): 49–66.

53. David Waters, *The Art and Science of Navigation in Elizabethan and Early Stuart Times* (Hollis and Carter, 1958), p. 120.

54. Ibid., pp. 243–244, 411.

55. Peter French, *John Dee: The World of an Elizabethan Magus* (Routledge & Kegan Paul, 1972), p. 163.

56. Michael S. Mahoney, "Christiaan Huygens: The Measurement of Time and of Longitude at Sea," in *Studies on Christiaan Huygens* (Swets & Zeitlinger, 1980), p. 234. It is curious that Constantine Huygens was deeply offended when Christiaan received a letter from a French minister that addressed him as "Mr. Huygens, Mathematician." The father responded, "I did not think that among my children I had artisans; he (i.e., the French minister) appears to think (Christiaan) is one of his fortification engineers." See *A History of Science in the Netherlands*, ed. K. van Berkel et al. (Brill, 1999), p. 55.

57. For a detailed description of Robins's analysis of the ballistic pendulum, see Steele, The Ballistics Revolution, pp. 87–91.

58. Lewis Mumford, *Technics and Civilization* (Harcourt, Brace, 1934), pp. 91–92. For a general discussion of the mechanistic nature of the army of Maurice of Nassau, including the use of standardized artillery with interchangeable parts, see Barry H. Nickle, The Military Reforms of Prince Maurice of Orange, Ph.D. dissertation, University of Delaware, 1975, pp. 132–136.

59. William Bourne, *The Art of Shooting in Great Ordnaunce, London 1587* (Da Capo, 1969), pp. 6–7.

60. Congreve is also known for his invention of the lightweight artillery carriage that the British deployed during the Napoleonic Wars.

61. Walter G. Vincenti, *What Engineers Know and How They Know It: Analytical Studies from Aeronautical History* (Johns Hopkins University Press, 1990), pp. 138–139.

62. Christopher Duffy, *The Fortress in the Age of Vauban and Frederick the Great, 1660–1789* (Routledge & Kegan Paul, 1985), pp. 182–197.

63. Charles Edward White, *The Enlightened Soldier: Scharnhorst and the Militarische Gesellschaft in Berlin, 1801–1805* (Praeger, 1989).

64. Howard Rosen, The Système Gribeauval: A Study of Technological Development and Institutional Change in Eighteenth Century France, Ph.D. thesis, University of Chicago, 1981. Also see Ken Alder, *Engineering the Revolution: Arms and Enlightenment in France, 1763–1815* (Princeton University Press, 1997). Both of these histories of the Gribeauval artillery regime, however, overlook the highly innovative artillery developments in Piedmont-Savoy and Austria that commenced during the War of the Austrian Succession, which Gribeauval effectively imitated—at least as far as the hardware is concerned. These included the use of interchangeable parts in gun carriages and the machining of cannon bores.

65. Simon Pepper and Nicholas Adams, *Firearms and Fortifications: Military Architecture and Siege Warfare in Sixteenth-Century Siena* (University of Chicago Press, 1986), pp. 136–137.

66. Steele, The Ballistics Revolution, 241.

67. Niyazi Berkes, *The Development of Secularism in Turkey* (Routledge, 1998), pp. 56–63. Also see Ekmeleddin Ihsanoglu, "Changes in Ottoman educational life and efforts towards modernization in the 18th-19th centuries," in *The Introduction of Modern Science and Technology to Turkey and Japan* (International Research Center for Japanese Studies, 1998), pp. 120–123.

I

The Global Development of Gunpowder Weaponry

1

Facing the New Technology: Gunpowder Defenses in Military Architecture before the *Trace Italienne*, 1350–1500

Kelly DeVries

In *The Medieval City under Siege* (1995), I published an article, "The Impact of Gunpowder Weaponry on Siege Warfare in the Hundred Years' War," which I would now like to subtitle "A Preliminary Study."[1] This essay is a follow-up.

The thesis of my earlier article was simple: Gunpowder weapons "had, by the end of the Middle Ages, completely altered siege warfare."[2] Yet my method for proving this was unorthodox. Whereas other historians have focused on the offensive capabilities of gunpowder weapons, most importantly on the speed of fortification capture,[3] I concentrated on the changes in fortification design and construction inspired by the use of gunpowder weapons against them during the same period. There is a precedent, however, in Renaissance history for my unconventional means of relying on military architecture as a gauge for broader military change. In *The Military Revolution*, Geoffrey Parker argues that the diffusion of *trace italienne* fortifications led to an increase in the size of armies during the sixteenth century. Because these were shorter and thicker fortifications that employed angular bastions, or gun platforms, their engineers sought to exploit fully both their structural resistance and their active artillery.[4] Rather than pursuing the origins for the *trace italienne* system described in Parker's thesis, my "Preliminary Study" established that other gunpowder weaponry systems predated the *trace italienne* and actively protected fourteenth- and fifteenth-century fortifications against new siege technology. Such gunpowder systems comprised not only gunports, boulevards, artillery towers, and other structural additions to existing fortifications but also entirely new fortification constructions. I therefore concluded that the move to the *trace italienne* signified an evolution from medieval military engineering, not the sixteenth-century military revolution that Parker and others have claimed.

In this essay, I will center on France and the southern Low Countries during the fifteenth century. Nowhere in Europe during this century was there more fierce or constant warfare, and nowhere were there more sieges undertaken with gunpowder weapons. (This continual use of gunpowder weapons at sieges and, to a lesser degree, in battles may explain why these weapons in Western Europe evolved into a military technology that was so much more sophisticated than that found in other civilizations.[5]) Other studies, including Parker's text and my earlier article, have focused on England and Italy, lands that did not suffer the degree of siege warfare experienced by France and the southern Low Countries during the fifteenth century. These studies have overlooked the fact that within those embattled western European lands one can find all of the evolutionary stages of pre- and non-*trace italienne* fortification.[6]

My "Preliminary Study" traced many of the origins of non-*trace italienne* gunpowder fortification to the fourteenth century and to England. I now conclude that this was the case for two reasons. First, a constant fear of invasion seems to have existed in England during the fourteenth century. This was not an irrational fear: French and French-allied ships often raided the southern and eastern coasts of England during the early part of the Hundred Years' War; and in the north, the Scots, also allies to the French, constantly crossed into England and pillaged lands as far south as York.[7] The second reason is that Edward III (1328–1377), like no one else of his time, foresaw the future of gunpowder weapons. As such, he was responsible for a large increase in the production and offensive use of gunpowder weapons, and he took them with him in his many conquests during the early stages of the Hundred Years' War; here they appeared both at sieges and in battles. Yet his concern was not entirely offensive. He also saw the need to protect his own fortifications against the increased use of guns in sieges. Edward added firearms to several vulnerable fortifications of England by piercing their walls with gunports, a practice continued by his grandson, King Richard II. By 1390, gunports had been installed in Quarr Abbey on the Isle of Wight, Queensborough Castle, Asseton's Tower at Porchester, Carisbrooke Castle, Canterbury's town wall, Cooling Castle, Southampton Castle and its town wall, Saltwood Castle, Norwich's town wall, Dover Castle, Bodiam Castle, and Winchester's town wall.[8] Edward III also may have been the first to construct a purposely built gunpowder artillery fortification with Queensborough Castle. This fortification, constructed in the south of England during the 1360s, had a completely round shape, making it more resistant to gunfire, and was outfitted with a large number of gunpowder weapons. (Unfortunately, Queensborough has not survived; therefore, it is impossible to ascertain the purpose of its anti-gunpowder artillery.[9])

The development of gunpowder weaponry fortifications in France and the southern Low Countries during the fourteenth century was much slower than that in England. A castle in Bioule, France, may have been the first to be defended by gunners who fired through arrowslits that had been widened to permit gunfire, as recorded to have occurred on March 18, 1347.[10] Such an innovation, however, did not effect a change in fortification construction policy both there and in the southern Low Countries. Why this was so is not known, nor is it known why France and the southern Low Countries did not follow the same pattern of gunpowder weaponry development or fortification adaptations that England did. Certainly neither area had the comparable leadership of an Edward III. On the other hand, both regions suffered far more military activity in the fourteenth century, especially in the form of sieges, than did England. Large cities, such as Cambrai, Tournai, Caen, Calais, Rennes, Reims, Ghent, Oudenaarde, and Ypres, all suffered sieges during this century, as did smaller fortified sites such as Barfleur, Cherbourg, Valognes, Carentan, Saint-Lô, Roche-Derrien, Auberoche, Evreux, Breteuil, La Rochelle, Saint-Saveur-le-Vicomte, and Odruik.[11] The fear of invasion should also have been present in France and the southern Low Countries, one would think, even more so than in England, based on the number of sieges which the inhabitants of those lands suffered. At most of these sieges there was also gunpowder weaponry. Although not effective enough to bring down a wall or gate early in the century, these weapons were succeeding in breaching fortification walls by the end of the fourteenth century.[12] Why then did France and the southern Low Countries not develop new gunpowder weaponry fortifications in the fourteenth century? I have contended that this may stem from the local control over gunpowder weapons and fortifications in these regions during the fourteenth century, by contrast with England's central, royal control[13]; however, more research is needed.

In terms of gunpowder weaponry fortifications during the fifteenth century, England exchanged places with France and the southern Low Countries. The people in England seemed to have lost some concern with the protection of their fortified sites against gunpowder weapons, while those in France and the southern Low Countries paid much more attention to outfitting existing fortifications with gunports, as well as to developing other gunpowder fortification elements. Why the English lost interest in adapting fortresses with guns is not well established. There was a lack of foresight among fifteenth-century English military leaders, however, especially after the death of Henry V in 1422. This resulted in the decline of English military successes not only during the Hundred Years' War but also during the Wars of the Roses, which were fought almost entirely on the battlefield and with relatively few gunpowder weapons.[14]

During the fifteenth century the threat of invasion in England apparently declined. The French navy had been effectively destroyed at the battles of Sluys in 1340 and La Rochelle in 1372, and elsewhere. It would not recover enough to pose even a raiding threat, let alone an invasion, until well into the sixteenth century. The Scots, too, reduced their border raids after 1400. Almost all Scots with any military ability chose to fight with the French armies on the continent rather than against the English on the British Isles.[15]

The shift from royal to local control of gunpowder weapons and fortifications may signify the most important reason for the lack of gunpowder fortifications in England during the fifteenth century.[16] In effect, the expensive acquisition of gunpowder weapons, as well as the construction of gunpowder fortification elements, had to be borne by the individual and not the state. Unlike the French and the Flemish, the English state apparently was unwilling to pay for such advanced fortification elements, and English individuals could not afford them. William Worcestre, who toured England from 1479 to 1480, observed very few fortifications that had altered their defense in response to gunpowder weapons; numerous traditional fortifications lay derelict or in ruin.[17] Only a man with the wealth of a William, Lord Hastings, was able to afford the gunpowder fortification construction found throughout France and the southern Low Countries. Yet his attempt at Kirby Muxloe Castle remained unfinished after Richard III dramatically murdered him.[18]

In contrast to the English royalty and nobility, the rulers in France and the southern Low Countries initiated an active policy for altering or constructing gunpowder fortifications during the fifteenth century. What prompted this change cannot be determined, although it might be traced to the end of the fourteenth century, when Charles V ruled France and Philip the Bold acquired the principalities of the southern Low Countries. Like the earlier Edward III, both of these wise rulers recognized the future of gunpowder weapons and sought to have such weapons built for offensive purposes. They may also have sought to fortify their existing fortifications against gunpowder weapons. Nonetheless, such changes did not occur until the reigns of their successors: Kings Charles VI and Charles VII of France and Dukes John the Fearless and Philip the Good of Burgundy.[19]

Structural changes to the fortifications of France and the southern Low Countries may also have been spurred on by the conquests of John the Fearless during the Armagnac-Burgundian civil war that ravaged a large part of France between 1407 and 1419, and by Henry V's invasion of France from 1415 to 1420. Both of these military leaders preferred sieges to battles, and both used a large number of gunpowder weapons. Because of this, many French fortifications fell quickly to Duke John and King Henry. For example, at the end of 1407 and the beginning of 1408, John the Fearless used his gunpowder weapons at the siege of Maastricht.[20] In September 1409, John besieged

Vellexon, and by January 22, 1410 its walls had been destroyed by mining and gunfire.[21] In 1411, with gunpowder artillery, the Duke both besieged Rougemont and achieved a rapid victory in the conquest of Ham.[22] In 1412, John besieged Bourges.[23] In 1414, John used artillery against the town of Arras,[24] and, in 1417 and 1418, he used his guns successfully against Paris.[25] Henry V experienced a similar success. His siege of Harfleur in 1415 is well known largely because of Shakespeare's celebration of it. This siege succeeded only when Henry abandoned his mining of this fortification, as his mines were continually being countermined; he relied instead on his guns to breach the walls.[26] In 1417, Henry started anew with his conquest of Normandy, a conquest achieved with unprecedented speed. His first target was the castle of Bonneville; unrelieved, its garrison surrendered on August 9. The nearby castle of Auvilliers also fell to him two days later. Henry V then moved towards Caen, capturing Lisieux and Bernay along the way. Caen was a well-fortified town, with strong walls, a strong castle, and a larger garrison than Henry had previously encountered. Yet, without a relief force to aid in its defense, it fell on September 20. On the heels of this conquest, Henry occupied Bayeux, Tilly, Villers Bocage, and other nearby towns and villages, all of which surrendered to him without resistance. Moving southwards, Argentan and Alençon also fell quickly and with little conflict. Falaise was stronger, given its castle, walls, and large garrison. Still, the town only persevered until January 2, 1418, with the castle surrendering a month later.[27]

In spite of the winter, Henry pressed on with his invasion. Targeting the Cotentin, before the winter was out, he had attacked Saint-Lô, Carentan, Valognes, Cherbourg, Coutances, Avranches, Domfront, and Saint-Saveur-le-Vicomte. Only Cherbourg, Domfront, and Mont-Saint-Michel did not fall before the spring; Cherbourg and Domfront were captured in the summer, while Mont-Saint-Michel successfully held out.[28] As the spring of 1418 arrived, Henry moved toward Rouen, the capital of Normandy. By April he had captured all of lower Normandy, and in June and July he besieged and reduced the fortified Seine bridge crossing at Pont-de-l'Arche. Rouen would last somewhat longer, enduring a six-month siege; but in the end it also fell to the English army.[29] The English entered Rouen on January 19, 1419. The fortifications of Arques (later known as Arques-la-Bataille), Lillebourne, Vernon, Mantes, Neufchâtel, Dieppe, Gournay, Eu, Fécamp, Tancarville, and Honfleur quickly followed, all of which fell to the English before the end of February 1419, as did those of Gisors, Ivry, La Roche-Guyon, Pontoise, Meulan, Poissy, Saint-Germain, and Château Gaillard. These held out a little longer but were still captured by the end of the year.[30] After 1420, Henry's progress slowed largely due to his new involvement in French political affairs: Paris fell to

Henry's ally, John the Fearless, in 1418; the Treaty of Troyes granted Henry a wife in Catherine of Valois, Charles VI's daughter, in 1420; and the inheritance of the French throne upon Charles' death was negotiated between Henry and Charles. There was also a reinvigoration of the Armagnac opposition, led by Henry's brother-in-law, the dauphin (later Charles VII).[31]

Gunpowder weapons undoubtedly enabled John the Fearless and Henry V to capture so many fortifications so quickly. Allow me to cite one example I have used before to describe how fast a fortification could fall to this new technology. At the siege of Ham, conducted by John the Fearless in 1411, only three shots were fired from the bombard known as "Griette." The first passed over the town and fell into the Somme; the second hit the ground in front of the fortification but still destroyed a tower and two adjacent walls; and the third, also falling short, made a breach in the wall itself. Before a fourth shot could be fired, the fortification capitulated.[32] (I will return to the siege of Ham below, as the townspeople and their new Burgundian lord, obviously impressed by this gunpowder onslaught, made considerable changes to their fortification's defensive capability in order to improve their resistance to a similar attack.) But not all French fortifications fell in such a manner. In fact, most were never attacked at all: they simply agreed to surrender should a relief army not arrive by a given date. The small size of a fortification garrison was never intended to withstand an army. The garrison only hoped to resist until relieved by an army whose numbers could effectively counter those of the besiegers. Without a French relief army, surrender came quickly.

It is true that the time required to capture a fortification was greatly diminished by gunpowder weapons between 1370 and 1420; however, in many sieges after 1420 the time required to reduce a fortification returned to that of the pre-gunpowder era.[33] Here we also find many fortifications holding out against a protracted siege until the besiegers had retreated. Defenders in Mont-Saint-Michel, Orléans, Tournai, Vaucouleurs, Calais, Neuss, and many other locations refused to surrender, persevering until the besiegers had run out of money, supplies, or patience, or until a relief army had freed them from their attackers, as at Orléans. Sometimes this was due to the valor of the defenders, as at Mont-Saint-Michel and Neuss, or the valor of both the defenders and the relievers, as at Orléans. It could also be due to the strategic or tactical shortsightedness of the besiegers. A prime example of the latter is a quote from Philippe Commynes concerning the siege of Beauvais in 1472: "My Lord of Cordes ... had two cannons which were fired only twice through the gate and made a large hole in it. If he had more stones to continue firing he would have certainly taken the town. However, he had not come with the intention of performing such an exploit and was therefore not well provided."[34]

Even with the assistance of gunpowder weapons, medieval sieges hardly involved formalized or systematic operations. As with the different styles of fortifications, fit to the local terrain, different tactics existed to resist them. It remains uncertain whether a causation exists between the increased length of time required to defeat some of these fortifications after 1420 and the new gunpowder defenses which France and the southern Low Countries had begun adding to fortifications mostly after that time. Nonetheless, two facts can be determined with certainty: first, these fortifications did begin to receive new gunpowder defenses in greater numbers from 1420 to 1480; second, many of those that did receive these defenses had fallen to siege armies outfitted with gunpowder weapons during the period 1370 –1420. In addition, Duke Philip the Good prophetically feared that his duchy of Burgundy, which had previously been secure, could be threatened by gunpowder weapons in the future.

In France and the southern Low Countries during the fifteenth century, four basic gunpowder fortification defenses were constructed: gunports, boulevards, artillery towers, and pre-*trace italienne* and non-*trace italienne* artillery fortifications. Although not necessarily chronological in their construction, these represent the four evolutionary stages of gunpowder defenses that preceded and perhaps inspired the *trace italienne*.

Gunports

The first of these was the gunport. Although a French castle may have had the first gunports, which were recorded to have been placed in Bioule Castle in 1347, these apparently had no influence on later gunport construction in France and the southern Low Countries. Only in the town wall of Mont-Saint-Michel and at the castles of Saint-Mâlo and Blanquefort are gunports confirmed before 1400.[35] Others were built after 1412, with Paris receiving them in 1415 and Rennes in 1418.[36] Still, prominent voices in French political and military society did call for this important gunpowder defense; for instance, Christine de Pisan, in her *Faits d'armes et de chevalerie* (c. 1410), acknowledged their need.[37]

After 1420, gunports proliferated throughout France. In a detailed study of gunports in extant French fortifications, Jean Mesqui established that 168 fortified sites had acquired gunports during the fifteenth century, and most before the end of the Hundred Years' War in 1453. Additionally, these gunport-equipped fortifications can be found throughout France, especially in areas plagued or threatened by warfare.[38] Although not as thorough as Mesqui's survey of France, work by Alain Salamagne shows a similar number and distribution of gunports throughout the southern Low Countries.[39] By comparison, a gazetteer of English gunports compiled by John Kenyon shows only 36 fortified sites containing gunports before the end of the fifteenth century.[40] Most of these contain only a small number of gunports in each fortification. Numbering fewer than seven, they protected the most vulnerable spots. For instance, York town walls have gunports principally at the river crossings, while Bodiam Castle concentrates them only at the main gate. Only the incomplete Kirby Muxloe Castle and the Southampton town walls provide exceptions.[41]

Gunports in France and the southern Low Countries appear as arrowslits adapted for the use of firearms, but more often as wholly new constructions. These new gunports penetrated both newly built and older fortifications. At one time it was believed that gunports adapted from arrowslits were older than those newly constructed as gunports alone.[42] But this hypothesis must now be discounted. Gunports were constructed in whatever way was easiest for the builder: if a gunport was to be built where a suitable arrowslit existed, that arrowslit would be adapted. A new gunport was installed if no arrowslit existed; if no arrowslit was found to be suitable in an older fortification, which was common given the different functions that a bow and a gun performed; or if the wall structure in which the gunport was to penetrate was itself newly constructed. New gunports were especially evident in the artillery towers of the fifteenth century; most had little or no resemblance to arrowslits.

The easiest way to adapt an arrowslit for firearm use is to enlarge the slit at the point where the gun was to protrude. Most often this occurred at the bottom of the slit, if the thickness of the wall or the nearby floor could be used for mounting the weapon. Less frequently, it occurred in the middle of the slit when a wooden or iron bar was installed to permit the mounting of a "hook gun."

For the newly constructed gunport, there was no fixed style of construction. Many diverse styles have been documented throughout France and the southern Low Countries, but little justification is given for any style; the technology of the guns or fortifications seems to have made little difference. In fact, at the castle of Tonquédec in France, built around 1472, different gunport styles exist side by side, all apparently constructed at the same time by the same builders.[43] By contrast, at the castle overlooking Dieppe, one tower constructed in the first half of the fifteenth century contains gunports of one style, while a later tower, built during the reign of Louis XI, contains gunports of an entirely different style.[44]

In the adapted arrowslits, the slit extending above the gunport served as a sighting device. Whereas some newly constructed gunports were built with a sighting slit, others were built as a port alone, with no such sighting device.[45]

Figure 1.1
Gunport (outside), Bruges Town Wall Gate Tower.

It should also be noted that all of the extant fifteenth-century gunports are small in size. There are no gunports made for large or even medium-sized gunpowder weapons. Thus, the guns used in these gunports would also have been small. To date, no one has conducted a formal study of the gunpowder weapons that would have been used in these early gunports. Likewise, no one has ever identified a gunpowder weapon as a "fortification piece," although Robert D. Smith and David Starley (both of the Royal Armouries Conservation Department) and I have recently inspected and weighed four of the Royal Armouries "handguns" dated to the fifteenth and sixteenth centuries. We discovered that two of these weapons, one bronze (Royal Armouries XII.959) and one iron (Royal Armouries XII.960), respectively weighed 16.45 and 16.30 kilograms without their stocks. By comparison, the other

Figure 1.2
Gunport (inside), Caen Castle Artillery Tower, shelf mount.

two weapons in the collection, again one bronze (Royal Armouries XII.1787) and a second iron (Royal Armouries XII.3748), weighed 4.75 and 5.85 kilograms. Clearly, the heavier pieces can be held in the hands of an operator; however, after prolonged use, their weight may cause undue fatigue. Their heaviness might be better explained by recognizing their application as fortification weapons.[46]

Although the type of guns used in these gunports has not been established, their mounts can be identified by archaeological remains. These remains indicate that the guns which filled these ports were mounted on beds that were then laid on a stone platform or sill built only a few centimeters below the gunport. This installation is exemplified at Crèvecour and Caen Castles, and at most other fortifications outfitted with gunports. In a few gunports, firearms

Figure 1.3
Gunport (inside), Beersel Castle, bar mount.

were mounted by a hook attached to a wooden or iron bar transfixing the port. This may have been the origin for the name "hook guns" (*hackenbüsches, hakkenbussen, hacquebutes*), which later became the common term for hand-held guns of the early modern age.[47]

But how effective were these gunports? Flanking fire was arranged for arrowslits as early as the late thirteenth century; the Welsh castles built under the direction of Edward I demonstrate this.[48] Peter Jones and Derek Renn have shown that arrowslits from the same period minimized, if not eliminated, those "dead zones" that rendered a medieval fortification vulnerable to attack.[49] Jean Mesqui has confirmed that arrowslits, and later gunports, provided similar protection for a French fortification at the end of the Middle Ages.[50] Above all, the most vivid proof of the effectiveness of gunports must be the number of late

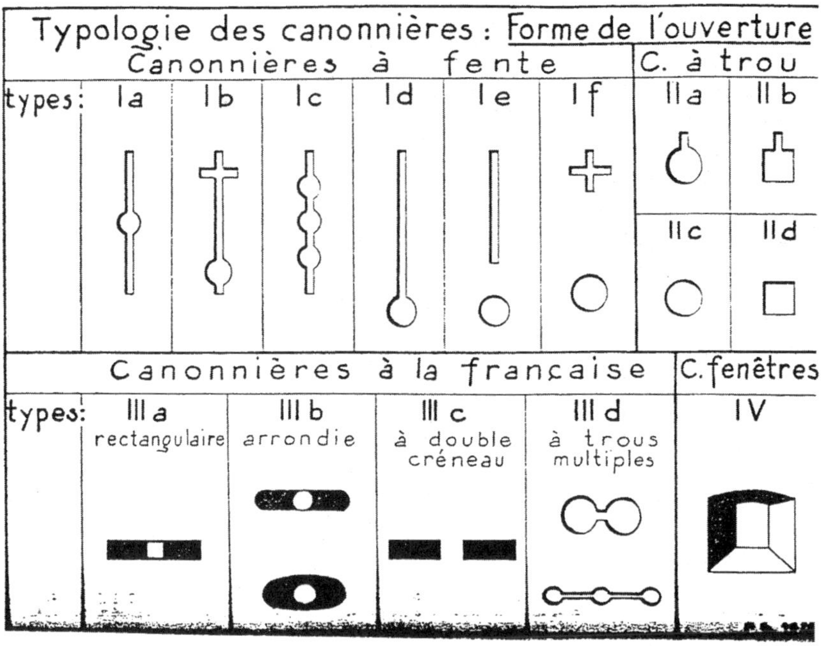

Figure 1.4
Typology of gunports. Source: Pierre Sailhan, "Typologie des archères et cannonières: Les archères des châteaux de Chauvigny," *Bulletin de la société des antiquaries de l'Ouest et des musées de Poitiers*, fourth series, 14 (1978), p. 522.

medieval fortifications in France and the southern Low Countries that feature gunports either newly constructed or adapted for gunpowder weapon use; where there is a supply there must be a demand, as economists like to say.

Boulevards

The boulevard represented the second gunpowder fortification defense constructed in France and the southern Low Countries during the fifteenth century. A low earthwork defense, it was generally placed before a vulnerable gate or wall. In essence, as a gunpowder artillery platform, its defense derived from a large number of guns, a low height (over which it was easier to fire), and earth and timber walls (which more readily absorbed the impact of any attackers' stone and metal cannonballs).[51] The boulevards were principally French, although the English knew how to build them and did so in some of their occupied sites in France, notably Poitiers, Orléans, and Cadillac.[52] Yet no evidence exists that the English transferred them to their home soil.

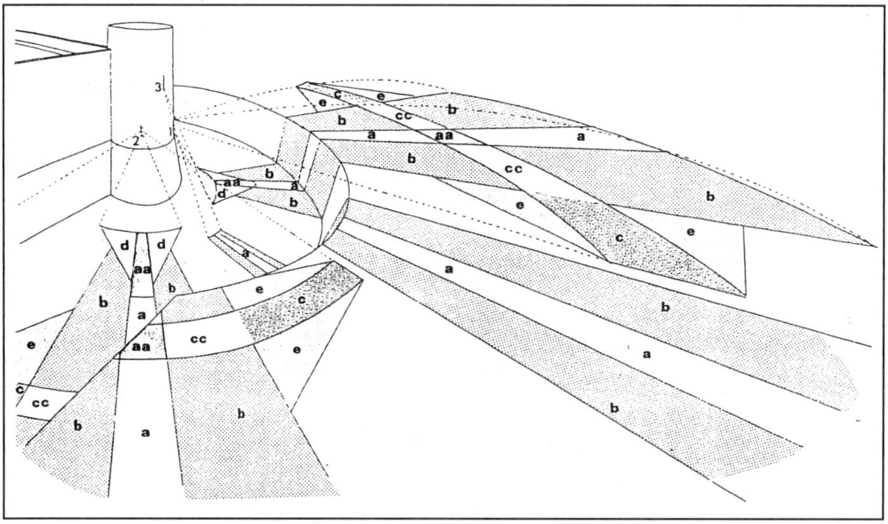

Figure 1.5
Areas of fire from gunports in artillery towers. Source: Jean Mesqui, *Châteaux et enceintes de la France médiévale: De la défence à la résidence* (Paris, 1991), volume II, p. 289.

In France and the southern Low Countries during the fifteenth century, boulevards were very popular, with many fortifications having one or more to augment their defense. The list of these fortifications is quite long, although not nearly as long as that of similar fortifications outfitted with either gunports or artillery towers. Additionally, most fortifications with boulevards in France and the southern Low Countries also had gunports.[53]

The boulevard also emerged as an early and relatively inexpensive gunpowder defense. Perhaps in France and the southern Low Countries it was even contemporaneous with the more famous gunport, which also may have derived from the same push for new defenses against gunpowder weapons. But exactly when and where the first boulevard was built is difficult to establish. Because the early boulevard was constructed solely from earth and wood, none survives. To date, little archaeological work has occurred at any boulevard sites. Written records provide only a bit more information. The earliest boulevard recorded was in 1407 at Liège in defense against the Burgundians.[54] In the southern Low Countries, others appear in the same year at Barbençon, Braine, Brussels, and Leuven.[55] Early boulevards are recorded at Quesnoy in 1414, at Douai from 1414 to 1415, and at Lille in 1416, with more in the 1420s and 1430s.[56] However, these dates should not establish the original boulevard, since the earliest records noting the existence of these boulevards imply that this was

Figure 1.6
Boulevard, Pierrefonds Castle (earth and wood later remade in masonry).

a standard defense. These dates correspond with the attacks made by the Burgundian duke, John the Fearless, in his military conquests, first against the southern Low Countries and then against France. As such, the issue is not the boulevard's inception but its diffusion as a gunpowder defense.

The most famous boulevards were those the English built (or assumed) in their siege of Orléans from 1428 to 1429. Joan of Arc would later attack and capture them for the French. Yet, because of a shortage of manpower (numbering fewer than 4,000), Thomas Montagu, the Earl of Salisbury and commander of the Anglo-Burgundian siege, could not surround Orléans.[57] Instead, he manned only a few boulevards located some 600 meters from the city walls along the western side of the town, one at the bridgehead in front of the Tourelles, a second one to the north, a third on an island in the Loire River to the west of the town, and a fourth (the boulevard of Saint Loup) along the eastern road from Orléans to Jargeau (this last one located some two kilometers from the Orléans walls).[58] These imposing structures generally contained relatively few soldiers; however, when occupied with gunners and longbowmen they could be taken only through costly frontal assaults.

The most formidable of these was the first boulevard, which the English built in front of the fortified bridgehead over the Loire. The capture of this Tourelles bridgehead had occupied much of the English effort, costing

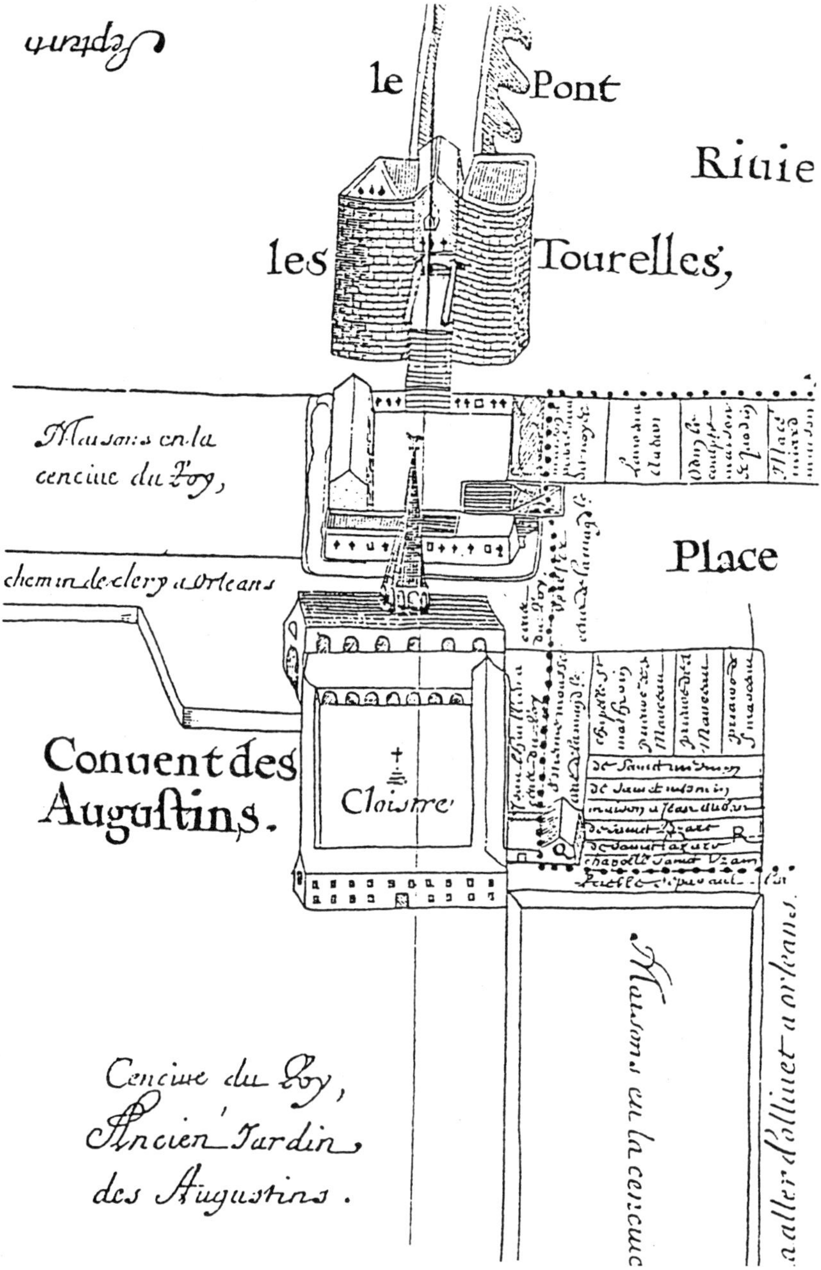

Figure 1.7
Boulevard of the Tourelles (across the river from Orléans) from a plan made in 1676. Source: Jacques Debal, "La topographie de l'enceinte fortifiée d'Orléans au temps de Jeanne d'Arc," in *Jeanne d'Arc: une époque, un rayonment*, Colloque d'histoire médiévale, Orléans—Octobre 1979 (Paris, 1982), p. 37.

Salisbury his life. Because it remained the focal point of the English siege, they were determined to protect it. Thus, upon its capture, the English constructed a large boulevard over the earthwork that they had earlier captured outside of the Tourelles.[59] This boulevard, called either the "boulevard of Tourelles" or the "boulevard of the Augustins" in contemporary sources, was imposing, despite its earth-and-wood construction. According to Régine Pernoud, it measured 20 meters in length and 26 meters in width, with a ditch surrounding it 8 meters in depth.[60] At its completion, the English commanders who succeeded Salisbury supplied the boulevard with a large number of men and gunpowder weapons, perhaps the majority of English resources at Orléans.[61]

For Joan of Arc to relieve the siege of Orléans, she had to seize the boulevard of Tourelles. This was not to be an easy task, but it would secure Joan's legacy. The following account of one of the bloodiest military engagements of the Hundred Years' War comes from the *Journal du siège d'Orléans*:

> Early in the morning on the day after, which was Saturday, the seventh day of May, the French attacked the Tourelles and the boulevard while the English were attempting to fortify it. And there was a spectacular assault during which there were performed many great feats of arms, both in the attack and in the defense, because the English had a large number of strong soldiers and had strengthened skillfully all of the defensible places. And also they fought well, notwithstanding that the French scaled the different places adeptly and attacked the angles at the highest of the strong and sturdy fortifications so that they seemed by this to be immortal. But the English repulsed them from many places and attacked with artillery both high and low, both with cannons and other weapons, such as axes, lances, pole-arms, lead hammers, and other personal arms, so that they killed and wounded many Frenchmen.[62]

In the midst of the battle, Joan was wounded, but this did not stop her from carrying on the battle. Nor did her wound stop her from intensifying the attack when other leaders, including the Bastard of Orléans, the commander of all French forces, became fatigued and wished to retreat from the fight to rest until the following day. The Bastard would later testify:

> The attack lasted from early morning until the eighth hour of vespers [eight o'clock in the evening], so that there was almost no hope of victory on this day. On account of this, this lord [the Bastard of Orléans] chose to break it off and wanted the army to retreat to the city. And then the Maid came to him and requested that he wait for a little while, and at that time she mounted her horse, and retired alone into a vineyard at a distance from the crowd of men. In this vineyard she was in prayer for a space of seven minutes. She

> returned from that place, immediately took her standard in her hands, and placed it on the side of the ditch. And instantly, once she was there, the English became afraid and trembled. The soldiers of the king regained their courage and began to climb [up the ramparts], making an attack on those against the boulevard, not finding any resistance. And then the boulevard was taken, and the English in it were put to flight.[63]

Joan collaborated this testimony when she testified that she "was the first to put her ladder on the boulevard of the Tourelles."[64]

The above testimonies illustrate the difficulty in conquering a boulevard, even when supported by gunpowder weapons, which Joan of Arc had in relative abundance at Orléans.[65] But these gunpowder defenses were only effective when occupied with men and guns. Most of those boulevards at Orléans were not. For example, Joan had assaulted and captured the undermanned and underarmed boulevard of Saint Loup a few days earlier (May 4) with only minimal French casualties.[66] Even the boulevard of Tourelles proved vulnerable against its most determined foe, Joan of Arc, who seemed unconcerned about how many lives such an attack would cost.[67]

By the 1440s and later, some boulevards had begun to be built with stone, thus acquiring a more permanent defensive capability. It cannot be ascertained when this commenced, although records show that two of the most threatened sites of the fifteenth century, Rennes and Mont-Saint-Michel, acquired stone boulevards by 1440, while similar structures were constructed at Loches in 1480 and Lassay from 1485 to 1489.[68] Could the stone boulevard be the origin of the late-fifteenth-century or early-sixteenth-century bastion? The similarity between these earlier defenses and the Italian bastions indicates an evolution from the medieval gunpowder defense toward the *trace italienne* fortress.[69]

Artillery Towers

The next chronological development in gunpowder defense was the artillery tower, which comprised very thick walls (usually more than 2 meters thick) of various shapes—circular, rectangular, polygonal, or spur/"U-shaped."[70] Artillery towers either remained free-standing or, more frequently, flanked the walls of previously constructed fortifications. Here their sole purpose was to support earlier fortifications that were deemed vulnerable to the new besieging gunpowder weapons. To perform this function, these towers were equipped with gunports—sometimes in large numbers and most often extending at least two storeys—and were armed with gunpowder weapons. Frequently they provided flanking fire.[71] According to Jean Mesqui, because these towers protruded from their fortifications, they forced besieging gunners to turn their gunfire away from the more vulnerable castle walls.[72] Unlike boulevards, artillery towers were also known in the British Isles, appearing in England as the Cow Tower on the Norwich town walls and at Raglan Castle, as well as in Scotland at Threave Castle.[73]

Because of a greater intensity of warfare, far more artillery towers were constructed with greater expertise in France and the southern Low Countries than in the British Isles during the fifteenth century. Although no chart exists like that composed by Jean Mesqui for French gunports, there were at least forty fortifications built in that kingdom with artillery towers, along with many more in the southern Low Countries.[74] Many of these are also on sites that had previously been attacked or captured during the Hundred Years' War, especially in urban areas. Such vulnerability helps explain this gunpowder defensive construction. For example, a number of fortifications captured by Henry V in his conquest of Normandy later acquired artillery towers. These included Falaise and Caen Castles, both captured in 1417; Rennes town wall, captured in 1418; and Arques-la-Bataille and Dieppe Castles, captured in 1419. In addition, the town wall of Chartres, seized by the Burgundian Duke John the Fearless also in 1419, later acquired an artillery tower. Finally, Mont-Saint-Michel, under siege almost constantly between 1417 and 1434, and Dinan, an equally threatened site, also acquired artillery towers. For captured fortifications, it is often difficult to determine who was responsible for the construction of the artillery towers because of the dearth of records. Some were built by captors, as at Falaise and Rennes, while others were erected after the fortifications had been liberated, as at Arques-la-Bataille and Dieppe. At Mont-Saint-Michel and Dinan, the artillery towers appear to have been constructed during the times of conflict by their defenders as a successful means of warding off gunpowder artillery attacks.[75] One of the most impressive artillery towers built in the fifteenth century was attached to the Castle at Ham in France. Ham, too, had been subject to conquest by John the Fearless in 1411. This castle had fallen to gunpowder weapons, or, more appropriately, it had surrendered to the threat of gunpowder weapons after only a small show of their offensive force. By a rather convoluted inheritance, Ham Castle eventually fell into the hands of the Burgundian lieutenant, Louis de Luxembourg, who determined that the site was strategically significant because of its proximity to Senlis, Compiègne, Paris, and other northern French towns. Louis vowed not to allow the castle to fall as easily as it had in 1411. In 1460, he set about rebuilding it with all the available "state-of-the-art" gunpowder fortification devices; this rebuilding would last intermittently until 1475. The reconstruction outfitted the castle with gunports, a boulevard, and, most impressively,

Figure 1.8
Artillery Tower, Falaise Castle.

Figure 1.9
Artillery Tower, Caen Castle.

Figure 1.10
Artillery Tower, Dinan Town Wall.

two artillery towers. Placed along the Somme River, and at both ends and across the river from the boulevard, the Ham Castle artillery towers, known as the *tour du connétable* and the *tour aux poudres*, were built to guard the fortification against another attack by gunpowder weapons. The *tour du connétable* was the largest and most formidable of the two artillery towers and perhaps the largest ever built in the fifteenth century. It was round in shape and measured 32 meters in diameter at the base. It was also tall, standing more than 28 meters above the water level of the Somme River, and divided inside into three storeys. It thus jutted out from the castle's rectangular shape and towered over its walls. (By comparison, the *tour aux poudres* was smaller in diameter and only as high as the castle walls.) According to studies made and photographs taken before this artillery tower was destroyed in World War I, each level of the *tour du connétable* featured gunports; yet those on the second and third storeys, with six on each level, protected the fortification from attack on the tower itself or along either side of the fortification running southeast (toward the *tour aux poudres*) and northwest from the tower.

Based on *Le livre des trahisons de France envers la maison de Bourgogne*, it was this northwest side of the castle wall which had been the object of the 1411 attack. So impressive was the completed Ham Castle, with its new gunpowder defenses, that the chronicler Phillippe de Commynes wrote that in 1475 Louis of Luxembourg, while facing defeat by either the French king, Louis XI, or the duke of Burgundy, Charles the Bold—mortal enemies themselves, "decided to stay on in his strong castle of Ham, which had cost him so much, for he had built it for the purpose of protecting himself in such a necessity; and he had it provided with everything, as well as any castle I have ever seen."[76] Ham Castle did not fall then, nor would it again fall to a siege until 1557. The circumstances of this engagement do not indicate a failure of the artillery towers.[77]

Ham Castle's endurance through the middle of the sixteenth century testifies to the effectiveness of its artillery towers. But does this represent the overall effectiveness of this particular gunpowder defense? This is impossible to answer definitively, although support can be found in the large number of artillery towers. While most often placed on previously vulnerable fortifications, they seem to have withstood any further attacks. In addition, these towers outnumbered the boulevards, while also having a higher unit-construction cost. This further suggests their relative effectiveness.

Pre-*Trace Italienne* Artillery Fortifications

There exists a final demonstration of the effectiveness of artillery towers as a gunpowder fortification. The *trace italienne* fortress was not the first fortification specifically built in response to gunpowder weapons. An earlier design was developed in France and the southern Low Countries during the middle of the fifteenth century, incorporating all of the innovations mentioned above but evolving principally from the artillery tower. Few of these fortresses were built, for their cost was high and their construction time long—a strategic hindrance in an era of almost continual warfare. Nevertheless, several were erected. I will highlight two of these fortifications, both of which were constructed in the middle of the fifteenth century.

The first of these, Posanges Castle, was built under the direction of Philip the Good, duke of Burgundy. Philip's well-known interest in the development and proliferation of gunpowder weapons in his duchy[78] inspired an interest in innovating gunpowder defenses. This can be seen in the "Mémoire de François de Surienne," written in 1461, which established that the Burgundians fully comprehended the utility of flanking fire from the artillery towers and boulevards.[79] But this was prior knowledge, as these same principles are evident in Posanges Castle.

Posanges Castle is a square fortification, with four large circular artillery towers built on its corners. The fortress is not very large, with its curtain walls measuring only 35 meters between the artillery towers. These towers, rising above the 8-meter curtain walls by more than 2 meters, easily flank the walls with gunports. Built in each of the two storeys, these gunports faced down the walls and away from the towers. To improve defenses further, the castle was also surrounded by a moat measuring 3.5–7 meters in width and filled with water.[80] Overall, Posanges proved to be militarily insignificant. It protected neither an important region of the Duchy nor a strategic waterway or road. Yet, in viewing its importance in terms of gunpowder fortification development, I must agree with Jean-Bernard de Vaivre when he argues that "the castle of Posanges constitutes one of the better examples of military architecture in Burgundy during the fifteenth century."[81] It may have been one of the first fortifications built wholly for the purpose of resisting gunpowder siege weapons.

This honor may also go to Rambures Castle.[82] The more famous of these two castles, Rambures could also be considered a Burgundian initiative that was built during the time of Philip the Good. While its construction date remains dubious, its reputed builder was André de Rambures, a member of the

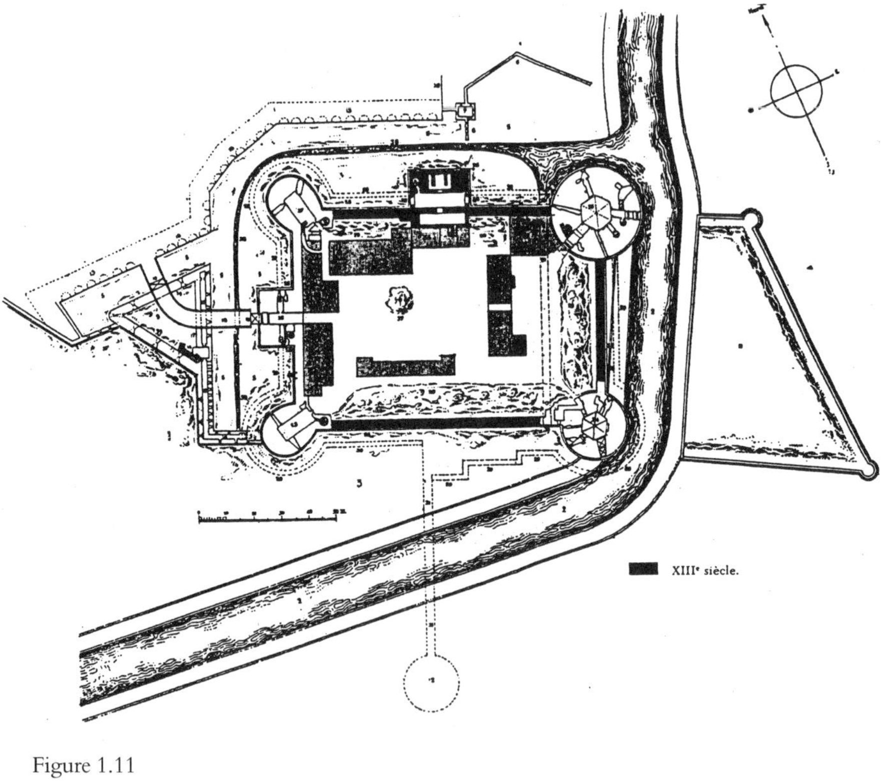

Figure 1.11
Plan of Ham Castle. Source: Phillippe Seydoux, *Fortresses médiévals du nord de la France* (Bellegarde, 1979), p. 215.

Figure 1.12
Artillery fortification, Posanges Castle.

Figure 1.13
Artillery fortification, Posanges Castle (detail of towers and gunports).

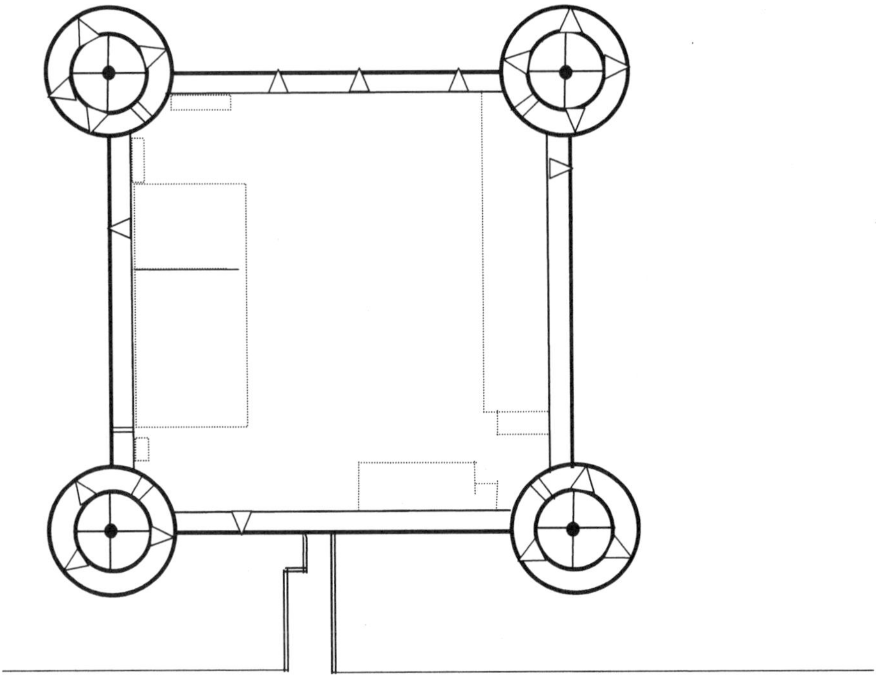

Figure 1.14
Plan of Posanges Castle. Source: Jean-Bernard de Vaivre, "Le château de Posanges," *Congrès archéologique de France* 144 (1986), p. 217.

Burgundian ducal court and son of the French King Charles VI's master of crossbowmen, David. This may have given him the unique ability to straddle those two French sides of the Hundred Years' War that had become enemies over the Armagnac-Burgundian controversies. But it was a hazardous ability that could have betrayed him suddenly. This may have prompted Rambures to investigate the new gunpowder defensive technologies and then to build a fortified residence for himself.

A castle is mentioned for the first time at Rambures in 1421; whether or not this is the castle which currently stands is unknown. If the two do correspond, this would make the extant castle the earliest fortification to have been built entirely as a gunpowder fortress. This could also imply that it was built over a long period of time and not completed until 1465, the date of André de Rambures' death. In this light, it could be inferred that Rambures added the new gunpowder defenses as the century progressed and as their popularity increased. An obvious rebuttal to this hypothesis is the apparently sudden construction of Rambures Castle. On the other hand, the castle could have been initially built in 1421, and then torn down and reconstructed as a complete gunpowder fortification later in the century; but no evidence substantiates this.

Rambures is a small castle, measuring only 29 meters from side to side; it comprises little more than eight towers without a curtain wall. Functioning as artillery towers, the four towers on the end are round, protrude slightly, and stand taller than the interior towers. The four interior towers are "U-shaped" and shallower than their corner counterparts; the northern and southern towers open onto a cylindrical hall that runs the length of the castle between them. The sole entrance is into the northern tower. Each corner artillery tower measures 12 meters in diameter and has a height of 20 meters. The interior of the castle is more difficult to describe. Some storeys open to the interior but lack even a gunport to the exterior; others only open to the interior by a staircase yet remain open to the exterior of the fortification. Those storeys that open to the exterior are penetrated by gunports—three in the corner towers and two in the interior towers. They face away from the castle; yet, because of this fortification's unique shape, all approaches of attack would have been amply covered by defensive gunfire. In addition, a wide, dry ditch surrounds the castle. The strongest gunpowder defense of Rambures Castle may be its brick construction. Masonry appears only on the top of the corner towers, forming the machicolations; this was added later in the seventeenth or eighteenth century. Only one or two brick fortifications of its kind were built at the end of the Middle Ages, with Lord Hastings' Kirby Muxloe Castle providing a later example.[83] While construction in brick would not catch on as a gunpowder defense in the *trace italienne*, its use at Rambures, Kirby Muxloe, and elsewhere was logical. Despite its expense, brick was a stronger building material than masonry, and it did not fracture as quickly under the ballistic impact of artillery fire. Perhaps such innovative builders as André de Rambures concluded that brick construction was as relevant as the gunport, the artillery tower, and the ditch. Neither the Rambures nor the Posanges Castle faced serious enemy threats—at least no assault at either site is documented. This may reflect their small size or insignificant strategic positioning. Then again, the manifest strengths of these castles may have discouraged direct assaults.

King Louis XI's artillery fortification at Dijon represents the culmination of fifteenth-century gunpowder defenses. After the defeat and death of Charles the Bold at the battle of Nancy in 1477, Louis XI occupied the ducal capital of Dijon. Wanting to consolidate his hold on this "vulnerable" site, Louis began building a new castle, which his son, Charles VIII, completed. Archaeological excavations have revealed that this fortress, which is no longer extant, included all the gunpowder defenses of the previous century: gunports that penetrated all walls and towers of the castle; boulevards that set out from the castle on both the

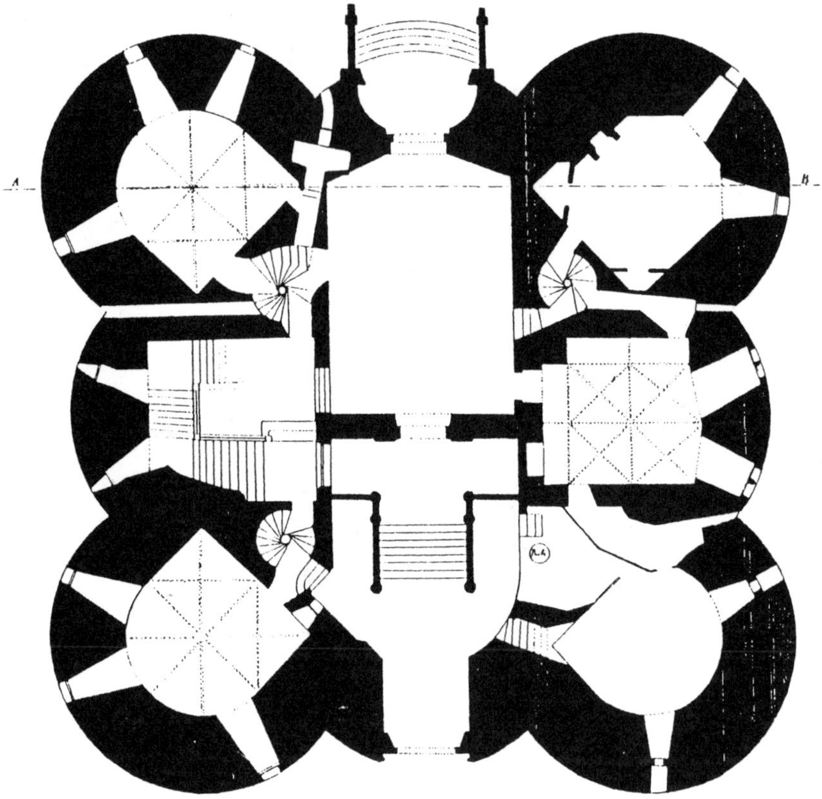

Figure 1.15
Plan of Rambures Castle. Source: Philipe des Forts, "Le château de Rambures (Somme)," *Bulletin monumental* 67 (1903), 250.

north and south sides; and artillery towers that stood at the four corners of the fortification. A wide moat also surrounded the fortress. Unfortunately, nothing more can be said about its construction or defensive purposes. Yet the fact that the King of France constructed a fortification on this vulnerable site suggests his confidence in the overall effectiveness of gunpowder defenses.[84] The purpose of this essay is not to diminish the importance of the *trace italienne* as a fortification system. As Christopher Duffy has shown, the endurance of that system throughout the early modern period, and throughout the world, secures its prestige as one of the most significant fortification systems ever devised.[85] But it was not a system that developed in a vacuum, and it did not signify a revolutionary overturning of medieval military architecture. Europeans had been employing gunpowder weaponry for nearly two centuries by that time. As a technology, it

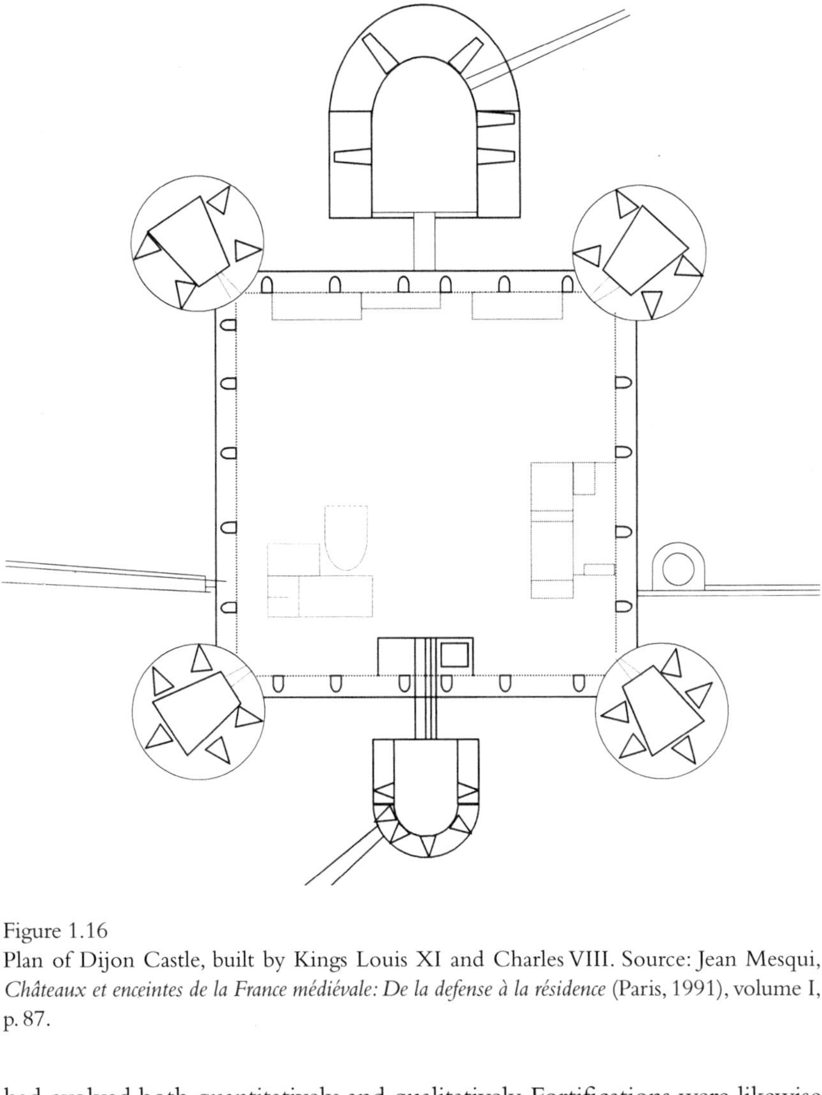

Figure 1.16
Plan of Dijon Castle, built by Kings Louis XI and Charles VIII. Source: Jean Mesqui, *Châteaux et enceintes de la France médiévale: De la defense à la résidence* (Paris, 1991), volume I, p. 87.

had evolved both quantitatively and qualitatively. Fortifications were likewise evolving in response. As demonstrated in France and the southern Low Countries (domains that endured numerous military engagements during the late fourteenth and fifteenth centuries), fortifications were being constructed, reconstructed, and adapted to both employ and resist gunpowder weaponry. Gunports, boulevards, and artillery towers were added to castles and town walls perceived to be vulnerable to siege guns, while by the middle of the fifteenth century entirely new pre-*trace italienne* artillery fortifications were constructed.

In most cases, these defensive technologies were successful in what they were designed to do: to shield their inhabitants from the emerging "military revolution" that would dominate the early modern era.

Notes

1. Kelly DeVries, "The Impact of Gunpowder Weaponry on Siege Warfare in the Hundred Years' War," in *The Medieval City under Siege*, ed. I. Corfis and M. Wolff (Boydell, 1995).

2. DeVries, "The Impact of Gunpowder Weaponry," p. 244.

3. Clifford J. Rogers, "The Military Revolutions of the Hundred Years' War," *Journal of Military History* 57 (1993): 241–278, and in *The Military Revolution Debate: Readings on the Military Transformation of Early Modern Europe*, ed. C. Rogers (Westview, 1995).

4. Geoffrey Parker, *The Military Revolution: Military Innovation and the Rise of the West, 1500–1800* (Cambridge University Press, 1988), pp. 7–24. He first mentioned this in "The 'Military Revolution,' 1560–1660—a Myth?" *Journal of Modern History* 48 (1976): 195–214. Parker has continued to insist on this proof as the cornerstone of his Military Revolution thesis. See Parker, "In Defense of *The Military Revolution*," in *The Military Revolution Debate: Readings on the Military Transformation of Early Modern Europe*, ed. C. Rogers (Westview, 1995); "The Gunpowder Revolution, 1300–1500," in *The Cambridge Illustrated History of Warfare: The Triumph of the West*, ed. G. Parker (Cambridge University Press, 1995), pp. 106–117. My definition of a *trace italienne* fortification follows John A. Lynn, "The *trace italienne* and the Growth of Armies: The French Case," *Journal of Military History* 55 (1991): 297.

5. Kelly DeVries, "Gunpowder Weaponry at the Siege of Constantinople, 1453," in *War, Army and Society in the Eastern Mediterranean, 7th–-16th Centuries*, ed. Y. Lev (Brill, 1996), pp. 346–350; "The Technology of Gunpowder Weaponry in Western Europe during the Hundred Years' War," in *XXII. Kongreß der Internationalen Kommission für Militärgeschichte* Acta 22: *Von Crécy bis Mohács Kriegswesen im späten Mittelalter (1346–1526)* (Vienna: Heeresgeschichtliches Museum, 1997).

6. Parker, *Military Revolution*; DeVries, "The Impact of Gunpowder Weaponry." See also M. W. Thompson, *The Decline of the Castle* (Cambridge University Press, 1987); B. H. St. J. O'Neil, *Castles and Cannon: A Study of Early Artillery Fortifications in England* (Oxford University Press, 1960); N. J. G. Pounds, *The Medieval Castle in England and Wales: A Social and Political History* (Cambridge University Press, 1990); D.J. Cathcart King, *The Castle in England and Wales* (Croom Helm, 1988); John R. Kenyon, "Early Artillery Fortifications in England and Wales: A Preliminary Survey," *Archaeological Journal* 138 (1981): 205–240; John R. Kenyon, "Early Artillery Fortifications in England and Wales," *Fort* 1 (1976; Rev. ed., 1993): 33–36.; John R. Hale, "The Early Development of the Bastion: An Italian Chronology," in *Europe in the Late Middle Ages*, ed. J. Hale (Northwestern University Press, 1965); John R. Hale, *Renaissance Fortification: Art or Engineering?* (Thames and Hudson, 1977); Simon Pepper, "Castles and Cannon in the Naples Campaign of 1494–95," in *The French Descent into Renaissance Italy, 1494–95: Antecedents and Effects*, ed. D. Abulafia (Ashgate Variorum, 1995); Simon Pepper and Nicholas Adams, *Firearms and Fortifications: Military Architecture and Siege Warfare in Sixteenth-Century Siena* (University of Chicago Press, 1986); Horst de la Croix, "The Literature on Fortification in Renaissance Italy," *Technology and Culture* 4 (1963): 30–50.

7. On the French raids, see Christopher Allmand, *The Hundred Years' War: England and France at War c. 1300–c. 1450* (Cambridge University Press, 1988), pp. 88–89. On the raids of the Scots, see J. A. Tuck, "War and Society in the Medieval North," *Northern History* 21 (1985): 33–52; Cynthia J. Neville, "Keeping the Peace on the Northern Marches in the Later Middle Ages," *English Historical Review* 109 (1994): 1–25.

8. See DeVries, "The Impact of Gunpowder Weaponry," pp. 233–236; Kelly DeVries, "Gunpowder Weaponry and the Rise of the Early Modern State," *War in History* 5 (1998): 127–145.

9. Those who acknowledge the possibly of Queensborough's anti-gunpowder weaponry purpose include DeVries ("The Impact of Gunpowder Weaponry," p. 240), Thompson (*The Decline of the Castle*, p. 36), and R. A. Brown, H. M. Colvin, and A. J. Taylor (*The History of the King's Works*, volume 2 (Her Majesty's Stationary Office, 1963), pp. 801–802. King (p. 154) disputes the link between Queensborough Castle and gunpowder artillery.

10. "Règlement pour la défense du château de Bioule, 18 mars 1347," *Bulletin archéologique* 4 (1846–1847): 490–495. See also Philippe Contamine, *War in the Middle Ages* (Blackwell, 1984), 202.

11. Jim Bradbury, *The Medieval Siege* (Boydell, 1992), pp. 156–163.

12. DeVries, "The Impact of Gunpowder Weaponry," p. 229.

13. DeVries, "Gunpowder Weaponry and the Rise of the Early Modern State," pp. 132–145.

14. On the use of gunpowder weapons during the Wars of the Roses, see Kelly DeVries, "The Use of Gunpowder Weapons in the Wars of the Roses," in *Traditions and Transformations in Late Medieval England*, ed. D. Biggs et al. (Brill, 2001).

15. On the lack of Scottish raids across the English border in the fifteenth century, see Neville. On the Scots serving in the French continental armies, see P. Jubault, *D'Azincourt à Jeanne d'Arc, 1415–1430* (Moulat, 1969), pp. 113–186; Bernard Chevalier, "Les écossais dans les armées de Charles VII jusqu'a la bataille de Verneuil," in *Jeanne d'Arc: Une époque, un rayonnement* (CNRS, 1982).

16. See DeVries, "Gunpowder Weaponry and the Rise of the Early Modern State," pp. 139–144.

17. William Worcestre, *Itineraries*, ed. J. Harvey (Clarendon, 1969), pp. 20–21, 140, 242–243, 262–263, 397–400. See also Thompson, p. 19.

18. Sir Charles Peers, *Kirby Muxloe Castle, Leistershire* (English Heritage, 1957), and Anthony Emery, "Kirby Muxloe Castle," *Nottingham Area Proceedings* (1989): 72–77.

19. On the gunpowder weaponry policies of Charles V, see DeVries, "Gunpowder Weaponry and the Rise of the Early Modern State," pp. 131–132. On those of Philip the Bold, see DeVries, "Gunpowder Weaponry and the Rise of the Early Modern State," pp. 133–135; Robert D. Smith and Kelly DeVries, *A History of Gunpowder Weaponry in the Middle Ages: The Artillery of the Valois Dukes of Burgundy, 1363–1477* (Boydell, 2005).

20. Contamine, pp. 200–201.

21. U. Plancher, *Histoire générale et particulaire de Bourgogne*, volume 3 (Dijon: Antoine de Fay, 1739–1781), pp. 291–297; Richard Vaughan, *John the Fearless: The Growth of Burgundian Power* (Longmans, 1966), pp. 175–176; Pierre Bertin, "Le siège du château de Vellexon dans l'hiver 1409–1410," *Revue historique des armées* 27 (1971): 7–18.

22. Enguerran de Monstrelet, *Chronique*, volume II, ed. L. Douet-d'Arcq (Société de l'histoire de France, 1857–1862), pp. 172–175; *Chronique des Pays-Bas, de France, d'Angleterre et de Tournai*, in *Corpus chronicorum Flandriae*, 3, ed. J. de Smet (Hayez, 1856), 342. See also L. de Laborde, *Les ducs de Bourgogne* (Plon freres, 1849–1852), volume I, p. 24.

23. Religieux de Saint-Denis, *Chronique*, volume IV, ed. L. Bellaguet (Crapelet, 1839–1852), p. 652.

24. Religieux de Saint-Denis, volume V, pp. 370–375; Monstrelet, volume III, pp. 22–31; Jean le Fevre, *Chronique*, volume I, ed. F. Morand (Libraire Renouard, 1876–1881), p. 184.

25. Religieux de Saint-Denis, volume VI, pp. 85, 127–129; Monstrelet, III: 216; and Jean Juvenal des Ursins, *Histoire de Charles VI*, in *Nouvelle collection des mémoires relatifs à l'histoire de France* 2, ed. M. Michaud (Editeur du Commentaure Analytique du Code Civil, 1857), pp. 537–538.

26. *Gesta Henrici quinti*, ed. F. Taylor and J. Roskell (Clarendon, 1975), pp. 23–28. See also Alfred H. Burne, *The Agincourt War: A Military History of the Latter Part of the Hundred Years' War from 1369 to 1453* (Eyre and Spottiswoode, 1956), pp. 42–46.

27. Burne, *Agincourt War*, pp. 115–126; Jubault, pp. 45–52; Christopher Allmand, *Henry V* (University of California Press, 1992), pp. 116–120; E. F. Jacob, *Henry V and the Invasion of France* (Hodder and Stoughton Limited, 1947), pp. 125–129; and Richard Ager Newhall, *The English Conquest of Normandy, 1416–24: A Study in Fifteenth Century Warfare* (Yale University Press, 1924), pp. 37–91. For the siege of Caen, see Léon Puiseaux, *Siège et prise de Caen par les anglais en 1417: Épisode de la guerre de cent ans* (Le Gost-Clérisse, 1868).

28. Burne, *Agincourt War*, pp. 126–127; Newhall, pp. 71–72, 92–97; Jacob, *Henry V*, pp. 129–130.

29. Burne, *Agincourt War*, pp. 129–133; Jubault, pp. 57–64; Allmand, *Henry V*, pp. 121–128; Newhall, pp. 97–105, 110–123; Jacob, *Henry V*, pp. 130–141; and Léon Puiseaux, *Siège et prise de Rouen par les anglais (1418–1419)* (Le Gost-Clérisse, 1867).

30. Burne, *Agincourt War*, pp. 133–134; Newhall, pp. 123–132; E. F. Jacob, "The Collapse of France, 1419–20," *Bulletin of the John Rylands Library* 26 (1941–42): 307–326.

31. Allmand, *Henry V*, pp. 151–182; Burne, pp. 139–180.

32. *Le Livre des trahisons de France envers la maison de Bourgogne*, in *Chroniques relatives à l'histoire de la Belgique sous la domination des ducs de Bourgogne (texts Français)*, ed. M. de Kervyn de Lettenhove (F. Hayez, 1873), 96.

33. This conclusion contradicts what I have said previously in *Medieval Military Technology* (Broadview, 1992), pp. 145–147, as well as what Clifford J. Rogers has written in "The Military Revolutions of the Hundred Years' War," in *The Military Revolution Debate: Readings*

on the Military Transformation of Early Modern Europe, ed. C. Rogers (Westview, 1995), pp. 66–68. Still, I now contend that after 1420, for every siege that was decided quickly by gunpowder weapons, there were many more in which gunpowder weapons did nothing to decrease the time of the siege, or which were decided by means other than gunpowder weapons. In support of this, see Alain Salamagne, "L'attaque des places-fortes au XVe siècle à travers l'exemple des guerres anglo et franco-bourguignonnes," *Revue historique* 289 (1993): 65–113.

34. Philippe de Commynes, *Mémoires*, ed. J. Calmette and G. Durville I (Libraire ancienne Honoré Champion, 1924), 235.

35. DeVries, "The Impact of Gunpowder Weaponry," p. 234. On the gunports at Mont-Saint-Michel, see Contamine, p. 202. On those at Saint-Mâlo, see Michael Jones, "The Defence of Medieval Brittany: A Survey of the Establishment of Fortified Towns, Castles and Frontiers from the Gallo-Roman Period to the End of the Middle Ages," *Archaeological Journal* 138 (1981): 174. And on gunports at Blanquefort Castle, see M. G. A. Vale, "Seigneurial Fortification and Private War in Late Medieval Gascony," in *Gentry and Lesser Nobility in Late Medieval Europe*, ed. M. Jones (Gloucester: A. Sutton, 1986), 141. Alain Salamagne ("A propos de l'adaptation de la fortification à l'artillerie vers les années 1400: quelques remarques sur les problèmes de vocabulaire, de typologie et de méthode," *Moyen Age* 75, 1993: 829–830) reports the dates of several purportedly fourteenth-century French and southern Low Countries' gunports, but then presents compelling arguments against accepting any of them.

36. DeVries, "The Impact of Gunpowder Weaponry," p. 234. On Paris, see *Choix de pièces inédites relatives au règne de Charles VI*, volume II ed. L. Douet-d'Arcq (Libraire Renouard, 1863–1864), pp. 32–33. On Rennes, see Jones, p. 175.

37. Christine de Pisan, *The Book of Deeds of Arms and of Chivalry*, ed. C. Cannon Willard (Pennsylvania State University Press, 1999), p. 105.

38. Jean Mesqui, *Châteaux et enceintes de la France médiévale: De la défense à la résidence*, volume 2 (Grands Manuels, 1991), pp. 319–322. Mesqui's chart shows that at least 49 French *departements* have fortifications with fifteenth-century gunports.

39. Alain Salamagne, "A propos de l'adaptation de la fortification à l'artillerie vers les années 1400: quelques remarques sur les problèmes de vocabulaire, de typologie et de méthode," *Moyen Age* 75 (1993): 809–846. See also his articles "Les années 1400: La genèse de l'architecture militaire bourguignonne ou la définition d'un nouvel espace urbain," *Revue Belge d'histoire militaire* 26 (1986): 325–344, 405–434; "L'attaque des places-fortes au XVe siècle à travers l'exemple des guerres anglo et franco-bourguignonnes," *Revue historique* 289 (1993): 65–113; "La défense des villes des Pays-Bas à la mort de Charles le Téméraire (1477)," in *La guerre, la violence et les gens au Moyen Âge. I: Guerre et violence*, ed. P. Contamine and O. Guyotjeannin (Editions du CTHS, 1996), pp. 295–307.

40. Kenyon, "Early Gunports," pp. 33–36. Kenyon actually lists 58 total English sites outfitted with gunports, but 22 of these were built in the sixteenth century.

41. See Kenyon, "Early Gunports," pp. 33–36; Peers; Emery; Catherine Morton, *Bodiam Castle, Sussex* (Plaistow: The National Trust, n.d.); Nigel Saul, "Bodiam Castle," *History Today* 45 (Jan 1995): 16–21; C. Taylor, P. Everson, and R. Wilson-North, "Bodiam Castle, Sussex,"

Medieval Archaeology 34 (1990): 155–157; D. J. Turner, "Bodiam, Sussex: True Castle or Old Soldier's Dream House?" in *England in the Fourteenth Century: Proceedings of the 1985 Harlaxton Symposium*, ed. W. Ormond (Boydell and Brewer, 1986), pp. 267–279; and John R. Kenyon, "Artillery and the Defences of Southampton circa 1360–1660," *Fort* 3 (spring 1977; rev. ed., 1993): 21–30.

42. This was O'Neil's belief.

43. Jean Mesqui, *Châteaux forts et fortifications en France* (Flammarion, 1997), pp. 378–380; Raymond Ritter, *L'architecture militaire du moyen âge* (Fayard, 1974), pp. 171–174.

44. M. Deshoulieres, "Dieppe," *Congrès archéologique de France* (1926), pp. 287–297; Daniel Queney, *Histoires d'un château: Promenade d'un "passe-muraille" et quelques anecdotes, suivies du catalogue de l'iconographie du Château de Dieppe, présentée en exposition du 20 November 1993 au 28 Février 1994* (Chateau-Musée de Dieppe, 1993).

45. See, for example, the typological charts found in Pierre Sailhan, "Typologie des archères et canonnières: Les archères des châteaux de Chauvigny," *Bulletin de la société des antiquaires de l'Ouest et des musées de Poitiers* (4th Series) 14 (1978): 522. For a comparison with arrowslit typologies, see Sailhan, p. 513. (Sailhan has categorized these arrowslits and gunports, but these typologies have not, to my knowledge, found general acceptance.)

46. This will be discussed further in Kelly DeVries, "The Early Use of Hand-held Gunpowder Weapons," and "The Royal Armouries' Handguns" (both forthcoming). See also Graeme Rimer, "Early Handguns," *Royal Armouries Yearbook* 1 (1996): 73–78, which contains illustrations of these four "handguns," but does not differentiate between the weights of the heavier and lighter weapons.

47. On these weapons, see Bert S. Hall, *Weapons and Warfare in Renaissance Europe: Gunpowder, Technology, and Tactics* (Johns Hopkins University Press, 1997), pp. 100, 176–177.

48. Quentin Hughes, "Medieval Firepower," *Fortress* 8 (1991), February: 31–43.

49. Peter N. Jones and Derek Renn, "The Military Effectiveness of Arrow Loops: Some Experiments at White Castle," *Château Gaillard* 9–10 (1982): 445–456.

50. Mesqui, *Châteaux et enceintes*, volume II, pp. 289, 318.

51. DeVries, "Impact of Gunpowder Weaponry," p. 237; Mesqui, *Châteaux et enceintes*, volume I, pp. 86–87, 356–361; Kelly DeVries, *Joan of Arc: A Military Leader* (Sutton, 1999), pp. 61–62.

52. DeVries, "Impact of Gunpowder Weaponry," p. 238.

53. Mesqui, *Châteaux et enceintes*, volume I, pp. 356–361; Salamagne, "Les années 1400," pp. 405–426; Nicolas Faucherre, "Barbacanes, boulevards, ravelins et sutres demi-lunes; inventaire incertain," in *Aux portes du chateau: Actes du troisième colloque de castellologie* (Flaran, 1987), pp. 105–115.

54. Salamagne, "Les années 1400," p. 407.

55. Ibid.

56. For Quesnoy, see Salamagne, "Les années 1400," pp. 407–408; Alain Salamagne, "Les fortifications médiévales de la ville du Quesnoy," *Revue du nord* 63 (1981): 1000–1001. For Douai, see Salamagne, "Les années 1400," p. 407. For Lille, see Salamagne, "Les années 1400," p. 408; Gilles Blieck and Laurence Vanderstraeten, "Recherches sur les fortifications de Lille au moyen âge," *Revue du Nord* 70 (1988): 112–113. For boulevards built in the 1420s and the 1430s, see Salamagne, "Les années 1400," pp. 408–414; Mesqui, *Châteaux et enceintes*, volume I, pp. 356–61; Faucherre, "Barbacanes, boulevards," pp. 108–115.

57. M. Boucher de Molandon and Adalbert de Beaucorps' tally is 4,365, including almost 900 pages (M. Boucher de Molandon and Adalbert de Beaucorps, *L'armée anglaise vaincue par Jeanne d'Arc sous les murs d'Orléans* [H. Herluison, 1892], especially pp. 134–139) while Louis Jarry's total is 3,189 (Louis Jarry, *Le compte de L'armée anglaise au siège d'Orléans, 1428–1429* [H. Herluison, 1892], especially pp. 58–65); both judgments are based on an English *compte* of Salisbury's army. Even when some 1,500 Burgundian soldiers later joined in the siege, there seems not to have been enough soldiers to surround the city, let alone to capture it by siege (Burne, *Agincourt War*, p. 229). See also DeVries, *Joan of Arc*, p. 59.

58. DeVries, *Joan of Arc*, pp. 59–60. See also Philippe Contamine, "Les armées française et anglaise à l'époque de Jeanne d'Arc," *Revue des sociétés savantes de haute-normandie, lettres et sciences humaines* 57 (1970), pp. 5–6.

59. See DeVries, *Joan of Arc*, p. 61. If Jacques Debal is to be believed, this was actually a reconstruction of the earlier earthwork, also called a boulevard by Debal, built by the French and occupied by Salisbury before his capture of the Tourelles (Jacques Debal, "Les fortifications et le pont d'Orléans au temps de Jeanne d'Arc," *Dossiers d'archéologie* 34 (May 1979): 88–90, and Jacques Debal, "La topographie de l'enceinte fortifiée d'Orléans au temps de Jeanne d'Arc," in *Jeanne d'Arc: une époque, un rayonment* [CNRS, 1982], 25–26). However, if the earlier structure was a boulevard, based on the later attack of this fortification by Joan, it appears that the English structure was far stronger than its French precursor.

60. Régine Pernoud, *La libération d'Orléans, 8 mai 1429* (Gallimard, 1969), 83. On whose calculations she bases these figures is unknown.

61. On the Tourelles boulevard, see DeVries, *Joan of Arc*, pp. 61–62; DeVries, "Impact of Gunpowder Weaponry on Siege Warfare," p. 238; Debal, "Les fortifications et le pont d'Orléans," pp. 88–90; Debal, "La topographie de l'enceinte fortifiée d'Orléans," pp. 25–26. This boulevard remained in place until at least 1676, as a drawing from that time (reproduced in Debal, "Les fortifications et le pont d'Orléans," p. 89, and Debal, "La topographie de l'enciente fortifiée d'Orléans," p. 37) shows it still in place. But this illustration shows the boulevard to be of stone construction, thus different from the boulevard which was constructed by the English in 1428. (This has led some to believe that the original boulevard was also in stone—including the model in the Musée de Jeanne d'Arc in Orléans—but that is clearly not the case, either from the original sources or from traditional fifteenth-century boulevard construction techniques.)

62. *Journal du siège d'Orléans* in *Procès de condamnation et de réhabilitation de Jeanne d'Arc dite la Pucelle*, volume IV, ed. J. Quicherat (Jules Renouard et Cie., 1841–1849), pp. 159–160 (hereafter Quicherat). Cagny (in Quicherat, IV: 8) reports that three or four assaults were made against the Tourelles. See also DeVries, *Joan of Arc*, pp. 87–88.

63. The Bastard of Orléans, then Lord Dunois, in *Procès en nullité de la condamnation de Jeanne d'Arc*, volume I, ed. P. Duparc, Société de l'histoire de France (C. Klincksieck, 1977), pp. 320–321. See also DeVries, *Joan of Arc*, pp. 88–89.

64. Joan, in Quicherat, volume I, p. 79.

65. On the gunpowder weapons used by Joan of Arc in her attack of the Tourelles boulevard and elsewhere, see Kelly DeVries, "The Use of Gunpowder Weaponry By and Against Joan of Arc During the Hundred Years' War," *War and Society* 14 (1996): 1–16.

66. DeVries, *Joan of Arc*, pp. 80–81.

67. On this as an aspect of Joan's military strategy, see DeVries, *Joan of Arc*, and Kelly DeVries, "The Military Strategy of Joan of Arc" (forthcoming).

68. On the stone boulevards at Rennes, see Jean-Pierre Leguay, *La ville de Rennes au XVme siècle à travers les comptes des Miseurs* (Rennes: Université de Rennes, Faculté des lettres et sciences humaines, 1968). On Mont-Saint-Michel, see Mesqui, *Châteaux et enceintes*, volume I: 304; Nicolas Faucherre, "Les défenses," in *Le Mont-Saint-Michel: Histoire et imaginaire* (Editions du Patrimoine/Anthèse, 1998), pp. 144–155. On Loches, see Faucherre, "Barbacanes, boulevards," p. 112. On Lassay, see Mesqui, *Châteaux forts*, pp. 210–211; Mesqui, *Châteaux et enceintes*, volume I, pp. 358–359; Faucherre, "Barbacanes, boulevards," p. 108.

69. Mesqui (*Châteaux et enceintes*, volume I, pp. 357–361) accepts this. On the late-fifteenth-century–sixteenth-century bastion, see Hale, "The Early Development of the Bastion," pp. 466–494; *Renaissance Fortification*; Christopher Duffy, *Siege Warfare: The Fortress in the Early Modern World, 1494–1660* (Routledge and Kegan Paul, 1979); Parker, *Military Revolution*, pp. 9–16.

70. Mesqui (*Châteaux et enceintes*, pp. 287, 309) lists fortresses with each of these shaped towers; most had circular or spur/"U-shaped" towers.

71. DeVries, "The Impact of Gunpowder Weaponry," pp. 239–240. See also Mesqui, *Châteaux et enceintes*, volume I, pp. 87–88, 273–285, 380–381.

72. Mesqui, *Châteaux et enceintes*, pp. 380–381.

73. On English artillery towers, see DeVries, "The Impact of Gunpowder Weaponry," p. 239; King, pp. 161–163; Brown, Colvin, and Taylor, eds., volume II, p. 606; Colin Platt, *The Castle in Medieval England and Wales* (1981; rpt. Barnes and Noble, 1996), 160; Kenyon, *Medieval Fortifications*, pp. 75–76; Hilary L. Turner, *Town Defences in England and Wales: An Architectural and Documentary Study, AD 900–1500* (John Baker, 1971), pp. 60, 165. On Threave Castle, see DeVries, "The Impact of Gunpowder Weaponry," p. 239; Kenyon, *Medieval Fortifications*, pp. 75–76, 172; Christopher J. Tabraham and George L. Good, "The Artillery Fortification at Threave Castle, Gallowey," in *Scottish Weapons and Fortifications, 1100–1800*, ed. D. Caldwell (J. Donald, 1981).

74. This is according to my own research, which is far from comprehensive. See also Mesqui, *Châteaux et enceintes*, volume I, pp. 87–88, 380–381.

75. On the artillery towers on Falaise Castle, see Régis Faucon, *Falaise* (Nouvelles Editions Latines, n.d.). On Caen Castle, see Michel de Boüard, *Le château de Caen* (Centre de

recherches archéologiques medievales, 1979). On Rennes town wall, see Jean-Pierre Leguay, *Un réseau urbain au moyen âge: les villes du duché de Bretagne aux XIVème et XVème siècles* (Maloine, 1988), and P. de Cortenson, "Les remparts de Rennes," *Bulletin monumental* (1907): 431–441. On Arques-la-Bataille Castle, see Mesqui, *Châteaux forts*, pp. 34–35, and Jean Achille Deville, *Histoire du château d'Arques* (Imprimé Chez Nicétas Periaux, 1839), and *Notice sur le château d'Arques* 8th ed. (Imprimère de H. Boissel, 1866). On Dieppe Castle, see Queney, and Deshoulieres, pp. 287–297. On Mont-Saint-Michel, see Marc Déceneux, *The Mont-Saint-Michel: Stone by Stone* (Rennes: Éditions Ouest-France, 1996). On Dinan, see Mesqui, *Châteaux forts*, pp. 150–152; Mesqui, *Châteaux et enceintes*, pp. 150, 280–281; Xavier Barral i Altet, "L'enceinte urbaine de Dinan," in *Dinan au moyen âge* (Pays de Dinan, 1986), pp. 73–100; Gilles Ollivier, "A propos du plan de Dinan à la fin du moyen âge," in *Dinan au moyen âge* (Dinan, 1986), pp. 343–351; Raymond Cornon, "Dinan," *Bulletin monumental* 107 (1949): 172–186.

76. Philippe de Commynes, *Mémoires*, volume II, ed. J. Calmette and G. Durville (Libraire Ancienne Honoré Champion, 1924–25), p. 84. I am using the translation of Isabelle Cazeaux (University of South Carolina Press, 1969), volume I, p. 290.

77. On Ham Castle see DeVries, "The Impact of Gunpowder Weaponry," p. 239; Mesqui, *Châteaux et enceintes*, pp. 174, 251, 277–278; Ritter, pp. 161, 167–168; Philippe Seydoux, *Fortresses médiévales du nord de la France* (Morande, 1979), pp. 212–230.

78. DeVries, "Gunpowder Weaponry and the Rise of the Early Modern State," pp. 135–138; DeVries and Smith, *The Gunpowder Artillery of the Dukes of Burgundy*.

79. See Jean Richard, "Quelques idées de François de Surienne sur la défense des villes à propos de la fortification de Dijon (1461)," *Annales de Bourgogne* 16 (1944): 36–43.

80. Mesqui, *Châteaux et enceintes*, p. 275; Mesqui, *Châteaux forts*, pp. 304–305; Jean-Bernard de Vaivre, "Le château de Posanges," *Congrès archéologique de France* 144 (1986): 211–234; and Henry Soulange-Bodin, *Les chateaux de Bourgogne* (Vanoest, 1942): 78–79.

81. Vaivre, p. 232.

82. DeVries, "The Impact of Gunpowder Weaponry," pp. 237, 240; Mesqui, *Châteaux forts*, pp. 315–316; Mesqui, *Châteaux enceintes*, volume I, pp. 210–211; Ritter, p. 165; Seydoux, pp. 47, 51–52, 267–275; Josiane Sartres, *Châteaux "brique et pierre" en Picardie* (Nouvelles éditions latines, 1973): 65–67; E. Prarond, "Rambures," *La Picardie* 4 (1858): 299–307, 406–413, 451–457, 513–520, 563–658; Philippe des Forts, "Le château de Rambures (Somme)," *Bulletin monumental* 67 (1903): 240–266; "Rambures," *Congrès archéologique de France* 99 (1936): 445–458.

83. See Peers and Emery.

84. Mesqui, *Châteaux enceintes*, volume I, p. 87; Nicolas Faucherre, *Muraille de Dijon: Le château de Dijon*, Catalogue de l'exposition du Musée Archéologique de Dijon, juillet-septembre 1989 (Musée archeologiques, Dijon, 1989).

85. See Duffy.

2

The French Reluctance to Adopt Firearms Technology in the Early Modern Period

Frederic J. Baumgartner

Any discussion of the Military Revolution, regardless of the precise period in which it is placed, accepts the proposition that it was directly linked to new technology, specifically gunpowder weaponry. What happened, however, when a major power was reluctant to adopt this new technology? The case of England is well known, where firearms did not replace the longbow until the end of the sixteenth century; but France also proved slow to adopt firearms. It was this delay that had the greater impact on history, because France played a major role in the European wars of the sixteenth century.

The modern progressive attitude deems new technology inherently superior to old technology. It suffices to say that a powerful imperative exists in contemporary American society to adopt a new technology well before the current one has become ineffectual. The attitude toward new technology in the early modern era differed greatly from the modern one. Change of any sort was regarded with great suspicion; thus, adoption of new technology came about slowly. I believe it was true of technology in general,[1] but I am prepared here to make the case for only one example of military technology. Until recently, the military traditionally maintained a proven military technology until a major defeat or institutional crisis demanded change; however, current armed forces have become major catalysts of innovation for new technology, much of which industry and commerce eventually adopts. The result has been a dramatic cultural change in response to new military technology—that is, relative to the sixteenth century.

During the Military Revolution, there was a military-industrial complex of sorts: numerous merchants and artisans depended upon the production and sales of arms, but they and their guilds were inherently conservative. They had little interest in change, because the production of a new type of weapon rarely improved their profit margins and usually disrupted the finely tuned competition system among their guilds. French towns were responsible for overseeing the production of firearms to suffice their needs. This reflected the French monarchy's indifference until the seventeenth century, although Francis I did organize a royal monopoly on saltpeter. By contrast, the Holy Roman Empire and Venice boasted arsenals that produced vast numbers of firearms by 1500. As late as the middle of the seventeenth century, the French produced weapons and gunpowder in small shops and homes, while rivals developed large-scale manufacturing facilities for their instruments of war.[2]

I wish to focus, however, on the attitude of the French military forces. Had the Gallic warriors been committed to change, their market demand could have compelled the artisans to produce new weaponry. The first suggestion of gunpowder weapons dates from 1326; Europeans gained significant expertise with them over the ensuing century. Unreliability, inaccuracy, heavy weight, and a painfully slow rate of fire initially made early gunpowder weapons—both firearms and artillery—scarcely superior to traditional projectile machinery.[3] These defects were mitigated in siege warfare, and early cannon were routinely used in sieges by 1400. Having been inspired by the "first arms race" between France and Burgundy during the fifteenth century, as well as the English occupation due to the Hundred Years' War, the French forged the classic system of more powerful yet lighter artillery coupled with more mobile gun carriages that also doubled as firing platforms. The French monarchy consequently fielded the best artillery train in Europe, which served it so well through the French Invasions of Italy.

However, relative to their rivals on the continent, the French did prove resistant to the deployment of firearms. The arquebus or matchlock was developed by 1460, primarily by the Germans. The word 'arquebus' probably comes from the German word for "hook gun." Originally, the word referred to a handcannon with an attached hook that fit over a wall to absorb its recoil when fired. Having a gradual impact on the battlefield, the arquebus first replaced the crossbow as a siege weapon. Early firearms appealed to the urban militiamen who guarded city walls because they required minimal training to fire effectively from walls. Although they cost more than crossbows, the artisans and merchants who composed the militia still purchased them. It probably was within the context of the siege that the arquebus was introduced to the field armies, which always doubled as siege forces.

No French arquebuses built before 1520 have survived, and French sources rarely mention new style firearms prior to that date. To be sure, the words for handguns in that era were used interchangeably; some of the *couleuvriniers*—the handgunners listed in 1471 as present in Louis XI's forces[4] and usually viewed as men armed with the by-then obsolete handcannon—may have carried newer weapons. When Louis XII celebrated the betrothal of his daughter Claude to Francis of Angoulême (the future Francis I) in 1506, he staged a large mock battle as part of the festivities; but there is no mention of handgunners among the many different types of participating infantrymen.[5] By the end of Louis' reign, the French were using the arquebus as a defensive siege weapon, as shown by an inventory of several forts and towns from 1513 that revealed 947 arquebuses. In contrast, Emperor Maximilian I reportedly stored 18,000 arquebuses and 22,000 hand culverins in his Innsbruck arsenal a decade earlier.[6]

Had Charles the Bold's mixed forces, including handgunners and crossbowmen, defeated the Swiss pike formations in their war of 1476 to 1477, the French might have developed a corps of arquebusiers. Instead, the French monarchy began recruiting the Swiss as mercenaries, who limited their use of firearms until the late sixteenth century. The few native French infantry units of the last years of the fifteenth century consisted mostly of Gascon pikemen and crossbowmen. As Bert Hall has noted, "French infantry for decades remained an uneasy combination of Swiss mercenaries and home-grown archers and crossbowmen. This combination, which seems so strange to the modern mind, would carry the House of Valois into the sixteenth century."[7] French gunmakers were so slow to manufacture the arquebus that the few in French hands by 1515 were Italian made. As late as 1568 Brantôme commented on the superior quality of Italian firearms, saying that none from France could attain the perfection of those from Milan.[8]

The army that Charles VIII led into Italy in 1494 had no handgunners on the muster rolls. Its strength lay in heavy cavalry and artillery. The French won their initial siege triumphs in Italy without small arms. Philippe de Commines' extensive memoirs of the Italian campaign mention firearms only once, when used by 300 German mercenaries recruited during Charles' return to France from Naples in 1495.[9] The same year at Seminara in the kingdom of Naples, the French defeated an overmatched Spanish army through the combined actions of French cavalry and heavy Swiss infantry.

A victorious army is typically cocksure about its tactics and weaponry. When the French and the Spanish fought again in 1503 at Cerignola, the French army was unchanged. The chronicler Jean d'Auton listed halberdiers, pikemen, and arbalesters in the army that Louis XII sent to Italy, but he failed to mention arquebusiers.[10] The Spanish forces, by contrast, had changed. Gonsalvo de Cordoba had substantially increased the number of arquebusiers in his forces and devised a technique to reduce their vulnerability. By digging trenches in front of his lines, he transformed the battlefield into a fortress, since arquebusiers had long demonstrated their effectiveness in that domain. The arquebusiers consequently raked the French *gens d'armes* with impunity as they approached the Spanish trenches, and even killed the French commander, the Duc de Nemours.[11] Gonsalvo rapidly increased the number of handgunners in his forces over the next three years and eventually drove the French out of southern Italy in a series of battles. The French government revamped its army after the defeat of 1503, but it focused on reducing the expense of their Swiss mercenaries rather than directly confronting the successful Spanish deployment of handguns. Marshal Pierre de Gié proposed to Louis XII a native infantry force of 20,000 men bearing pikes and crossbows. The proposal excluded arquebusiers.[12] Gié's disgrace a year later put a halt to its implementation, and the French relied on foreign mercenaries for another half-century.

To succeed with the arquebus, the Spaniards relied on two conditions: a sufficient amount of time to dig field fortifications and the impulsivity of French knights and Swiss pikemen, who traditionally charged upon sighting the enemy. When the Spaniards failed to entrench themselves or when the French commander recognized their tactics, they could be defeated. For instance, at the Battle of Ravenna in 1512, Gaston de Foix held his men back until the Spaniards, out of frustration, charged into his heavy artillery fire. Unfortunately, de Foix was killed late in the battle, and his successors resumed their impulsive frontal assaults against entrenched Spanish forces bristling with firearms.

By 1513 the French finally attempted to introduce arquebusiers into their forces. The army Louis XII sent to recover Milan, which the Swiss had captured the previous year, included 500 French arquebusiers. Nevertheless, the army still endured a devastating defeat at Novara. Documents indicate that the French commander, Louis de la Trémoille, deployed the arquebusiers incompetently: he left them as a block in his line without shock-troop support. The Swiss pikemen steadfastly charged through their fire and slaughtered them.[13] Claude de Seyssel responded to this blow by proposing in *The Monarchy of France* (1515) that prizes should be offered to induce men to become proficient with the bow and arquebus.[14]

In 1515, when Francis I sought to reverse the French defeats of the previous three years, he commissioned Pedro Navarro, the Spanish commander at Ravenna and now in French service, to raise a French infantry force. Drawing men mostly from Gascony and Picardy, areas under foreign control in previous centuries, Navarro raised 8,000 men carrying pikes, halberds, and crossbows. Based on his experience in the Spanish army, he also sought to recruit arquebusiers but found very few in France. Since the French intended to fight the Swiss for control of Milan, Navarro looked to the Germans. He hired 2,000 German arquebusiers, representing a tenth of the mercenary foot soldiers who signed on.[15] The sources for the ensuing Battle of Marignano mentioned handgunners, but they excluded them in the various drawings of the battle. The

French *gens d'armes* showed their contempt for infantry troops by taunting the retreating Swiss with "Go back and eat cheese in your mountains."[16] If they felt that way about the Swiss troops, who had crushed the French in several previous battles, imagine how they responded to the manifestly inferior French infantrymen.

Because of their victory at Marignano, which was won largely without handgunners, the French secured permanent access to Swiss mercenaries. The victory also set in motion a chain of events that culminated in two major battles near Milan a decade later—La Biccoca and Pavia. At both battles the French army included numerous Swiss mercenaries; consequently, it had few arquebusiers. In his preparations for the war with Emperor Charles V, Francis I had issued an edict on the French infantry that established its weaponry: two-thirds of the infantrymen were to carry pikes and the other third, a combination of halberds, crossbows, and arquebuses.[17] Charles' army, on the other hand, was strong in arquebusiers, who played a major role in the devastating French defeat in 1525 at Pavia, where Francis was captured. Even the French were now forced to reexamine their tactics; no longer were they willing to assault frontally entrenched enemy gunfire. In turn, Charles' army had to revise its tactics to bring the French to battle. The solution lay in creating a mobile wall of pikes to defend the arquebusiers while the infantry line moved forward to assault the enemy. Thus was created the Spanish square system, formally instituted by an edict of Charles V in 1534, which established an infantry system that combined pikemen and handgunners in a ratio of about three pikes to one firearm.

The French response to the Spanish square was immediate. Francis I also issued an edict in 1534 that created seven provincial legions of 6,000 men each. The conscious imitation of the ancient Roman army extended to giving a gold ring to each legionnaire who showed special valor. The ratio of pikemen and halberdmen to arquebusiers was 4 to 1, with the latter serving as auxiliaries. The smaller ratio of handgunners relative to that of the Spanish infantry indicated either how little the French king valued arquebusiers or how difficult it was to recruit them in France. This scarcity was probably reflected in their higher pay, six *livres* per month during war compared to the pikemen's five *livres*. (The arquebusiers were not paid in peacetime but were exempt from the *taille*.) It may also have been a consequence of the higher cost of their weapons.[18] The actual weapon ratio of the legions varied considerably, with the southern units having a higher proportion of arquebusiers. In 1535 Francis I made his famous review of the Legion of Picardy at Amiens, and the next year he took 12,000 legionnaires with him to Italy. The innovation failed, however. The legionnaires never rose above 19,000 in number and proved ineffectual in combat. The army that the Dauphin Henry led against Charles V's forces in Champagne in 1544 included the legions, but it also included Italian arquebusiers, who composed up to 80 percent of the Italian infantry fighting for France.[19] By Francis' death in 1547, the legionnaires were used only as garrison troops. The first representation of firearms in French art hence appeared in the bas relief on Francis' tomb at St.-Denis, sculpted by Pierre Bontemps circa 1550.

The most complete description of the French army at the time of Francis' death came from the sieur de Fourquevaux, whose *Instructions sur le faict de la guerre* (1548) only referenced events through 1543. While influenced by the classical Roman art of war, Fourquevaux focused on contemporaneous warfare. In his passages on the infantry, he nearly always grouped archers, arbalesters, and arquebusiers together and treated them interchangeably. He preferred the reliability of the bow to the handgun, especially in damp weather, and argued that a well-shot arrow was deadlier from a greater distance than an arquebus ball.[20]

In 1552, when Henry II made his "promenade to the Rhine," arquebusiers constituted about a third of his infantry, including two to three thousand native troops. Thus, a half century later than Spain, France finally achieved approximately the same proportion of handgunners in its army. Nonetheless, the anonymous author of a French military manual of 1559 recommended that the army return to the crossbow, because firearms could not be used in the rain, and they took too long to be loaded in the event of a surprise attack.[21] It was only in 1568 that the king ordered the royal army to replace the halberd with the pike and the arquebus.[22]

Being relatively cheap and light, the arquebus remained the basic infantry firearm throughout the sixteenth century. The Spanish musket, a heavier weapon, appeared shortly after 1520. It fired a larger ball that could reliably penetrate the plate armor, and it was more effective against cavalry. Yet the original musket was so heavy that the musketeer required a forked rest to avoid dropping the barrel and firing into the ground. The fork further complicated and slowed the process of reloading and firing. The French again were tardy in adopting this weapon, although its potential was well demonstrated to them when Marshal Piero de Strozzi, the noted military engineer in the French service, was killed at the siege of Thionville in 1558 by a musket ball fired from 500 paces.[23] According to Brantôme, the musket first appeared in the French army in 1568, although it was already widely used in Germany, Italy, and Spain. Brantôme said that Strozzi's son Philippe demonstrated its great range by using one to kill a horse, also from 500 yards; this feat persuaded several infantry captains to adopt it for their companies.[24]

The pistol represented the second major weapon development in the early 1500s. The first datable illustration of its wheel-lock mechanism appeared in a German manuscript of 1505.[25] The technical demands for machining this mechanism were very high, since it required unusually close tolerances relative to the manufacturing standards of that era. While inherently delicate, the wheel lock still had to fulfill the demands of heavy combat. More than the arquebus, the wheel-lock pistol was labor- and technology-intensive and hence expensive to produce. As with the arquebus and musket, the French artisans appear to have had difficulty acquiring the technical skills needed to manufacture wheel-lock mechanisms.

Because pistoliers were required to bear two or more pistols, only the nobility could afford this armament. Although its price dropped significantly after 1560, the pistol remained well out of the price range of the common soldier, for whom the arquebus was better suited. The arquebus offered greater reliability and range while also delivering a heavier shot (although it could not be concealed as well, much to the relief of civil authorities). For cavalrymen, by contrast, the absence of the glowing match, which frightened their horses, was an important advantage. Armies of the early 1500s had mounted arquebusiers, whom the French called *arquebusiers à cheval,* but they usually had to dismount to fire. The arquebus required two hands to use, whereas the pistol left one hand free for the reins. Because the pistol-carrying cavalryman could ride a smaller and cheaper horse, even a lower aristocrat could afford the financial burden of carrying several pistols.

The first fighting men to use the pistols were German *ritters*, whom the French called *reitres*. With a loaded pistol in their right hand and others in their boots and saddles, they would approach their foe and fire their pistols at close range—perhaps as close as 10 yards. This was just beyond the reach of an infantryman's pike or a cavalryman's lance. The *reitres* would fire as they wheeled about to the left, return to the rear of their company, reload, and wait their turn to come forward again. A smoothbore pistol fired from a moving horse is highly inaccurate, but at close range it can hit its target often enough and with sufficient impact velocity to unhorse an armored opponent.[26] The first mention of pistols on the battlefield was in 1544 during the war between Charles V and the French. In 1552 in Lorraine, a company of *reitres* routed a company of French heavy lancers of nearly the same size.[27] King Henry II of France was so surprised by the outcome that he ordered his forces to begin to recruit *reitres*. Few if any had been added to the French army before the battle of St.-Quentin in 1557, when the German *reitres* in Philip II's army played a major role in crushing the French. Since the battle was largely a pursuit of the fleeing French, the speed of the *reitres* was a major factor in the heavy casualties the French suffered. In a muster a year later, the French army added some 8,200 pistoliers to its manpower, but few were natives.

During the French Wars of Religion that followed Henry II's death in 1559, pistoliers represented a majority of the French mounted troops. They were stronger in the Huguenot army because of the large number of German Lutheran *reitres* deployed. At the Battle of Dreux in 1562, the Germans in the Huguenot army pioneered the pistolier maneuver called the *caracole*. The *caracole* enhanced the performance of the *reitres*, which accelerated the changeover from lancer to pistolier among the French nobility. By 1568 French pistoliers had become so common that only the name of the captain indicated whether a documented *reitre* company consisted of Frenchmen or Germans.

By the time of Henry IV most French nobles carried the pistol, as did the King himself. However, an episode at the Battle of Ivry in March 1590 against the army of the Catholic League revealed its drawback. According to the poet Du Bartas, Henry IV led 600 horsemen in an assault on 2,000 enemy horsemen. When he reached the enemy line, he raised his pistol to shoot, but it misfired. In an angry voice he shouted "O treacherous weapon! I'm done with you. The sword is the man-at-arms' true glory." Taking his sword, he "unleashed a deluge of blood."[28] For the next two centuries the French nobles would maintain that cold steel, the *arme blanche,* was the most effective and honorable military technique.

Thus far, our discussion has centered on the timeline for the adoption of firearms in France and the technical problems involved. Nevertheless, there were important cultural reasons for the French reluctance to embrace these new arms. We are speaking of the Renaissance, after all, a period when the admiration, or obsession, with the ancients was dominant. With respect to war, the great military manual from the ancient world was Vegetius' *Military Institutions of the Romans*. One of the first books to be printed, it was translated into all the major languages of Europe and printed in dozens of editions. *The Art of War* (1520), by Niccolo Machiavelli, who spent three terms as the Florentine ambassador to France, was largely an effort to update Vegetius for the early sixteenth century.[29] Machiavelli, for whom the study of war was a major concern, had witnessed plenty of military conflicts in his tenure with the Florentine government and was an unabashed admirer of the Roman military system. (What was good enough for Julius Caesar was good enough for Machiavelli.) Having little interest in new technology, he was convinced that military operations would remain indifferent to firearms: they were good only for frightening peasants. Cannon also failed to impress him, at least as field weapons. Cannon balls too often passed over the heads of the enemy. Once cannon had been fired, they took so long to reload that the enemy, especially cavalry, would overrun the gunners before they could fire again. Machiavelli argued that quality infantrymen could avoid cannon fire similar to the way the

Roman legionaries avoided Hannibal's elephants. He saw the arquebusiers as the equivalent of the Roman light infantry auxiliaries—archers and slingers—whose impact on battle was limited; therefore, he applied the Latin word *veliti* to them, a term that Fourquevaux would also use three decades later. Machiavelli regarded the Swiss with their phalanxes of sturdy pikemen as the true heirs of the Roman legions. The Swiss used handgunners, after all, in much the same fashion as the Romans had used their light auxiliaries.

In the sixteenth century Machiavelli's *Art of War* was more popular in France than *The Prince*, based on the number of French editions, which numbered at least six. Machiavelli set the tone for the other humanists—admittedly few—who wrote knowledgeably about war. The works of classical authors were extensively studied and considered the best sources for military instruction. For instance, François Rabelais, who served as physician to the noted captain and diplomat Guillaume Du Bellay, was exposed to military technology. Nonetheless, he still saw shock weapons as decisive armaments in battle, and he rarely mentioned handguns.[30] Michel de Montaigne, who knew something of war but probably never experienced battle, was capable of describing the Battle of Dreux (1562) only in terms of several clashes in antiquity. He wrote that a sword depends only on the courage of the wielder, while a pistol requires the successful performance of the powder, wheel, and stone; if one of these fails, "it endangers your fortune."[31] Montaigne also contrasted the reputed accuracy of ancient slingers with the poor rate of hits from firearms. He concluded that the only value of a gunpowder weapon was its noise; and since the fright that it caused soon diminished with familiarity, "I look upon it as a weapon of very little effect and hope we shall one day lay it aside."

The Project for American Research on the Treasury of the French Language further documents the French ambivalence toward firearms.[32] Using its digital database that allows word searches in French literary texts, I searched for weapon terminology in works written mostly by humanists from 1520 to 1620. Words such as *arquebus*, *pistol*, or *musket* were uncommon by comparison with *sword* or *pike*. Although most of the authors were hardly military men, they frequently used words dealing with shock weapons and gunpowder artillery, which hardly suggests an ignorance of warfare. According to J. R. Hale's extensive survey of both positive and negative remarks about gunpowder weapons during the Renaissance, there are fewer citations from France than from England, Germany, Italy, or Spain.[33] While none of this proves the French rejected firearms, it does suggest that such weapons failed to capture the imagination of French intellectuals.

The nobles had their own reasons for rejecting individual gunpowder weapons. Firearms were unchivalrous because they allowed common footsoldiers to kill mounted nobles from a distance—a humiliating death for warriors who cherished the opportunity to display their skill at fighting hand to hand and face to face. The French noble Bayard, the "knight without fear and reproach," was reputed to have slaughtered any arquebusier and cannonier who fell into his hands. Ironically, a vengeful arquebus ball killed him in 1524.[34] Although less brutal and more pragmatic, many noblemen shared his sentiments.

The memoirs of Blaise de Monluc date to about 1570, the year he was wounded in the face by an arquebus ball. The tough old French warrior reflected on 50 years of military experience, during which he had frequently commanded infantry companies. Probably thinking of his three sons who died from gunshots, he wrote a stinging denunciation of firearms:

> Would to God that this accursed engine had never been devised, . . . and so many brave and valiant men had never been slain at the hands of those who are often the most pitiful fellows and the greatest cowards, poltroons who had not dared look in the face of those men which at distance they lay low with their wretched bullets. But it was the devil's invention to make us murder one another.[35]

Despite his attitude, Monluc sought handgunners for his unit in 1523 but found none among his fellow countrymen. He was reduced to recruiting some Spanish deserters as his arquebusiers. Monluc made another comment about their scarcity in France in the year of 1527. By 1544, at the Battle of Ceresole, Monluc nevertheless had managed to bring 700 to 800 arquebusiers under his command. They served as the *enfants perdus* who opened the battle.[36]

Other mid-sixteenth-century works reveal similar disdain for firearms. Lamenting the many deaths of French nobles from arquebus shot, a veteran of Pavia demanded that firearms be restricted to fighting the infidel. In 1544 Michel d'Amboise, a nephew of the famed Cardinal Georges d'Amboise, declared that a nobleman who carried a pistol forfeited his honor.[37] A Venetian ambassador reported to his government in 1559 that the French nobles insisted on fighting in the long thin line, *en haie,* with their lances and swords, because to fight in any other way was cowardly.[38] François de La Noue, writing three decades later, still retained that attitude. He remarked that only the lance matched the spirit of the *gens d'armes*, while firearms were diabolical instruments invented to reduce kingdoms to sepulchers. He asserted that 700 or 800 *gens d'armes* were sufficient to rout 18,000 arquebusiers. By his time, however, the pistol had become the primary weapon of the cavalry, and La Noue conceded that properly handled pistols were effective against mounted lances.[39]

The French did eventually adopt the new firearms, albeit considerably later than their enemies. It is difficult to explain the apparent contradiction between France's slow response to firearms and its innovative achievements with artillery. I have tentative answers only. It appears that French artisans were slow in learning the techniques for producing firearms. We can assume, however, that had the military created a demand for those weapons, the artisans would have responded with a supply. Surely another factor was the inbred fear among French nobles of putting more effective weapons in the hands of the peasants, who could use them for rebelling or hunting; it is difficult to say which was more threatening. Perhaps the most important element was the French nobleman's class identity, if not his manliness, that was linked with fighting on horseback with shock weapons. The French monarchs of the sixteenth century were fully in tune with this view, making it difficult for them to mandate the adoption of new weaponry. (Henry II, after all, was killed in a traditional jousting tournament.) It long remained a tenet of French military doctrine that French soldiers possessed a great flair for shock combat with edged weapons, which partially explains why the Revolutionary government sought to restore the pike in 1793.[40]

We cannot ascertain whether or not these cultural factors explain the French reluctance to adopt firearms on a broad scale when the technology and tactics were available. The consequences of this resistance in the sixteenth century, however, are obvious enough. The French royal treasury was plundered to contract foreign handgunners. Upon Henry II's death in 1559, about half of his infantry forces were mercenaries, and the majority of them were arquebusiers. Largely because of the greater expense of mercenaries relative to that of native troops, the royal debt reached about 43,500,000 *livres,* about two and a half times the annual income of the monarchy.[41] We can only speculate how many French defeats in the sixteenth century might have gone the other way had the French deployed adequate numbers of both affordable and capable arquebusiers. With its agricultural wealth, vast population, and relatively cohesive government, France should have been the most powerful state in Europe from 1477 to 1559. It is true that Spain's success in acquiring that supremacy came from its new-found colonial wealth and religious unity; but its success also owed much to the adoption of a formidable firearms culture and, in turn, to France's ambivalence towards relinquishing its medieval material heritage.

Notes

1. See H. Heller, *Labour, Science and Technology in France* (Cambridge, 1996), for a number of non-military examples. The best example of relatively rapid technology adoption in the early modern period was the printing press, but even that took more than 20 years to reach Paris from Cologne.

2. J. Nef, *War and Human Progress* (New York, 1968), pp. 67–69.

3. B. Hall, *Weapons and Warfare in Renaissance Europe* (Baltimore, 1997), pp. 40–57

4. P. Contamine, *Histoire militaire de France*, volume 1 (Paris, 1992–1994), p. 132.

5. J. D'Auton, *Chroniques de Louis XII*, volume 4, ed. R. de Maulde (Paris, 1889–1895), pp. 48–51.

6. Contamine, *Histoire militaire*, pp. 149–150.

7. Hall, *Weapons and Warfare*, p. 123.

8. Baumgartner, *Louis XII* (New York, 1994), p. 111; Brantôme, *Discours sur les colonels de l'infanterie de France*, ed. E. Vaucheret (Paris, 1973), p. 164.

9. S. Kinser, ed. *The Memoirs of Philippe de Commynes*, volume 2 (Columbia, SC, 1973), p. 541.

10. D'Auton, *Chroniques de Louis XII*, volume 1, p. 113; volume 3, p. 172.

11. Nemours was hit three times by arquebus shot. Ibid., III, p. 173.

12. R. de Maulde de Clavière, ed., *Procedures politiques du règne de Louis XII* (Paris, 1885), pp. 87–97.

13. D. Godefroy, *Lettres de Louis XII*, volume 4 (Amsterdam, 1712), p. 248.

14. C. de Seyssel, *The Monarchy of France* (New Haven, 1991), p. 114.

15. F. Lot, *Recherches sur les effectifs des armées français des guerres d'Italie aux guerres de religion* (Paris, 1962), p. 42.

16. Brantôme, cited in G. Gaier, "L'Opinion des chefs de guerre français sur les progrès de l'art militaire," *Revue internationale d'histoire militaire* 29 (1970): 734.

17. *Ordonnances des roys de France de la troisième race*, 22 vols. (Paris, 1723–1846): François Ier, III, no. 299.

18. Ibid., VII, no. 666.

19. Lot, *Recherches*, p. 103.

20. Sieur de Fourquevaux, *Instructions sur le faict de la guerre*, ed. G. Dickinson (London, 1954), p. 12r.

21. *Institution de la discipline militaire du Royaume de France* II (Lyon, 1559), p. 46.

22. J. Wood, *The King's Army* (Cambridge, 1996), p. 91.

23. I. Cloulas, *Henri II* (Paris, 1985), p. 494. The French source used the term *arquebuse à croc*, which referred to a heavy firearm that used a fork. See Hall, *Weapons and Warfare*, p. 171.

24. Brantôme, *Discours*, p. 165.

25. Concerning the debate over where the wheel-lock mechanism was first developed, see V. Foley et al., "Leonardo, the Wheel Lock, and the Milling Process," *Technology and Culture* 24 (1983): 399–427. The first known examples of the wheel-lock pistol come from Germany.

26. F. de La Noue, *Discours politiques et militaires*, ed. F. Sutcliffe (Geneva, 1967), pp. 360–367. See also Hall, *Weapons and Warfare*, pp. 193–194.

27. The following is taken largely from my "The Final Demise of the Medieval Knight in France," in *Regnum, Religio, et Ratio*, ed. J. Friedman (St. Louis, 1988), pp. 9–17.

28. DuBartas, *The Works of Guillaume de Salluste, sieur DuBartas*, volume 2 (Chapel Hill, 1940), p. 496.

29. Machiavelli, *The art of war . . . in seven books* (Albany, 1815).

30. S. Gigon, "L'art militaire dans Rabelais," *Revue des Etudes Rabelaisiennes* 5 (1907): 3–23. Rabelais' only mention of handguns in *Gargantua* is the use of two variants of *arquebusier*, despite the fact that his hero is a king who fights a major war.

31. Montaigne, *Essais*, Book I, Number 48; ed. J. Plattard (Paris, 1960), II, p. 208.

32. http://humanities.uchicago.edu/ARTFL

33. Hale, "Gunpowder and the Reniassance: An Essay in the History of Ideas," *Renaissance War Studies* (London, 1983): 389–420.

34. *La très joyeuse, plaisante et récréative Hystoire de Seignenur de Bayard*, in M. Petitot, *Collection complête des mémoires relatifs à l'histoire de France* (Paris, 1819). This work makes clear both now many prominent French nobles were killed by arquebus fire and how little use the French made of the weapon up to 1524.

35. Monluc, *Military Memoirs* (Hamden, 1972), p. 41.

36. Ibid., p. 111.

37. M. d'Amboise, *Le guidon des gens de guerre* (Paris, 1544), p. 2.

38. In M. Gachard, *Relations des ambassadeurs vénétiens sur Charles-Quint et Philipppe II* (Brussels, 1856), p. 309.

39. La Noue, *Discours politiques et militaires*, pp. 352–355. For an excellent discussion of the attitudes and military styles of the mid-sixteenth-century French cavalry, see T. Tucker, "Eminence over Efficacy: Social Status and Cavalry Service in Sixteenth-Century France," *Sixteenth Century Journal* 32 (2001): 1057–1095.

40. J. Lynn, "The Military Resurrection of the Pike," *Military Affairs* 41 (1977): 1–7.

41. See Baumgartner, *Henry II King of France* (Durham, 1988), especially pp. 247–248.

3

Gunpowder and the Changing Military Order: The Islamic Gunpowder Empires, ca. 1450–ca. 1650

Barton C. Hacker

Gunpowder weapons altered the ancient military balance between the steppe and the sown across Eurasia. Ultimately, firearms completed the painful and much-prolonged decline of cavalry, though hardly overnight. Through most of Eurasia, guns also weakened most forms of resistance to central authority. Gunpowder artillery rendered most existing fortifications obsolescent. Everywhere in Eurasia during the early gunpowder era, great guns seemed to inspire military imaginations. Artillery improved rapidly, but every increase in power meant heavier, clumsier, and costlier guns. In the attack and defense of fixed positions, such shortcomings mattered less than the enormous weight thrown by big guns, and their great expense meant that central governments ordinarily enjoyed near monopolies on their use.

Marshall Hodgson coined the term "gunpowder empire" for the three large military-patrimonial-bureaucratic states—Ottoman in the Near East, Safavid in Iran, Mughal in India—that took advantage of the new weapons technology. Nomadic conquerors, so armed, divided the Moslem world among themselves between the middle of the fifteenth century and the middle of the sixteenth.[1] This essay focuses only on these three Islamic states, although further east the later Ming dynasty and its Manchu conquerors were roughly their contemporaries and reflected some similar if not strictly comparable trends.[2] Gunpowder weapons also played significant roles in the reorganization of polities on the periphery, in Europe, in Russia, in Southeast Asia, and in Japan.[3]

Central Asian Military Institutions

Firearms did not appear soon enough to counter the last great irruption of steppe nomads, the Mongol conquests of the thirteenth century. The Mongol Empire rose on a foundation laid during the preceding two centuries. Powerful but relatively short-lived nomad states along China's northern frontiers had elaborated and improved administrative and command structures modeled on those of their more settled neighbors. Mongol success owed much to this reorganization of steppe military and political institutions.[4]

Although other steppe dwellers swelled their ranks, Mongol armies seldom matched their foes in size. Sophisticated organization, adroit leadership, rapid movement, and skillful tactics—rather than the vast hordes imagined by stunned victims—accounted for the extraordinary Mongol victories.[5] Like other nomad conquerors over nearly two millennia, the Mongols were horse archers. Men bred to the saddle and armed with compound bows combined mobility and striking power to a degree rarely matched by the armed forces of any sedentary state.[6]

From their central Asian homeland, Mongol armies swept across Eurasia, defeating every army they faced from northern China through western Asia to eastern Europe.[7] Geography alone seemed to limit their advance, as in the desert approaches to Egypt, the mountains of Korea, or the islands of Japan.[8] Although the Mongol Empire soon fragmented, its successors dominated much of Eurasia well into the fourteenth century. The Yüan dynasty in China, the Ilkhanids in Iran, and the Golden Horde in Russia were only the most notable.[9] The Mongol legacy remained alive in memory far longer, from the conquests of Tamerlane to modern Mongolia.[10]

Paradoxically, steppe arms enjoyed these great triumphs just as the conditions that made such feats possible faded away. Gunpowder had been little more than a battlefield curiosity in making its way westward from China early in the second millennium. But coupled with the gun in the West, gunpowder became an era-defining commodity from the thirteenth century onward. Spreading eastward, firearms undercut the ancient bases of nomad military superiority and the independence of the steppe peoples. No longer could steppe warriors rely on their own skills and resources to arm themselves, and the very arms they were compelled to seek in settled lands would in time drive horses from the battlefield. Ultimately, properly equipped standing armies freed civilized society from the age-old threat of barbarian incursion. As Adam Smith observed in the late eighteenth century, "The invention of fire-arms, an invention which at first sight appears to be so pernicious, is certainly favourable both to the permanency and to the extension of civilization."[11]

But the day of the horseman did not close quickly. In the fifteenth and sixteenth centuries, three conquerors originating from the steppes—Ottomans, Safavids, and Mughals—divided the Islamic world among themselves. They won lasting empires by melding traditional steppe cavalry with centrally controlled gunpowder weapons: small arms, field guns, and siege artillery. Mounted archers remained the military foundation for all the Islamic gunpowder empires—and for most of their rivals. This traditional warrior elite normally disdained gunpowder weapons. Imperial success hinged on finding ways to incorporate the new weapons despite the horsemen's cultural resistance.

Ottomans and Mughals solved the problem, as did the Safavids with more difficulty; others, like the Mamluks of Egypt, did not. Yielding to European and Ottoman pressure, the Mamluks adopted gunpowder weapons reluctantly. Not only would warriors have to master new weapons, they would have to do it on foot. This they largely refused to do themselves, and they also strongly resisted raising arquebus units manned by low-class recruits. The threatened devaluation of their traditional skill with the bow on horseback challenged their honor as much as their status. In the end, the Ottoman conquest of Egypt did look like a triumph of overwhelming firepower, in spite of the Mamluk's belated efforts to catch up.[12]

Ottoman Military Institutions and Empire

"The Ottoman government had been an army before it was anything else," observed Albert Howe Lybyer in his classic study of the Ottoman polity under Suleiman the Magnificent. "Fighting was originally the first business of the state, governing the second."[13] Government followed the army on campaign, the two scarcely distinguishable. To a degree, the officers who commanded the army were the officials who administered the government. Ultimately, imperial expansion and the growth of civil society loosened the tight linkage between government and army. For the several centuries it lasted, however, that unity helped ensure the well-organized administration and support services that underpinned Ottoman military success.

During the centuries of imperial expansion, horse archers normally furnished the bulk of Ottoman armies. Freelance light cavalry, the *akinci* forces, predominated until the fifteenth century, trading their military service for the lion's share of war booty. Early Ottoman armies also mustered provincial infantry and auxiliary troops. The sixteenth century was marked by the growing importance of the provincial cavalry, *sipahis* or *timariots*, supported by land grants (*timar*). Fewer in number than the *timariots*, but no less important, were the sultan's paid household troops, the *kapu kullari* (literally, servitors of the [palace] gate).[14] It was also a slave army, a traditional Islamic institution; those who accepted the sultan's pay became members of his slave household.[15]

Although including some cavalry, the *kapu kullari* chiefly comprised infantry—namely, the janissary corps (*yeni cheri*, new troops). As a body of troops dependent on, and personally loyal to, the sultan, the janissary corps provided a military counterweight to the sometimes fractious provincial forces. The corps drew its personnel from a regularized levy of Christian boys (*devshirme*). Taken from their Slavic families in the conquered Balkans, they trained in palace schools as soldiers; the brightest of them were reserved for the bureaucracy and had the potential to rise to the apex of Ottoman administration, second only to the sultan himself.[16]

Originating in the fourteenth century, the janissary corps assumed new importance with the spreading use of firearms in the following century. European craftsmen, some captured but many hired, became the chief agents of technology transfer.[17] Ottoman forces first met firearms in the Balkans in the middle of the fourteenth century, although no reliable evidence of Ottoman use of firearms or cannon antedates 1400. Firearms had certainly been introduced by the reign of Mehmed I (1413–1421). Despite firmly attested siege guns by 1422, effective Ottoman siege artillery did not appear much before its 1453 success against Constantinople. Under the well-reasoned leadership of Mehmed II the Conqueror (1451–1481), big guns played a major role in deciding the city's fate. The Ottomans were widely believed to have created the most powerful siege artillery in Europe.[18]

Field guns were in Ottoman service by the 1440s, about the same time the Ottomans most likely first encountered arquebuses (though it might have been as much as two decades earlier). Ambiguous nomenclature clouds the early history of firearms anywhere in the Islamic world, but the janissaries clearly were using the arquebus by the middle of the fifteenth century.[19] Ottoman artillery has been widely credited for the decisive Ottoman victory at Bashkent in 1473 that effectively halted the expansion from Iran of the Aq Qoyunlu (White Sheep) Turcoman confederation. But after defeats at the hands of the Mamluks in the 1480s, Bayazid II (1481–1512) augmented the janissaries and generally upgraded all gunpowder weapons and techniques.[20]

Renewed success followed these reforms. Ottoman armies inflicted crushing defeats on the Persian Safavids at Chaldiran in 1514, the Egyptian Mamluks at Marj Dabiq in 1516 and Raydaniyah in 1517, and the Hungarians at Mohacs in 1526. Once widely regarded as victories "of modern military technology over the outdated steppe ways of warfare" and credited largely to superior Ottoman field artillery and matchlock units, they more recently appear as battles decided by better tactics and organization.[21]

Adapting the old steppe practice of linking wagons as improvised field fortresses, the Ottomans added light guns and matchlocks. The immediate inspiration for this innovation, the Ottoman *tabur*, or *destur-I Rumi*, may well have been John Hunyadi's updated version of the Hussite *Wagenburg*, which the Ottomans met on campaign in Hungary during the early 1440s. Heavy wagons chained together anchored the battle line. Shielded (at least temporarily) against charging cavalry, gunners and janissaries could fire and reload and fire again, holding the center and freeing the Ottoman cavalry to envelop the enemy's flanks.[22]

Ottoman success against rival states owed a great debt to its effective use of siege artillery, field guns, and small arms. But gunpowder weapons could as easily point inward as outward. The adoption of firearms, as Djudjica Petrovic has observed, "coincided clearly with trends towards the creation of a strong centralized authority and of a centralized army standing close to the sultan."[23] With an effective monopoly on gunpowder weapons, the *kapu kullari* effectively counterbalanced the much more numerous provincial forces whose interests did not always coincide with the sultan's.

Safavid Military Institutions

The Ottoman Empire's great rival in the east was the revived Persian Empire under the Safavids.[24] Whatever uncertainty persists about the early history of firearms in Iran, Rudi Matthee concludes "that Iran took to firearms with the same eagerness exhibited by the Ottomans, the Mughals and the Russians."[25] Restricted access to suppliers of firearms presented a major problem. The first reliable evidence of gunpowder weapons in Iran comes from the late fifteenth century, with the establishment of Aq Qoyunlu rule. European efforts to arm Iranians against Ottomans started in 1471 but met little success before the end of the century.[26]

When the Safavids succeeded the Aq Qoyunlu at the beginning of the sixteenth century, firearms played no apparent role. The power of the Safavid founder, Shah Isma'il (1502–1524), was part feudal, part theocratic. He relied on tribal Turcoman levies, chiefly traditional steppe warriors.[27] But cannon and matchlock seemed decisive in the major Ottoman victory over the Safavids at Chaldiran in 1514. To compete with the Ottomans, Shah Isma'il sought to create matchlock and artillery units of his own. He tried hard to acquire European firearms, although, rather surprisingly, much material came from Ottoman sources. In 1528, the Safavids crushed the Uzbeks at the battle of Hashhad using the same Ottoman *tabur* tactics that had contributed to their own downfall at Chaldiran fourteen years earlier.[28]

For the most part, however, the rearmament effort may have promoted more propaganda than progress. Shah Isma'il's Qizilbash (Red Turban) supporters resisted gunpowder weapons with considerable effect. In a cavalry-based army, matchlock handguns posed practical problems, but they also conflicted with ideals of manhood, in Qizilbash eyes no less than Mamluk, Russian, or Japanese.[29] Recruitment of gun-carrying infantry from slaves and peasants did little to enhance the prestige of firearms, which remained despised into the nineteenth century. Although handguns became more common as the sixteenth century advanced, Qizilbash antipathy checked their military significance. Artillery held even less attraction for warriors who still relied on rapid movement and traditional Central Asian cavalry methods. Only in siege warfare did guns become increasingly important.[30]

Shah Abbas I (1587–1629) established a standing army as counterweight to the increasingly unreliable Qizilbash tribesmen that had been the chief Safavid armed force. Circassian military slaves (*ghulam*) formed the new permanent army, paid for by the shah and answerable only to him. The process began early in his reign with a *ghulam* cavalry unit. Its members, called *qullar*, eventually numbered 10,000. Later the shah formed a 3,000-*qullar* personal bodyguard; an artillery corps of 12,000 men (*tupchis*) and 500 guns; and 12,000 infantry armed with muskets. Paying such a force required radical changes in Safavid administration.[31]

The reforms of Shah Abbas had two purposes. Internally, they aimed to reduce the significance of the Qizilbash element in the army. Externally, they provided new troops equipped with firearms who could stand up to Ottoman forces.[32] Like Ottoman janissaries and Russian *strel'tsy*, Safavid *ghulam* represented a government attempt to reduce its dependence on tribal forces and to break the connection between land and service. In contrast to Ottoman and Russian practice, however, fully a third of the *ghulam* retained traditional arms and tactics.[33]

Difficult terrain, lack of resources for making both guns and gunpowder, and Ottoman blockade, all may have contributed to the well-attested absence of Safavid interest or competence in field artillery and under-utilization of siege artillery. Yet during the Safavid period the cities of Persia were largely unwalled, reflecting the empire's physical environment, internal stability, and municipal disorganization. Above all, Safavids neglected artillery because it served little purpose against their most immediate threat—nomadic horsemen from the north and east.[34]

Military Foundations of the Mughal Empire

The establishment of the Mughal Empire in northern India owed much, though certainly not everything, to the new gunpowder technology.[35] Reliable evidence of firearms or guns anywhere in the Indian subcontinent dates from the middle of the fifteenth century.[36] European expansion into the Near East and the Indian Ocean brought firearms to the attention of rulers, some of whom soon became eager to acquire these new weapons for themselves. The Ottomans provided them with guns to resist Portuguese expansion in the Indian Ocean and to curb the Russian advance into central Asia. In fact, they played a major role in introducing firearms throughout Islam, often as direct suppliers—to khanates in Turkestan and the Crimea, to the Gujeratis in India, to the Sultan of Aceh in Sumatra—sometimes provoking such rivals as the Aq Qoyunlu and Safavids in Iran and the Mamluks in Egypt to seek gunpowder weaponry from Christian Europeans.[37]

Babur, founder of the Mughal Empire, benefited greatly from two Ottoman firearm specialists, Usted Ali-Qulï and Mustafa Rumi. Not only a skilled marksman with the matchlock, Ali-Qulï could cast large guns.[38] Mustafa knew how to construct Ottoman-style wagons armed with guns, thereby enabling Babur to employ the tactics of *destur-I Rumi*, the Ottoman version of the *Wagenburg*.[39] But firearms were not the sole basis of Mughal victory. For the decisive battle of Panipat in 1526, Babur used the same tactics as the Ottomans at Chaldiran (1514). In the words of his memoir, he "ordered the whole army . . . to bring carts, which numbered about seven hundred altogether, . . . [and] to tie them together with ox-harness ropes instead of chains, after the Anatolian manner. . . . The matchlockmen could then stand behind the fortification to fire their guns."[40] Yet mounted archers proved decisive in a bold flanking action reminiscent of Alexander at the battle of the Hydaspes.[41]

A year later, Babur faced a much larger Rajput Confederacy army, but the battle of Khanua reprised the battle of Panipat.[42] "Babur's guns and his long-practiced use of the enveloping tactics of Central Asian cavalry," comments John F. Richards, proved no less effective against Rajputs than Afghans.[43] Rajput aristocrats, like all who prized the skills of mounted warriors, disliked guns. A few years after the battle of Khanua, "Maharana Vikramaditya of Mewar (c. 1536) took more interest in his foot soldiers (*paeks*) and their firearms than was acceptable to the equestrian Rajput aristocracy," as historian Robert Elgood recently observed. Consequently, he "suffered for his perceptiveness as innovatory rulers in Mamluk Egypt had done."[44]

Although Babur's son failed to consolidate his father's conquests, his grandson, Akbar, regained them. At the second battle of Panipat in 1556, the Mughal army under Bayram Khan, acting for the young Akbar, defeated a joint Hindu-Afghan army and restored Mughal rule of North India.[45] Maintaining rule over a mighty empire, Akbar (1556–1605) still took a personal interest in matchlocks.[46] His chronicler, Abu al-Fazl, noted the importance of artillery to the Mughals: "Guns are wonderful locks for protecting the august edifice of the state, and befitting keys for the doors of conquest. With the exception of the Ottoman empire, there is perhaps no country which in its guns has more means of securing the government than this."[47]

Yet gunpowder weapons had become widely available in mid-sixteenth-century India. While Akbar used them more effectively than any of his opponents, his success owed more to organizational prowess than technological innovation.[48] Although more Timurid than Mongol (from which Mughal derives), the dynasty was strongly influenced by patrimonial aspects of the Mongol Empire. It also drew on the highly bureaucratic model of ancient Sasanid rule in Persia and adopted aspects of the widespread *iqta* system of land tenure and other traditional methods of maintaining armed forces. For all their borrowings, the Mughals carried centralization and systematization much farther than any earlier Islamic system, and that became a major foundation of Mughal longevity.[49]

Abu al-Fazl compiled a comprehensive manual of Akbar's system of statecraft, the *A'in-I Akbari* (Regulations of Akbar). Akbar's *mansab* system was the major administrative innovation that sustained the Mughal state. The government had three branches—household, army, and empire—administered on military lines. All officials were designated *mansabdar* (Persian for "officeholder"). The basic *mansab* unit was called *suwar*, horseman; it was the military contingent that its possessor was obliged to maintain. Each of the 33 offices required the holder to provide a specified number of horsemen to the imperial army. In practice, compliance was limited; personal rank (*zat*) became the grade of the holder within the imperial service, while the *suwar* rank designated the horse contingent that was actually provided.[50]

Mughal tactical superiority rested on a unique (in the Indian context) combination of mounted archers, artillerists, and musketeers. Their enemies increasingly retreated behind fortress walls, which the Mughals demonstrated they could breach, but not always easily and cheaply. As a result, Mughal expansion involved a good deal of negotiation. The Mughals mostly absorbed, rather than displaced, existing local and regional elites, each of which tended to maintain its own military contingent.[51] For much the same reasons, the Mughal's own "fortresses served as bulwarks against revolt, because the Mughals alone had the ability to take them, except in unusual circumstances."[52] The military importance of towns further benefited from their central role in manufacturing guns and muskets. As towns increasingly dominated the countryside, the Mughal ruling class became urbanized.[53] Whether attacking enemy cities or defending their own, the Mughal control of fortress-breaching artillery made a decisive difference.

Guns and Government

From Abbasids to Mughals, the pattern of Islamic state building began with the conquest of settled lands by confederated nomads. Conquerors then sought to impose centralized administrations on low-surplus agrarian economies ill-suited for the purpose.[54] The usual expedient was some form of the system traditionally termed *iqta*—or *timar* by the Ottomans, *tiyul* by the Safavids, *tiyul* or *jagir* by the Mughals—which traded land revenue for service. Providing the state with civil and military officers, members of the service class in essence derived their salaries directly from the rents or taxes they collected on lands assigned to them. Provincial levies provided the bulk of army manpower. An elite core of troops, however, were paid from the royal treasury and remained loyal to the ruler only.[55]

Many land grants nevertheless went to local leaders whose new resources reinforced the loyalty of tribal warriors to them rather than to some distant ruler. The recurrent bane of the *iqta* system was subordinates grown too powerful. Although the army's royal core might well overpower the force of any single warlord, it was always vastly outnumbered by the sum of tribal warriors. When provincial interests clashed with imperial desires, as they inevitably did, the numbers weighed against the ruler.[56]

By the sixteenth century, gunpowder weapons sharply altered the balance between central government and provincial forces. Steppe warriors did not, for the most part, welcome firearms, which neither suited their traditional ethos nor promoted their skills in mounted warfare. Confronted by foes armed with the new weapons, Ottoman sultans in the fifteenth century, as well as Safavid shahs and Mughal emperors in the sixteenth, bypassed the resistance of mounted archers by making use of paid slave soldiers, an old Islamic military institution. Under direct central authority, they created from such troops new infantry and artillery units to exploit gunpowder weapons.[57]

These military changes not only improved battlefield performance but also encouraged state centralization and contributed greatly to the enhanced stability of the Islamic gunpowder empires. The competitive use of gunpowder weapons required a military transformation, which inevitably called into question other aspects of the social order that were linked to military organization. Accordingly, change extended beyond purely military adjustments. The high cost of artillery and the effort of keeping up with changing technology favored a well-organized central authority, at the expense of local power centers with more limited resources. Although the introduction of gunpowder weapons may not stand as the deterministic cause of the early modern changes, it certainly emerges as a major factor.

Notes

1. Marshall G. S. Hodgson, *The Gunpowder Empires and Modern Times*, volume 3 of *The Venture of Islam: Conscience and History in a World Civilization* (University of Chicago Press, 1974). See also William H. McNeill, "The Gunpowder Revolution and the Rise of Atlantic Europe," in *The Pursuit of Power: Technology, Armed Force, and Society since A.D. 1000* (University

of Chicago Press, 1982), pp. 79–102; McNeill, *The Age of Gunpowder Empires, 1450–1800* (American Historical Association, 1989); Arnold Pacey, "Gunpowder Empires, 1450–1650," in *Technology in World Civilization: A Thousand-Year History* (MIT Press, 1990).

2. Peter Lorge, "War and Warfare in China, 1450–1815," in *War in the Early Modern World, 1450–1815*, ed. J. Black (Westview, 1999); C. J. Peers, *Medieval Chinese Armies, 1260–1520*, Men-at-Arms Series 251 (Osprey, 1992); Albert Chan, *The Glory and Fall of the Ming Dynasty* (University of Oklahoma Press, 1982); Chia-chü Liu, "The Creation of the Chinese Banners in the Early Ch'ing," *Chinese Studies in History* 14 (summer 1981): 47–75; Frederic Wakeman Jr., *The Great Enterprise: The Manchu Reconstruction of Imperial Order in Seventeenth-Century China*, 2 vols. (University of California Press, 1985.

3. Geoffrey Parker, *The Military Revolution: Military Innovation and the Rise of the West, 1500–1800*, second edition (Cambridge University Press, 1996); Clifford J. Rogers, ed., *The Military Revolution Debate: Readings on the Military Transformation of Early Modern Europe* (Westview, 1995); Richard Hellie, "Warfare, Changing Military Technology, and the Evolution of Muscovite Society," in *Tools of War: Instruments, Ideas, and Institutions of Warfare, 1445–1871*, ed. J. Lynn (University of Illinois Press, 1990); Anthony Reid, "The Military Revolution," in *Southeast Asia in the Age of Commerce, 1450–1680*, volume 2 (Yale University Press, 1993); Stephen Morillo, "Guns and Government: A Comparative Study of Europe and Japan," *Journal of World History* 6 (Spring 1995): 75–106; Paul Varley, "Warfare in Japan, 1467–1600," in Black, *War in the Early Modern World.*

4. Thomas J. Barfield, *The Perilous Frontier: Nomad Empires and China* (Blackwell, 1989); Sechin Jagchid and Van Jay Symons, *Peace, War, And Trade along the Great Wall: Nomadic-Chinese Interaction through Two Millennia* (Indiana University Press, 1989); Gary Seaman and Daniel Marks, eds., *Rulers from the Steppe: State Formation on the Eurasian Periphery* (University of Southern California, Center for Visual Anthropology, Ethnographics Press, 1991).

5. Leo de Hartog, "The Mongol Army," in *Genghis Khan* (St. Martin's Press, 1989); David Nicolle, *The Mongol Warlords* (Firebird , 1990); David O. Morgan, "The Mongol Army," in *The Mongols* (Blackwell, 1986); S. R. Turnbull, *The Mongols* (Osprey, 1980).

6. Thomas J. Barfield, "The Devil's Horsemen: Steppe Nomadic Warfare in Historical Perspective," in *Studying War: Anthropological Perspectives*, ed. S. Reyna and R. Downs (Gordon & Breach, 1994); J. D. Latham, "The Archers of the Middle East: The Turco-Iranian Background," *Iran* 8 (1970): 97–103; Erik Hildinger, "Horse and Bow," in *Warriors of the Steppe* (Sarpedon, 1997).

7. Robert Marshall, *Storm from the East: From Genghis Khan to Khubilai Khan* (University of California Press, 1993); David O. Morgan, "The Mongol Armies in Persia," *Der Islam* 56 (1979): 81–96; Claude Cahen, "The Mongols and the Near East," in *A History of the Crusades*, second edition, volume 2, ed. R. Wolff (University of Pennsylvania Press, 1969); James Chambers, *The Devil's Horsemen: The Mongol Invasion of Europe* (Atheneum, 1979).

8. Reuven Amitai-Preiss, *Mongols and Mamluks: The Mamluk-llkhanid War, 1260–1281* (Cambridge University Press, 1995); John Masson Smith Jr., "Ayn Jalut: Mamluk Success or Mongol Failure?" *Harvard Journal of Asiatic Studies* 44 (1984): 307–345; William E. Henthorn, *Korea: The Mongol Invasions* (Leiden: E. J. Brill, 1963); Kyotsu Hori, "The Economic and

Political Effects of the Mongol Wars," in *Medieval Japan: Essays in Institutional History*, ed. J. Hall and Jeffrey P. Mass (Stanford: Stanford University Press, 1988; reprint of 1974 edition).

9. Elizabeth Endicott-West, "Imperial Governance in Yüan Times," *Harvard Journal of Asiatic Studies* 46 (1986): 523–549; Ch'i-ch'ing Hsiao, *The Military Establishment of the Yüan Dynasty* (Harvard University, Council on East Asian Studies, 1978); J. Boyle, ed., *The Cambridge History of Iran*, volume 5, *The Saljuq and Mongol Periods* (Cambridge University Press, 1993); Bertold Spuler, *History of the Muslim World: the Mongol Period* (Markus Wiener, 1994); Charles J. Halperin, *Russia and the Golden Horde: The Mongol Impact on Medieval Russian History* (I. B. Tauris, 1987); Donald Ostrowski, *Muscovy and the Mongols: Cross-Cultural Influences on the Steppe Frontier, 1304–1589* (Cambridge University Press, 1998).

10. Beatrice Forbes Manz, *The Rise and Rule of Tamerlane* (Cambridge University Press, 1989); David Nicolle, *The Age of Tamerlane: Warfare in the Middle East, c. 1350–1500*, Men-at-Arms Series 222 (Osprey, 1990); Sechin Jagchid and Paul Hyer, *Mongolia's Culture and Society* (Westview; Folkestone: Dawson, 1979); Reuven Amitai-Preiss and David O. Morgan, eds., *The Mongol Empire and Its Legacy* (Leiden: Brill, 1999).

11. Adam Smith, *An Inquiry into the Nature and Causes of the Wealth of Nations*, ed. E. Cannan (Modern Library, 1937), p. 669. For a review of the literature on early gunpowder weaponry, see Kelly DeVries, "Early Modern Military Technology: New Trends and Old Ideas (A Bibliographical Essay)," *Techniek* 8 (June 1992): 73–88.

12. David Ayalon, *Gunpowder and Firearms in the Mamluk Kingdom: A Challenge to a Medieval Society* (Vallentine, Mitchell, 1956); David C. Nicolle, *The Mamluks, 1250–1517* (Osprey, 1993).

13. Albert Howe Lybyer, *The Government of the Ottoman Empire: In the Time of Suleiman the Magnificent*, Harvard Historical Studies 18 (Cambridge, MA: Harvard University Press, 1913), 90. See also Virginia Aksan, "Ottoman War and Warfare, 1453–1812, in Black (ed.), *War in the Early Modern World*, p. 150.

14. Rhoads Murphey, *Ottoman Warfare, 1500–1700* (UCL Press, 1999), 36; Aksan, "Ottoman War and Warfare," pp. 150–151.

15. Daniel Pipes, *Slave Soldiers and Islam: The Genesis of a Military System* (Yale University Press, 1981); Patricia Crone, *Slaves on Horses: The Evolution of the Islamic Polity* (Cambridge, MA: Harvard University Press, 1980); Jere L. Bacharach, "African Military Slaves in the Medieval Middle East: The Case of Iraq (869–955) and Egypt (808–1171)," *International Journal of Middle Easter Studies* 13 (1981): 471–495.

16. Lybyer, *Government of the Ottoman Empire*, pp. 90–91; Godfrey Goodwin, "The Ottoman Armed Forces," in *The Janissaries* (Saqi Books, 1994); David C. Nicolle, *The Janissaries*, (Osprey, 1995); Nicolle, *Armies of the Ottoman Turks, 1300–1774* (Osprey, 1983); Özer Ergenç, "The Qualifications and Functions of Ottoman Central Soldiers," *Revue Internationale d'Histoire Militaire* no. 67 (1988): 45–56.

17. Gábor Ágoston, "Ottoman Artillery and European Military Technology in the Fifteenth [to] Seventeenth Centuries," *Acta Orientalia Academiae Scientiarum Hungaricae* 47 (1994): 15–48; Wayne S. Vucinich, *The Ottoman Empire: Its Record and Legacy* (D. Van Nostrand, 1965),

31; Silah Özbaran, "The Ottomans' Role in the Diffusion of Firearms and Military Technology in Asia and Africa in the Sixteenth Century," *Revue International d'Histoire Militaire,* no. 67 (1988): 77–84.

18. Kelly DeVries, "Gunpowder Weapons at the Siege of Constantinople, 1453," in *War and Society in the Eastern Mediterranean, 7th-15th Centuries*, ed. Y. Lev (Leiden: E. J. Brill, 1997); Robert Elgood, *Firearms of the Islamic World: In the Tareq Rajab Museum, Kuwait* (I. B. Tauris, 1995), pp. 31–32; Ágoston, "Ottoman Artillery and European Military Technology," pp. 24–25.

19. Witnesses are often unclear in distinguishing between cannon and handguns. Nomenclature is the historian's constant problem, and not only in the Ottoman realm. Firearms moved readily through the Moslem world, the technology often accompanied by the words to describe it. See especially Elgood, *Firearms of the Islamic World*, pp. 32, 136; Ágoston, "Ottoman Artillery and European Military Technology," pp. 32–44.

20. David Morgan, *Medieval Persia, 1040–1797* (Longman, 1988), p. 105; Halil Inalcik, "The Socio-Political Effects of the Diffusion of Fire-Arms in the Middle East," in *War, Technology and Society in the Middle East*, ed. V. Parry and M. Yapp (Oxford University Press, 1975), p. 207.

21. Morgan, *Medieval Persia*, p. 117.

22. Inalcik, "Socio-Political Effects of the Diffusion of Fire-Arms," 204; Elgood, *Firearms of the Islamic World*, 33; David Nicolle, *Hungary and the Fall of Eastern Europe, 1000–1568* (Osprey, 1988), pp. 12–13.

23. Djudjica Petrovic, "Fire-Arms in the Balkans on the Eve of and after the Ottoman Conquests of the Fourteenth and Fifteenth Centuries," in Parry and Yapp, eds., *War, Technology and Society in the Middle East*, p. 175.

24. Adel Allouche, *The Origins and Development of the Ottoman-Safavid Conflict (906–962/1500–1555)* (Berlin: Schwarz, 1983); Peter Jackson and Laurence Lockhart, eds., *The Timurid and Safavid Periods*, volume 6 of *The Cambridge History of Iran* (Cambridge University Press, 1986).

25. Rudi Matthee, "Unwalled Cities and Restless Nomads: Firearms and Artillery in Safavid Iran," in *Safavid Persia: The History and Politics of an Islamic Society*, ed. C. Melville (I. B. Tauris, with the Centre of Middle Eastern Studies, University of Cambridge, 1996), p. 393.

26. Elgood, *Firearms of the Islamic World*, pp. 113–114.

27. Laurence Lockhart, "The Persian Army in the Safavi Period," *Der Islam* 34 (1959), p. 89.

28. Elgood, *Firearms of the Islamic World*, 116; Inalcik, "Socio-Political Effects of the Diffusion of Fire-Arms," 207.

29. Matthee, "Unwalled Cities and Restless Nomads," pp. 392–393. Cf. Ayalon, *Gunpowder and Firearms in the Mamluk Kingdom*; Richard Hellie, "Warfare, Changing Military Technology, and the Evolution of Muscovite Society," in Lynn (ed.), *Tools of War*, Noel Perrin, *Giving up the Gun: Japan's Reversion to the Sword, 1545–1879* (Boston: David R. Godine, 1979).

30. Matthee, "Unwalled Cities and Restless Nomads," p. 393; Elgood, *Firearms of the Islamic World*, p. 117.

31. Morgan, *Medieval Persia,* 135.

32. Masashi Haneda, "The Evolution of the Safavid Royal Guard," *Iranian Studies* 22, no. 2–3 (1989), p. 59.

33. Matthee, "Unwalled Cities and Restless Nomads," 394.

34. Matthee, "Unwalled Cities and Restless Nomads," pp. 394–408.

35. Jos Gommans, "Warhorses and Gunpowder in India, c. 1000–1850," in Black (ed.), *War in the Early Modern World*; M. K. Zaman, "The Use of Artillery in Mughal Warfare," *Islamic Culture* 57 (1983): 297–304; S. P. Verma, "Fire-Arms in Sixteenth Century India (A Study Based on Mughal Paintings of Akbar's Period)," *Islamic Culture* 57 (1983): 63–69.

36. Iqtidar Alam Khan, "Early Use of Cannon and Musket in India, A.D. 1442–1526," *Journal of the Economic and Social History of the Orient* 24 (1981): 146–164.

37. Inalcik, "Socio-Political Effects of the Diffusion of Fire-Arms," 202. 210.

38. *Baburnama*, 326, 372. See also Jagadish Narayan Sarkar, *The Art of War in Medieval India* (New Delhi: Munshiram Manoharlal, 1984), 135; Inalcik, "Socio-Political Effects of the Diffusion of Fire-Arms," 204.

39. *The Baburnama: Memoirs of Babur, Prince and Emperor*, ed. W. Thackston (Freer Gallery of Art, Arthur M. Sackler Gallery, Oxford University Press, 1996), pp. 270, 363–364.

40. *Baburnama*, p. 323.

41. *Baburnama*, pp. 325–326. See also Vincent A. Smith, "India in the Muslim Period," in *The Oxford History of India*, third edition, ed. P. Spear (corrected reprint; Clarendon, 1961), p. 321.

42. *Baburnama*, pp. 372–386. See also Stanley Wolpert, *A New History of India*, 4th ed. (Oxford University Press, 1993), pp. 122–123.

43. John F. Richards, *The Mughal Empire*, volume 5, pt. 1, of *The New Cambridge History of India* (New York: Cambridge University Press, 1993), 8.

44. Elgood, *Firearms of the Islamic World*, 134.

45. Wolpert, *New History of India*, pp. 126–127.

46. Elgood, *Firearms of the Islamic World*, p. 135.

47. Douglas E. Streusand, "The Process of Expansion," chap. 3 in *The Formation of the Mughal Empire* (Delhi: Oxford University Press, 1989), p. 67.

48. Richards, *Mughal Empire*, 57; M. Athar Ali, "Towards an Interpretation of the Mughal Empire," in *The State in India, 1000–1700*, ed. H. Kulke (Oxford University Press, 1995), pp. 266–267.

49. Ali, "Towards an Interpretation of the Mughal Empire," p. 274; Richards, *Mughal Empire*, p. 7; Stephen P. Blake, "The Patrimonial-Bureaucratic Empire of the Mughals," in *The State in India, 1000–1700*, ed. H. Kulke (Oxford University Press, 1995), pp. 284–285.

50. Smith, "India in the Muslim Period," p. 359; Richards, *Mughal Empire*, pp. 142–143; Gommons, "Warhorse and Gunpowder in India," pp. 114–115. See also R. K. Phul, *Armies of the Great Mughals, 1526–1707* (Oriental, 1978); David Nicolle, *Mughul India, 1504–1761* (Osprey, 1993).

51. Streusand, "Process of Expansion," pp. 80–81.

52. Streusand, "Process of Expansion," pp. 66–67.

53. Ali, "Towards an Interpretation of the Mughal Empire," p. 275.

54. Ira M. Lapidus, "Tribes and State Formation in Islamic History," in *Tribes and State Formation in the Middle East*, ed. P. Khoury and J. Kostiner (University of California Press, 1990), pp. 25–47.

55. Streusand, "Process of Expansion," pp. 67–68; Blake, "Patrimonial-Bureaucratic Empire of the Mughals," pp. 281–284.

56. Streusand, "Process of Expansion," pp. 67–68. See also Nicoar Beldiceanu, *Le timar dans l'état Ottoman (debut XIVe-debut XVIe siècle)* (Wiesbaden: Harrassowitz, 1980); Ann K. S. Lambton, "The *Iqta*: State Land and Crown Land," in *Continuity and Change in Medieval Persia* (State University of New York Press, for Bibliotheca Persica, 1988).

57. Inalcik, "Socio-Political Effects of the Diffusion of Fire-Arms," p. 211.

4

Behind the Turkish War Machine: Gunpowder Technology and War Industry in the Ottoman Empire, 1450–1700

Gábor Ágoston

Despite the slow incorporation of Ottoman history into world history during the latter half of the twentieth century, and despite the growing number of comparative studies on Ottoman and Islamic military history in Western historiography, some of the old fallacies about Ottoman military technology are still with us. Historians of the Eurocentric and Orientalist schools alike have a tendency to present a fixed and facile picture of Islamic backwardness in the early modern period. Too much emphasis is placed on the alleged inability of Islamic civilizations to adopt Western innovations in general and military technology and know-how in particular. Kenneth Setton, E. L. Jones, and Paul Kennedy fault the "extreme conservatism of Islam,"[1] the "military despotism" that "militated against the borrowing of western techniques and against native inventiveness,"[2] and "cultural and technological conservatism"[3] for the failure of Islamic civilizations to keep pace with Western military technology. The popularity of these generalizations is particularly surprising since it is well documented that the Ottomans, the Mamluks, the Mughals, and even the Timurids, the Akkoyunlus, and the Safavids have systematically used gunpowder and firearms. Because the available evidence, originally published in specialized scholarly journals and monographs, has recently been incorporated into some basic monographs on general military history,[4] there is no need to repeat them here.

Recent studies on military history offer a more global, though Eurocentric, interpretation. Acknowledging that the Ottomans successfully adopted Western military technology and know-how during the fifteenth and the sixteenth centuries, Geoffrey Parker argues that Western techniques of the so-called military revolution were only "imperfectly practiced" by even the most developed empires of the Islamic world. As a consequence, Ottoman military technology soon became relatively inferior as early as the late sixteenth or the early seventeenth century. Parker claims that the Ottomans "experienced difficulty in mass-producing" and that their indigenous arms industry was weak and failed to meet the requirements of the empire from both quantitative and qualitative points of view. This technological inferiority was responsible for their military failures along the European frontier, and eventually led to their defeat and to Western military hegemony.[5] According to another hypothesis put forward by Keith Krause, the Ottoman Empire was a "third-tier producer" and "relied heavily on imported weapons and technologies."[6] Krause's Eurocentric views are repeated in Jonathan Grant's most recent study. Grant's main aim is to question the theory of Ottoman decline and to argue that the Ottoman military technology remained competitive in the regional context. Unfortunately, he had no access to Ottoman production data; that alone might have provided us with the necessary basis to challenge the received views about the supposed inferiority of the Ottoman arms industry. Basing his article on Krause's questionable model, Grant likewise argued that the Ottomans "remained a third-tier producer" throughout the period under discussion. He claimed that the Ottomans possessed capabilities comparable to "third-tier" rivals such as Hungary, Poland, and the medieval Balkan states.[7] In recent literature, only Jeremy Black significantly counters the above Eurocentric approach in his global and comparative narrative of warfare, which treats non-European armies and societies as autonomous participants in regional conflicts and questions the importance of technology for "the fate of the continents."[8]

In sharp contrast to Eurocentric historians, William McNeill and Marshall Hodgson emphasize the importance of gunpowder weapons in some of the Islamic civilizations, characterizing the Ottoman, Safavid, and Mughal states as "gunpowder empires."[9] Later historians of the Middle East have surpassed this characterization and have suggested that gunpowder weaponry played a crucial, if not determining, role in the military success of these empires.[10] Yet their interpretation exaggerates the importance of military technology. Neglecting other, equally important, factors of organized violence, they have displayed too much confidence in the decisiveness of gunpowder weaponry. Currently, however, there is a growing concern among European and Middle Eastern historians about technological determinism.[11] Rhoads Murphey has recently challenged the casual link between the rise of the Ottoman state and the use of gunpowder technology, on the one hand, and the decline of the empire and technological atrophy, on the other.[12]

The aim of this essay is not to present a simple counter-thesis to the Eurocentric views of Ottoman technological inferiority. Rather, it is to demonstrate the need for a more balanced and cautious approach in studying their military technology by broadening the scope of examination. The first section will comment on some of the questionable biases against Ottoman technology;

the second will concentrate on the Ottoman war industry and the supply of weaponry and ammunition.

The Introduction of Gunpowder Weapons in the Ottoman Empire

The date of the introduction of firearms into the Ottoman Empire remains debatable. The main problem, as with early Ottoman history in general, is the scarcity and questionable reliability of the sources. We hardly have any Ottoman sources contemporaneous with or close to the supposed time when firearms were introduced into the empire. It would be risky to base the argument solely on Ottoman chronicles dating from the late fifteenth and even the sixteenth centuries that are not confirmed by other independent sources, since chroniclers might have projected the terminology of their own times when referring to earlier events. Thus, in writing about fourteenth-century sieges and battles involving only mechanical artillery (e.g., trebuchet) or personal missile weapons (e.g., crossbows), Ottoman chroniclers might erroneously mention firearms, which were used regularly at the time they wrote their annals.

Terminology constitutes a major problem for the student of Islamic military technology. As with many European languages,[13] old terms were applied to new weapons. It is likely that to the east of the Ottomans, in the initially Samarkand-based Timurid Empire, the Arabic word *ra`d* (meaning "thunder") was used for a kind of mechanical artillery or missile weapon that hurled incendiaries during the reign of Timur (1370–1405). However, during the reign of Timur's son and heir, Shah Rukh (1405–1447), this very term was used for large stone-shooting cannon made of metal.[14] Similarly, in our fifteenth-century Ottoman sources the Turkish word "top" was used for both the shots of cannon and the cannon itself, and it is not always obvious which of these meanings we should apply. Furthermore, the mere lack of the word "cannon" or "gun" in the sources does not necessarily indicate that the weapon itself did not exist. (While the word "saltpeter," for example, first appears in English sources as late as the sixteenth century, saltpeter had been known and used 200 years earlier.) Lastly, we should not overvalue the importance of these "first references" to gunpowder weaponry. It took decades after the first appearance of firearms for soldiers to employ them regularly and in large enough quantities to be tactically significant. Likewise, it was not until the sixteenth century that these weapons proved to be central to military strategies.

We should bear this in mind since Turkish historians tend to ascribe too much importance to the first, though dubious, references to firearms cited in the earliest Ottoman chronicles. According to the historical chronology of the

Ottoman Empire, the Ottomans cast cannon as early as 1364 and used them, together with hand firearms, against the Karamanids in 1386.[15] Although these dates have found their way into Western scholarly literature through Carlo Cipolla's influential work on European artillery,[16] we should not forget that these dates derived from the sixteenth-century chronicle written by Şikari, who died in 1584—that is, 200 years after the events in question. When referring to the famous fifteenth-century Ottoman chronicler Neşri (deceased before 1520), other Turkish historians claim that the Ottomans used cannon at the battle of Kosovo in 1389.[17] A recent study argues that they used cannon as early as 1354 during the siege of Gallipoli. Yet again, this was based on a much later source, the early sixteenth-century chronicle of Kemal Paşazade (deceased 1534).[18] While we know that Kemal Paşazade used earlier sources and relied on an oral tradition based on eyewitnesses of the original events, the problem persists: he might have projected the terminology of his own time.

In view of these obstacles, it is hardly surprising that some European Ottomanists—trained originally in either classical philology or medieval and early modern European history, and thus well aware of the methods of source criticism—expressed their concerns about the first references to firearms in Ottoman chronicles. When Paul Wittek re-examined the earliest references to the Ottomans' use of firearms he expressed his doubts about the validity of the sources in question, and was "inclined to think that before 1400 the Ottomans had no knowledge of firearms."[19] Wittek mentioned that the first trustworthy references appearing in separate sources refer to the siege of Constantinople in 1422 and that of Adalia (Antalya) in southern Asia Minor in 1424.[20] His conclusions were later modified by Halil İnalcık, who referred to an Ottoman tax register from 1431. This alluded to a certain Ali, the son of a cannonier. İnalcık therefore claimed that the Ottomans might have used guns during the reign of Mehmed I (who ruled 1413–1421) "*and perhaps even earlier.*"[21] This chronology, established by Wittek and modified by İnalcık, was also accepted by Vernon J. Parry when writing his *Encyclopedia of Islam* article on 'Bārūd'.[22]

We can minimize these ambiguities by taking a closer look at the early history of gunpowder weaponry in the Balkans, Byzantium, and the Islamdom, possible regions through which firearms might have reached the Ottomans. Sources indicate that bombards were employed in the siege of Zara in 1346. In 1351, the Ragusan senate ordered a certain Nicola Teutonicus to make a *spingarda*. Although Nikola never completed his work, from 1362 to 1363 a local smith in Ragusa manufactured several *spingardas*.[23] By 1378, firearms had become regular weapons in the defense of the city. In August of the same year, the defenders of Kotor employed bombards against Venetian warships, and, by 1380, firearms were fairly commonly used in Bosnia. The Serbians imported bombards from Venice in the 1380s before manufacturing their own weapons in the 1390s.[24] Thus, during their raids, sieges, and battles on the peninsula, the Ottomans had the opportunity to capture some of these new weapons along with their manufacturers and gunners. In view of the small size of these early cannon, smuggling or transporting them should not have been a problem (the smallest bombards manufactured in Ragusa were sixteen to forty centimeters in length, and the smallest *spingarda*s weighed only 14 kilograms).[25] Djurdjica Petrović, who examined archival sources from Venice and Ragusa (Dubrovnik) as well as near-contemporary narratives from Serbia, Bulgaria, and Western Europe, also drew attention to the fact that some of his Western and Slavic sources, unexplored before him, mention firearms during the first battle of Kosovo (1389) and the siege of Constantinople (1394–1402). These near-contemporary sources substantiate Petrović's conclusion that the Ottomans had used firearms before the close of the fourteenth century.[26]

Gunpowder weapons might also have reached the Ottomans through either Byzantium or Islamic states. Firearms had been used in Byzantium in 1390, and from 1396 to 1397. It seems, however, that most of these weapons were of Genoese origin.[27] The first references for Muslim use of firearms (1204, 1248, 1258–1260, and 1303) go back as far as the thirteenth century, though they must be taken with caution since the terminology used for gunpowder and firearms in contemporary Arabic sources is very confused. As with references to early Ottoman firearms, most of these testimonies were provided by later chroniclers of the fifteenth and sixteenth centuries. It is without doubt, however, that the Mamluks used artillery from 1366 to 1368.[28] These references also corroborate the assumption that the Ottomans might have been familiar with gunpowder weapons before the end of the fourteenth century.

Ottoman and European narrative sources alike affirm that firearms in the Ottoman Empire gained tactical significance only in the latter part of the fifteenth century. As Vernon Parry and Colin Heywood have suggested, the Hungarian-Ottoman wars in the Balkans of 1443–44 proved to be crucial for the evolution of Ottoman military technology and warfare, because the Ottomans were forced to match their opponents' weaponry and tactics. As for siege artillery, the Ottomans successfully used their cannon during the siege of Thessalonoiki in 1430, and in demolishing the eight-kilometer defensives of the Isthmus of Corinth in 1446. By 1444 they possessed cannon at Silistre and in the fortress of Nicopolis. At this time, their cannon "gave a hard time" to the Burgundian ships at the Dardanelles, indicating that Ottoman gunners deployed their weapons not only against city walls but also against highly maneuverable galleys. In these instances, the artillery was cast in situ from material the Ottoman forces had transported.[29] Yet we should not overvalue the role of cannon. Just as in contemporary Europe, they were not the only means of siege warfare. Even after the successful conquest of Constantinople, where artillery proved to be crucial, other types of siege weaponry continued to be employed. Furthermore, long-lasting blockades remained a regular practice that complemented or assisted siege artillery. The change in military technique was slow to take root, and new weapons competed with old ones for decades everywhere in Europe.

Channels of Military Acculturation: The Human Factor

Fifteenth- and sixteenth-century Western historical accounts and travel books have revealed that large numbers of European renegades, cannon founders, and artillerists worked for the Ottomans. European historians have relied on this evidence to demonstrate the Ottomans' dependence on Western military technology. They have paid special attention to such European renegades as Master Orban, most probably a Hungarian or German from Transylvania, or Jörg of Nürnberg.[30] Though European cannon founders and artillerymen might have played some role in the early history of the Ottoman artillery, their contribution should not be exaggerated. Our sources make as many references to Turkish as to European artillerists. During the siege of Constantinople in 1453, Mehmed II is known to have had Turkish cannon founders and technicians working independently of Master Orban. One of them, a Turkish founder named Saruca, also cast a large cannon.[31] Nicolo Barbaro reported that in 1453 the Ottomans used at least twelve cannon from four positions to besiege the walls of the Byzantine capital. Other sources claim that Mehmed II brought sixty-two cannon against Constantinople. Yet only one cannon had been made by Master Orban.[32] In 1456, during the siege of Belgrade, Mehmed II used at least twenty-seven large bombards, seven mortars, and more than 100 smaller artillery pieces. Giovanni da Capistrano, who was among the defenders of Belgrade, allegedly counted 300 guns after the siege. Even if we trust contemporary sources claiming that most of these cannon were operated by Germans, Italians, Hungarians, and other Europeans, there should have been dozens, if not hundreds, of Turkish gunners present.[33]

The employment of foreign military technicians and artisans does not necessarily indicate that the Ottomans were technologically inferior, since this was a well-established practice all over Europe. Most of the gunners serving in medieval Hungary or in Ragusa were of German and Italian origin. In the sixteenth century, Spanish monarchs repeatedly employed Italian, German, and Flemish cannon founders. At that time, the lack of skilled workers was a significant problem for the Portuguese war industry as well, and even England lacked sufficient native gun founders. Both countries therefore recruited foreigners.[34] In short, the casting of ordnance was an international business: technicians who worked for a certain ruler in one year could serve his enemy in another.

Both contemporary European sources and later historical works overemphasize the role of Marranos and Jews in the transmission of Western military technology to the "infidel enemy" of Christendom. Although the Marranos and Jews played some role in the transfer of European military know-how and weaponry to the Ottomans, the exaggeration of this role might have fueled the already hostile attitudes towards them in many parts of Europe, and might have helped to justify their expulsion.[35] We should note that besides the Marranos and Jews, there were numerous French, Venetian, Genoese, Spanish, Sicilian, English, German, Hungarian, and Slavic experts working at the Ottoman cannon foundries. Professional miners from Novo Brdo in Serbia were used by Mehmed II in 1453 to dig mines under the walls of Constantinople. Ottoman pay registers also show that until the middle of the sixteenth century a considerable number of Christian smiths, stone carvers, masons, caulkers, and shipbuilders served in Ottoman Balkan fortresses. We should not forget, however, that the situation was similar on the other side of the borders. Slavic and Gypsy artisans and technicians from the peninsula who escaped from the Ottoman rule served the Hungarian kings and such towns of his realm as Buda, Pécs, Brassó, and Szeben. They built ships and made swords, guns, projectiles, and gunpowder, thereby fostering military acculturation and homogenized weaponry.[36]

Historical sources on the Barbary corsairs also suggest that renegades and adventurers played an important role in the transmission of European maritime and gunpowder technology to the Ottomans. In his famous work on the history of Algiers, Diego de Hadeo, a Spanish Benedictine and a captive in Algiers from 1579 to 1582, listed 35 corsairs who owned galliots in Algiers in 1581. Of the 35 shipowners, 22 were renegades and three were sons of renegades. The renegades comprised six Genoese, three Greeks, two Spaniards, two Venetians, two Albanians, one Hungarian, one French, one *judeo de naçion*, one Corsican, one Calabrian, one Sicilian, and one Neapolitan. Only ten of them were Turks.[37] Given such a cultural variety among the Barbary ship owners, it is hardly surprising that the corsairs provided the Ottomans with an invaluable reservoir from which the Sultan's naval empire drew its best human capital. The employment of hundreds of these renegades in the Mediterranean facilitated military acculturation and resulted in a common military and nautical knowledge of the region. The Turkish naval vocabulary of Italian and Greek origin mirrors this "cultural unity" of the Mediterranean.[38]

On land, the Ottomans found unexpected opportunities for recruiting highly skilled European military technicians at the end of the sixteenth century. During the Long Habsburg-Ottoman War fought in Hungary between 1593 and 1606, mutinies of European mercenaries provided a pool of European musketeers skilled in the use of up-to-date weaponry and tactics. In 1596, during the siege of Eger, some 250 Christian soldiers defected from the garrison and "fled to the sultan's camp and became Turk."[39] In the summer of 1600, at a critical juncture of the war, the unpaid French and Walloon mercenaries of the garrison of Pápa seized the garrison and, after careful negotiations with the Ottomans, offered their services to the Sultan. To stop this betrayal, imperial forces besieged Pápa and killed several mutineers. Nevertheless, some 400 to 500 mercenaries escaped to the Ottoman garrison of İstolni Belgrad (Székesfehérvár).[40] In 1601, during the siege of the Ottoman fortress of Kanije (Kanizsa), which had been captured by the Sultan's forces just a year before, several Italians deserted the camp and fled to the Ottomans. Italian mercenaries also served in the Ottoman army during the siege of Székesfehérvár in 1602 and the siege of Buda by the imperial forces in 1603.[41]

The Ottomans clearly valued the skills of the Pápa mutineers. They offered them generous terms and increased their pay by four times what they had received at the service of the Emperor.[42] This is hardly surprising, since by this time both sides had realized the firepower and tactical superiority of the imperial forces. As early as 1577, at a military conference held in Vienna, Habsburg military counselors were of the opinion that "for the time being, hand firearms are the main advantage of Your Majesty's military over this enemy [i.e., the Ottomans]."[43] An Ottoman observer from Bosnia, who participated in the major battle of the Long War at Mezőkeresztes in 1596, complained that the imperial forces had gained an edge over the soldiers of Islam "in their use of certain newly invented weapons. They invented hand guns and cannon, that is to say several types of hand guns and cannon, and they used them excessively."[44] Later, in May 1603, Yemişçi Hasan Pasha, Grand Vezier and commander-in-chief on the Hungarian front, reported to the Sultan that "in the field or during a siege we are in a distressed position, because the greater part of enemy forces are infantry armed with muskets, while the majority of our forces are horsemen and we have very few specialists skilled in the musket."[45] A recent study based on recruitment contracts has convincingly demonstrated that the Habsburg legions employed in Hungary during the Long War were dominated by infantry soldiers carrying hand firearms, and that the tactics of these troops were based on firepower.[46]

The prevalence of firearms in infantry tactics explains why the Ottomans welcomed the French and other European musketeers; however, given the contradictory nature of our sources, it is difficult to determine the effects of their employment during the latter part of the war. Some 400 French and Walloon mercenaries from the Pápa garrison received their first assignment right after they defected in 1600. At the siege of Kanizsa, the most important Hungarian fortress in Trans-Danubia, they were appointed to oversee the Ottoman siege cannon. After the successful siege, they were rewarded for their useful service. The Pápa renegades fought with the Ottomans during the remaining years of the war: in 1601, they were employed at the defense of İstolni Belgrad; and in 1602, they helped to retake the very same garrison. When the war ended on the Hungarian frontier in 1606, they were assigned to other campaigns. In 1607, they were sent against the Cossacks at the mouth of the Danube; in 1610, those who were still alive fought against the Safavids in the East. The Sultan himself, who saw the renegades first when they entered Istanbul after the Hungarian war, was impressed by their muskets and arquebuses. He was especially amused when they fired a salute in an unfamiliar manner.[47]

During the seventeenth century, along the Mediterranean and the Danubian frontiers, the Ottoman military mastery was supposedly in decay. Yet the Ottomans continued to rely on military technicians similar to their Venetian or Habsburg adversaries; consequently, they shared technology. The Ottoman admiral in 1645—Yusuf Pasha, alias Yosef Masković—was a renegade from the Veneto-Ottoman frontier of Dalmatia. He successfully commanded the Sultan's fleet during the first landing in Crete in the summer of 1645, which ended with the surrender of Canea (Hanya).[48] When in 1669 an English ship captured a small vessel from Algiers, off the southern Mediterranean coast of Spain, the English found that her captain was a certain Lübeck renegade called Ali Reis. Ottoman chronicles and personal accounts also mention some celebrated defectors and Christian renegades in the service of the sultans.[49] In addition, the Ottomans seized Christian captives during major battles at sea, as well as during raids against Spanish and Venetian coastal cities and Spanish presidios in North Africa. These provided the Ottoman navy with thousands of European oarsmen if not experts. After the Venetian victory at the battle of the Dardanelles in 1656, the galley slaves on board the eleven captured Ottoman vessels comprised 194 Poles, 60 Germans, 51 Spaniards, 92 French, 182 Italians, 43 Sicilians, 26 Neapolitans, 106 Greeks, 143 Hungarians, 119 Muscovites, and 1,087 Ukrainians.[50] Records on 9,500 Spanish war prisoners freed by the Spanish redemptionist orders reveal that, although the North African subjects of the Sultan did not encourage conversion, they did welcome captives who possessed special technical skills. Furthermore, when North African Muslim rulers needed soldiers with adequate military skills, they offered deals to their

Christian captives; records suggest that in certain cases large numbers of Spaniards accepted conversion and service.[51]

No revolutionary innovations occurred in European firearms technology "between the middle decades of the sixteenth century and the widespread use of the bayonet late in the seventeenth." After the introduction of "controlled-grain corning" and the wheel-lock pistol, further innovations in the production of powder and weapons had no "important ballistic effects."[52] Along with the Ottomans' indigenous inventiveness, the small number of European renegades were thus able to keep pace with the Christian forces. Furthermore, Ottoman siege technique matched that of the Hungarians and Habsburgs at least until the second half of the seventeenth century. While at the end of the seventeenth century the Ottomans lacked experienced artillerists, the Sultan's sappers are reported to have done a better job than their colleagues in the imperial army. Marsigli praised the Sultan's Armenian sappers and miners who were from Istanbul and who were "especially skilled in wooden-works and in laying mines." They were not only more experienced and diligent than the Christians but more effective too. Unlike the sappers in the Habsburg army, these Ottoman-Armenian miners and sappers "were working in a sitting position," and "consequently they carried out the same work in just half of the time and with half of the effort" of their Christian counterparts.[53]

Although Marsigli is viewed as one of the foremost experts on the late seventeenth-century Ottoman army, his observations need to be cross-referenced with other accounts of Ottoman sieges.[54] His comments on Ottoman mastery of siegecraft are supported by numerous examples taken from the Hungarian and Cretan wars—that is, from the two major fronts where the Ottomans encountered more or less up-to-date fortifications and defense tactics. The medieval fortifications of the Hungarian Kingdom could not resist Ottoman artillery. A statistical analysis of the Ottoman sieges in Hungary during the reign of Süleyman the Magnificent, who ruled from 1520 to 1566, shows that between 1521 and 1566 only thirteen Hungarian castles were able to resist Ottoman sieges for more than ten days, and only nine for more than twenty days. In this period, only four fortresses were able to withstand Ottoman assaults: Kőszeg/Güns in 1532, Temesvár/Timişoara in 1551, Eger in 1552, and Szigetvár in 1556. However, only Kőszeg was besieged by the Sultan's army. Temesvár and Eger were attacked by the troops of the Grand Vizier, and Szigetvár by the governor-general of Buda. Three of these fortresses were nonetheless captured some years later: Temesvár in 1552, Szigetvár in 1566, and Eger in 1596.

During the latter part of the sixteenth century, the Habsburgs initiated a large-scale modernization project in Hungary. The key fortresses were redesigned in the *trace italienne* style, according to the plans of such Italian military engineers as Pietro Ferabosco and Carlo Theti. In the case of Győr (Raab), Komárom (Komron/Komárno), Érsekújvár (Neuhäusel/Nové Zámky), Kassa (Kaschau/Košice), Várad (Oradea), and Szatmár (Satu Mare), not only the fortress but also the entire town was fortified. Thus, the process gave way to the development of the fortified town, the *Festungstadt*. However, during the Long War at the end of the sixteenth century, the Ottomans captured most of these fortresses: Győr, in 1594, Eger in 1596, and Kanizsa in 1600. Other *trace italienne* fortresses were taken by the Köprülü Grand Veziers in 1660 and 1663.[55]

Nonetheless, by the end of the seventeenth century Ottoman cannoniers could not match the knowledge of their European counterparts. Paul Rycaut observed that few of the *topçu*s (cannoniers) "are expert in their art, and are ill practised in the Proportions and Mathematical part of the Gunners Mystery. . . . And herefore knowing their own imperfections in this exercise, when Christian Gunners are taken in the War, they entertain them with better usage than other Captives, quartering them in the Chambers appropriate to that Profession, alloting them with others a pay from 8 to 12 Aspers a day; but because this is too considerable a maintenance to allure men who are otherwise principled, most of them as occasion offers, desert the service of the Turk, and fly to their own Country."[56]

The Ottomans were also hit hard when their competent gunners who had gained experience in Candia were dead by the end of the seventeenth century. The head of the gunners, the *topçubaşı*, repeatedly complained to Marsigli about the lack of experienced cannoniers; therefore, he decided to fetch Christian experts to inform him of the latest achievements of European military engineering.[57] European observers also noted that Ottoman gunners were careless and often overloaded their cannon, firing a variety of cannon balls from the same piece. The retention of the elevating wedge also militated against accuracy.[58]

"The Great Equalizer of Civilizations": War as a Means of Military Acculturation

It is well established that constant wars, "the great equalizer of civilizations,"[59] played an important role in the transmission of Western military technology from Europe to the Ottomans.[60] We have already mentioned that Ottoman raids and campaigns were responsible for the transmission of firearms from the West to the Ottomans as early as the waning years of the fourteenth century. Somewhat later, in the 1440s, the Hungarian-Ottoman wars of János Hunyadi were of crucial importance in diffusing European military technology and know-how to the Ottomans. It was during these wars that the Ottomans became acquainted with the *Wagenburg* (wagon fortress) system, a defensive arrangement of "war wagons" chained together, wheel to wheel, and protected by heavy wooden shielding. Manned with crossbowmen and handgunners, it protected against cavalry assault. Hunyadi learned the use of war wagons during his wars against the Hussites in Bohemia, when he served as a commander for Sigismund of Luxemburg, King of Hungary (1387–1437) and Bohemia (1419–1437) and Holy Roman Emperor (1411–1439). When Hunyadi was preparing against the Ottomans in March 1443, he relied on the well-developed industry of the Saxon cities of Transylvania and ordered the artisans of Brassó to send "war wagons furnished with guns, arquebuses and other war-machines, made according to the instructions of a certain Bohemian artisan."

Brassó was not the only city whose artisans furnished Hunyadi's army with war wagons made according to the "Bohemian manner" (*currus Bohemico more instructos*). Hunyadi spent a great amount of money on the construction of war wagons, and his Czech mercenaries also brought several war wagons to his camp. In all, 600 wagons, operated almost entirely by Czech mercenaries, were reported to have been employed in his "winter campaign" of 1443–44. It was probably the first opportunity for the Ottomans to see the war wagons in operation. In November 1444, in the Battle of Varna, war wagons again played an important role. Learning of the arrival of the Ottoman forces led by the Sultan himself, Cardinal Giuliano Casarini, the papal legate and the guiding spirit of the anti-Ottoman crusade, suggested in a war council that the army take up a position behind the "wagon camp."[61] This strategy is confirmed by the well-informed anonymous author of the contemporary Ottoman chronicle, the Gazavatname. The same source indicates that by then the Ottomans knew how to besiege the *tabur,* or the Christian wagon camp named so in Ottoman sources after the Hungarian szekér (wagon) tábor (camp).[62] In this battle, the Ottomans defeated the Crusaders' army and captured the Christian war wagons and weapons. During these crusading wars, the Ottomans acquired a solid knowledge of the *tabur* system and the *tabur cengi* (the camp battle), the new technique which relied on war wagons armed with firearms. Later, Ottoman experts introduced the *tabur cengi* to the Safavids and Mughals, who called it *destur-i Rumi*, the Ottoman order of battle.[63] In the latter part of the sixteenth century, Lazarus von Schwendi, the commander of the imperial forces in Hungary from 1564 to 1568, observed that the Ottomans could use the *Wagenburg* system very successfully against the imperial forces and that they owed their military success to the *tabur.* Consequently, he urged the Emperor's troops to use war wagons armed with double arquebuses and light cannon.[64]

Hungary was the major theatre of land warfare between the Ottomans and their Christian adversaries during the sixteenth and seventeenth centuries.

There, wars offered a continuous opportunity for both sides to learn about the latest developments of the enemy's military techniques and skills. During the siege of Székesfehérvár of 1543, Ottoman soldiers confiscated several wheel-lock pistols from German horsemen. Introduced first in Italy about 1520, the wheel-lock was unfamiliar to the Ottomans serving in Hungary until this siege.[65] A report dated 1594, however, states that the Turkish soldiers had not yet adopted the pistol.[66] Other Western weapons had better luck with the Ottomans. After the Ottoman conquest of the Hungarian fortresses in the middle of the sixteenth century, most of the artillery pieces cast in Europe and employed in these strongholds by the Hungarians fell into Ottoman hands; they remained in use until the end of Ottoman rule. According to one member of the Habsburg embassy to the Porte in 1572, the majority of the cannon in the Ottoman garrison of Varadin (Pétervárad), near Belgrade on the Danube, were of European origin. He counted some thirty pieces, of which only one was a Turkish falcon, a gun somewhat longer than the others. There were pieces originally made in 1496 for Wladislaus II, King of Hungary (1490–1516), and in 1511 for Johannes Ország de Guthi, Bishop of Sirem.[67] After the reconquest of major Ottoman fortresses in Hungary at the end of the seventeenth century, imperial forces compiled inventories that also listed several European cannon in the possession of the Ottomans. Guns found at the fortress of Gyula in 1695 had been cast in 1547, 1548, and 1559 for Ferdinand I of Habsburg, King of Hungary (1525–1564) and Holy Roman Emperor (1556–1564). Some estimate that 80–90 percent of the cannon found in certain reconquered Ottoman fortresses in Hungary in the 1680s were of Western origin.[68] Official Ottoman registers, nevertheless, indicate a more significant presence of their cannon cast either in the Imperial Cannon Foundry in Istanbul or in the Balkans.

Merces Prohibitae: Trade in Weaponry and the Dependence Theory[69]

Trade also played an important role in the diffusion of military technology. Although both European rulers and Ottoman sultans forbade the export of arms and war materials and declared them prohibited goods (*merces prohibitae* and *memnu eşya*, respectively), they never stopped this lucrative trade. It seems that the illicit exchange of war material was part of the everyday pattern on the frontiers, and it constituted a significant component of East-West trade. While it is obvious from the sources at our disposal that this diffusion was never one-sided and involved Western exports as well as imports of strategic material, Western historiography has nonetheless focused on Western exports to the Ottoman realm and has advanced the dependency theory.

Quick to acquire Western military technology in the fifteenth and early sixteenth centuries, the Ottomans proved to be "expert imitators, but poor innovators," according to Geoffrey Parker. Relative to their Christian counterparts, the Ottomans were not only inferior designers of gunpowder technology but they also "experienced difficulty in mass-producing" it.[70] Consequently, both Ottomanists and European historians claim that the Ottomans became dependent on imported weaponry and war materials.[71] From the 1580s, other historians suggest, the English became the main supplier of the Ottomans, who, after their naval defeat at Lepanto (1571), were eager to replace their losses through imports from the West.[72] The two main commodities were tin, "without which they [i.e. the Ottomans] cannot cast their cannon,"[73] and gunpowder. According to an English captive in Istanbul from 1603 to 1606, the janissaries had "not one corne of good powder but whyche they get from overthrone Christians, or else is broughte them out of England."[74] However, by the end of the seventeenth century, claims Rhoads Murphey, the channels of this trade had clogged up. The decline of Western ammunition imports "correspond rather closely with the beginning of decline in Ottoman naval fortunes after 1669 and of military fortunes after 1683." Murphey argues that "it was neither inferior technology nor inferior tactics which brought about the lessening in the Ottomans' ability to wage war, but their supply situation."[75]

The main problem with the above arguments is that they are based on random evidence, often atypical narrative data. Ottoman archival evidence unearthed so far concerning domestic production of Ottoman weaponry and ammunition disproves the dependency theory and supports the Ottomans' military self-sufficiency as first suggested by Vernon Parry. To help resolve this question we must estimate at least the ratio of imports to domestic production on the one hand, and the ratio of imports to domestic consumption on the other. Yet, given the nature of the contraband trade, the exact volume of imports will remain unknown. Nevertheless, this essay will provide estimates on Ottoman gunpowder imports and domestic production based on published ambassadorial reports and official production data, respectively.

Venetian ambassadors report that between 1579 and 1610, a period of constant warfare, only eleven English ships reached the Ottoman capital. It was only in 1605 that an English ship actually carried gunpowder, of which 700 barrels were confiscated. If we add to this a similar quantity of Dutch powder and count the whole as an annual import, we still only get a modest quantity of imported powder relative to that of domestic sources.

In the sixteenth century, important gunpowder works operated in Istanbul, Cairo, Baghdad, Aleppo, Yemen, Buda, Belgrade, and Temeşvar. Secondary production centers were located in Estergon, Erzurum, Diyarbekir, Oltu, and Van. In the seventeenth century, the Ottomans established major gunpowder works in Bor (in the province of Karaman), Selanik (Thessalonica), Gelibolu (Gallipoli), and Izmir. In addition to these gunpowder works, smaller mills driven by either animals or manpower operated in several fortresses of the empire. Tables 4.1 and 4.2 show the actual and estimated production data of some of the major Ottoman gunpowder works and the stock of powder. Based on the data summarized in table 4.1, I estimate the quantity of gunpowder produced annually in the major gunpowder works of Egypt, Baghdad,

Table 4.1
Gunpowder production in the Ottoman Empire in the second half of the sixteenth century.

Gunpowder works	Date of production and/or inventory of stocks	Production (kantars/year)	Stock (kantars)
Istanbul/Kağıthane	1571	1,800–3,600*	1,600
Istanbul/Şehremini	1571	900–1,800*	
Istanbul	1594–95		4,460
Buda		1,000–2,000*	3,000–4,000
Temeşvár		800–1,200*	1,500–2,000
Baghdad	1570		3,000–4,000
Baghdad	1574–75	2,500	
Baghdad	1575–76	5,000	
Egypt	1574	4,000	
Egypt	1593		7,000
Egypt	1595		7,000
Egypt	1599		4,000
Aleppo	1570	1,000	1,000
Erzurum	1579		2,000
Total		11,100–18,101.2	16,100–24,460

Sources: Istanbul 1571: Başbakanlık Osmanlı Arşivi, Istanbul (Ottoman Archives of the Prime Ministry, henceforth BOA), Mühimme Defterleri (MD) 16, p. 375, no. 656; Istanbul 1594–95: ibid., Maliyeden Müdevver Defterleri (MAD) 383; Buda and Temeşvar: Gábor Ágoston, "Ottoman Gunpowder Production in Hungary in the Sixteenth Century: the *Baruthane* of Buda," in G. Dávid and P. Fodor, eds., *Hungarian–Ottoman Military and Diplomatic Relations in the Age of Süleyman the Magnificent* (Budapest, 1994), 149–59; Egypt 1593 and 1599: Topçular Katibi Abdülkadir Efendi, *Tarih-i Al-i Osman*. Vienna, Nationalbibliothek Handschriftensammlung, Codex Vindobonensis Palatinus Mxt. 130, ff. 7/a and 113/b, 1595: MD 73 p, 221. no. 518. and p. 353, no. 775; Baghdad: Turgut Işıksal, "Gunpowder in Ottoman Documents of the Last Half of the 16th Century," *International Journal of Turkish Studies* 2. 2. (1981–82), p. 85; Aleppo: MD 9, p. 46; Erzurum: MD 32, no. 579. *: estimates, based on figures of stocks or of imperial orders. Lower figures for Istanbul (1571) take into consideration the fact that the powder works operated only for 6 months a year.

Table 4.2
Gunpowder production and gunpowder stocks in Ottoman Empire, 1663–1800.

Gunpowder works	Dates of production	Actual or estimated production (kantars)	Stock (kantars)	Source
Istanbul	III.25.1663–VI.1664		11,211	MAD 3279[a]
Istanbul	III.30.1683–VI.10.1686	6,275		DBŞM 449
Istanbul	XI.07.1687–XI.23.1688		7,183	MAD 15758[b]
Istanbul	VI.05.1689–XII.02.1690	1,750		DBŞM 598
Istanbul	XII.03.1690–VI.15.1692	1,989		DBŞM 642
Istanbul	I.08.1693–I.07.1694	2,004		MAD 3620, pp. 80–81
Istanbul	XI.26.1696–III.23.1697	571		DBŞM 844
Istanbul	III.11.1697–X.24.1699	4,581		DBŞM 19085
Istanbul	XI.03–1701–X.12.1703	2,430		MAD 7488, pp. 2–13
Istanbul	IV.15.1706–IV.03.1707	2,500		MAD 2652
Istanbul	1793–94	1,500		DBŞM BR< 18319
Istanbul	1794–95	1,500		DBŞM BR< 18321
Istanbul	1799–1800	10,000		Cevdet Askeriye 9756
Gelibolu	VII.31.1696–VII.19.1697	1,000		MAD 3127, p. 45
Gelibolu	VII.20.1697–VII.09.1698	1,000		MAD 3127, p. 45
Gelibolu	VII.10.1698–VI.28.1699	1,000		MAD 3127, p. 54
Gelibolu	1747–54	1,000/year		KK 6691
Gelibolu	1782	2,000/year		Cevdet Askeriye 9594
Izmir	1685–87	3,144		MAD 885.10–14
Izmir	I.17.1694–II.19.1695[c]	2,248.5		MAD 3620, p. 70.
Izmir	VII.31.1696–VII.9.1698	3,534.4		MAD 6880, p. 26–27
Izmir	VII.10.1698–VI.28.1699	2,081		MAD 6880, p. 17
Selanik	XI.17.1686–X.25.1688	4,970		DMKF 27627/189–A
Selanik	1695–96	2,520	3,081	MAD 3620, p. 27
Selanik	1696–97	2,035.5	3,231	MAD 3620, p. 87
Selanik	1697–98	3,078.5	3,306	MAD 3620, p. 37
Selanik	1716–17	3,000		MAD 10312
Selanik	1717–18	3,000		MAD 10312
Selanik	1718–19	1,500		MAD 10312
Selanik	1719–20	2,000		MAD 10312
Selanik	1720–21	1,500		MAD 10312
Selanik	1741–42	1,200		KK 6691

Table 4.2
(continued)

Gunpowder works	Dates of production	Actual or estimated production (kantars)	Stock (kantars)	Source
Selanik	1742–51	1,800/year		KK 6691
Selanik	1751–62[d]	1,500/year		KK 6691
Selanik	1765–66	1,500		Cevdet Askeriye 9814
Selanik	1777	2,000/year		Cevdet Askeriye 9595
Karaman	1637–38	3,300		MAD 5472
Karaman	Second half of 17th century	1,800–2,000/year[e]		MAD 3279, 5685
Karaman	1644–45	7,392.8		MAD 7512
Temeşvar	1672	1,000[f]		MAD 1497
Temeşvar	1679–80	1,380		KK 2682
Egypt	1663–64	2,000		MAD 3279[g]
Buda	1684		10,000	MAD 177

a. Total gunpowder income of imperial ammunition stores.
b. Total amount of gunpowder to be found in the imperial armory in this period.
c. According to BOA, MAD 6880, p. 26, the production in H 1106 (VIII.22.1694-VIII.11.1695) was 1,887.6 kantars.
d. In H.1168 (X.18.1754-X.06.1755) only 750 kantars.
e. MAD 5685 and MAD 3279. The annual obligation of Karaman in this period was 80,000 okka (1,818 kantars). However, during campaigns such as the 1663–64 Hungarian campaign, the officials in Karaman had to send more gunpowder than their annual obligation.
f. In July 1672, 800 kantars of gunpowder were delivered to Varad (MAD 1497, p. 9).
g. On May 6, 1663, 818 kantars. On June 25 a further shipment of 613 kantars arrived from Egypt for the 1663–64 Hungarian campaign. On October 15, again 613. On April 17, another shipment of 818 kantars arrived from Egypt for the 1664 campaign.

Istanbul, Buda, and Temeşvar at the end of the sixteenth century to be 11,000–18,000 kantars (594–972 metric tons). Needless to say, this estimate does not include the production output of such gunpowder works as Belgrade, Yemen, or Aleppo, which would considerably increase the figure for domestic production.

It is not until the end of the seventeenth century that official Ottoman registers document gunpowder purchases from European merchants. In 1688 the Imperial Armory purchased 295 kantars of gunpowder from a European merchant and a Christian subject of the Sultan. However, this amount was less than 5 percent of the Armory's total income of gunpowder (6,003 kantars) in that year.[76] Exhausted from the long-lasting wars fought against the Holy League since 1683, the Porte purchased 818 kantars of powder from an English merchant for its navy between 1697 and 1698. This amount was again less than 10 percent of the 8,550 kantars of gunpowder that the admiral of the fleet received from various sources.[77] It was only in January 1700, after the peace treaty of Karlóca (Karlowitz), that the registers cite a considerable amount of imported gunpowder. At this time, the allotment of the Imperial Armory was for 2,108 kantars of gunpowder, of which 1,208 kantars (57 percent) were bought from an Englishman.[78] Data presented in table 4.2, nonetheless, suggest that domestic gunpowder production was still significant in the late seventeenth century. We may estimate total annual gunpowder production in the 1680s in the *baruthane*s of Istanbul, Buda, Temesvar, Selanik, Gelibolu, Izmir, Bor, and Cairo at 14,000–19,000 kantars (756–1,026 metric tons). In other words, there was no decline in the output relative to the sixteenth century, and we may assume that domestic gunpowder production still met the vast majority of the Ottoman's needs—at least until the middle of the eighteenth century.

Secondary literature maintains that by the second half of the eighteenth century the Ottomans produced only about 3,000 kantars of poor-quality powder annually.[79] It is obvious from the available sources that the domestic supply did not meet the demand. In the 1770s and the 1780s, the powdermills at Selanik and Gelibolu were supposed to produce 2,000 kantars of gunpowder each; however, both mills had serious difficulties in fulfilling these expectations.[80] Consequently, the Porte had to import ever-larger quantities of gunpowder; therefore, new suppliers appeared. In 1778, with the assistance of the Swedish ambassador to Istanbul, the Ottomans bought 1,500 kantars of gunpowder from Sweden.[81] It seems that the Ottomans purchased further shipments of gunpowder from Sweden. In November 1782, the Imperial Armory received 1,693 kantars of gunpowder from this source.[82] By the end of the eighteenth century the gunpowder works at Azadlı, which had been modernized by French assistance, were able to produce sufficient quantities of gunpowder of a much better quality.[83] In 1800 the gunpowder works at Azadlı supposedly produced 10,000 kantars of high-quality gunpowder.[84] Although there were some difficulties with providing sufficient saltpeter, this amount shows that by the end of the eighteenth century the empire had again become largely self-sufficient in the production of gunpowder.

The situation with firearms seems to be somewhat different, even in the seventeenth century. In November 1605, an English vessel, destined for Istanbul, was held up by the ships of the Duke of Savoy and the Knights of

Malta, who confiscated the cargo. Apart from the 700 barrels of gunpowder mentioned earlier, the cargo consisted of 1,000 arquebus barrels and 500 arquebuses for horsemen.[85] Data suggest that "this ugly business"—as Ottaviano Bon, Venetian ambassador to Istanbul called it[86]—continued. It is quite possible that other cargoes reached the Ottomans. From the point of view of technology diffusion, these imported weapons had some significance, although we need more research to understand fully their importance.

We should not overvalue the significance of such trade. We have to remember that the import of weaponry and ammunition was a widespread practice in Europe during the period under discussion. Studies concerning contemporary European arms industry and ammunition supply suggest that countries that supposedly were "second-tier producers," according to Krause's model, were more dependent on imported weaponry and ammunition than were the Ottomans. In 1570, Castile imported 6,000 arquebuses from the Netherlands and 20,000 arquebuses from Italy, and from the 1580s onward large quantities of gunpowder had to be obtained regularly from outside sources (including Naples, Flanders, Germany, and Liege) because Spanish domestic production failed to meet requirements.[87] On the other hand, in the sixteenth and seventeenth centuries the Ottoman Empire, a supposedly "third-tier producer," according to Krause, remained largely self-sufficient in the production of cannon, firearms, and gunpowder, with the proportion of imported weapons and ammunition never reaching a significant percentage of domestic production.[88] Unfortunately, available evidence unearthed so far does not offer similar estimates for pistols and flintlock mechanisms. Sporadic sources, however, indicate that they might have had a different story.

Similarities in Military Hardware and the Question of the Ottomans' Technological Inferiority

As we have seen, the employment of foreign experts ensured that new technology and know-how was disseminated relatively quickly within and outside Europe. As a consequence, it became virtually impossible to gain any significant technological superiority in the long run. However, echoing Carlo Cipolla's notion, recent Western historiography claims that the Ottoman artillery hardware missed out on developments in European artillery from the middle of the fifteenth century onwards. Whereas European ordnance increasingly emphasized the design of artillery for both sieges and battles, the Ottoman ordnance remained infatuated with giant cannon.[89]

As this essay has demonstrated, there was an intensive military acculturation through the common pool of military experts, direct military conflicts, and the prohibited trade in weaponry and ammunition during the sixteenth and seventeenth centuries. This not only questions the Ottomans' supposed technological inferiority but also suggests close similarities in military hardware. It has already been noted that such similarities are reflected in military terminology: names of the common Ottoman cannon, such as *balyemez, bacaluşka, darbzen, kolunburna,* and *prangi*, are distorted versions of Italian, Spanish, Catalan, and Portuguese designations of well-known European firearms. These mirror the common material culture of the Mediterranean.[90] Other weapons, such as the small *şakaloz* (which acquired its name from a Hungarian hand cannon, the *szakállas puska*), reflect the Central European component of Ottoman weaponry.[91] The *szakállas* (puska) is the Hungarian equivalent of the Latin (*pixis*) *barbata*, or "hook gun," referring to a large-caliber hand firearm with a hook. The hook served to fix the weapon firmly to the rampart in order to absorb the firearm's heavy recoil. A considerable part of the artillery in both Hungarian and Ottoman fortresses consisted of Hungarian *szakállas*es and Ottoman *şakaloz*es. Unlike the Hungarian weapons, the latter seem to have been set on stock (*kundak*) and transported by gun carriages.

Accounts of production output of the Istanbul State Cannon Foundry, as well as inventories of such strategically important Ottoman fortresses as Baghdad, Belgrade, and Buda, convincingly demonstrate that an overwhelming majority of Ottoman cannon included light and medium-weight guns, and that the Ottomans had a long tradition of deploying smaller artillery pieces. For example, 97 percent of the 1,027 guns cast at the Imperial State Cannon Foundry in Istanbul in the four years before the battle of Mohács (1526) consisted of light and medium-weight cannon. Seventy-two percent of the 300 cannon cast at the foundry in 1685 and 1686 fired cannon balls of 1.28, 0.64, and 0.32 kilograms. Contemporary narrative descriptions of the Ottoman weaponry in operation reveal that the use of small-caliber cannon on battlefields was a common practice in the Ottoman army.[92] The available evidence shows that, even though Ottoman artillery pieces were heavier than some of the European ones of the same caliber, this seldom constituted serious logistic difficulties for the Ottomans—except in the Siege of Vienna in 1683, for example.

We must also handle with caution the bewildering confusion of Ottoman artillery and the lack of standardization—observed by such contemporaneous Europeans as Luiggi Ferdinando Marsigli and presented in the secondary literature—as further proof of the Ottomans' technological inferiority. Contemporary descriptions and weaponry inventories of Spanish, Austrian, and French artillery reveal similar variety and confusion.[93] Research concerning European artillery has demonstrated that attempts to standardize weaponry in the sixteenth and seventeenth centuries largely failed, for these were no more than attempts.[94] The Spanish Army of Flanders and the Dutch forces under Maurice of Nassau were the notable exceptions.

The supposed "metallurgical inferiority" of the Ottoman artillery[95] is based on a superficial assessment of insufficient and random evidence. Chemical analysis has shown an Ottoman gun barrel cast in 1464 for Mehmed II to be composed of excellent bronze containing 10.15 percent of tin and 89.58 percent of copper.[96] The bronze of almost the same composition was recommended by Vanoccio Biringuccio (1480–1539), the famous Italian expert of the sixteenth century, and was used in contemporary Europe.[97] A study of the technology of cannon-casting, used by Mehmed II's technicians at the time of the Siege of Constantinople in 1453 and described in detail by Kritoboulus, has revealed that the Ottoman technology of gun-casting was the same as the one applied in contemporary Europe.[98] Given this fact, it is hardly surprising that the Spanish artillerist Collado, the author of one of the most popular treatises on gunnery in the sixteenth and seventeenth centuries,[99] described Ottoman cannon to be of good metal, though ill-proportioned.[100] Ottoman production data at our disposal suggest that, at least until the end of the seventeenth century, their cannon founders used the typical tin bronze, which contained 8.6–11.3 percent tin and 89.5–91.4 percent copper. (See table 4.3.)

Table 4.3
Compositions of Ottoman bronze cannon.

	Copper	Tin
1464	89.58%	10.15%
1517–1523	91%	9%
1522–1526	90.5%	9.5%
1604	90.8%	9.2%
1685–86	91.4%	8.6%
1693–94	89.5%	10.5%
1704–1706	89.6%	10.4%
1704–1706	89.5%	10.5%
1704–1706	88.7%	11.3%
1706–07	89.5%	10.5%

Sources: Piaskowski, "The Technology of Gun Casting," 168 (for 1464); Istanbul, BOA, MAD 7668 and Heywood, "The Activities of the State Cannon-Foundry," and İdris Bostan, "XVI Yüzyıl Başlarında Tophane-i Amirede Top Döküm Faaliyetleri" manuscript to be published in the İnalcık Festschrift (for 1517–1523 and 1522–1526); MAD 2515 (for 1604); MAD 4028 and DBSM TPH 18597, 18598 (for 1685–86); MAD 5432 (for 1693–94); MAD 2652 (for 1704–1706) and MAD 2679 (for 1706–07).

The above data ought to be handled cautiously. In spite of similar composition, sloppy foundry techniques or impurities in the metal all might have caused significant porosity. It is obvious that, apart from further archival research, we need more metallurgical examinations of extant artillery pieces. However, the results of these examinations should not be taken at face value. Such results always reflect the composition or purity of the metal of the individual cannon examined, and can hardly be applied to other pieces cast elsewhere from different raw material and by another cannon founder. More importantly, slight technological inferiority alone did not matter in the sixteenth and seventeenth centuries, especially when the Ottomans could easily compensate for it with their enormous resources.

Apart from the technology of gun-casting, Ottoman gunpowder production also shows close similarities to that of the Europeans. In Turkish archival documents from the end of the seventeenth century, we frequently come across a certain type of gunpowder called "English powder" *(İngiliz perdahtı)*. This was not an imported powder from England, but a specially refined gunpowder produced locally in Ottoman powder-works according to the new mixture (*be ayar-i cedid*) or the so-called English proportion (*be ayar-i perdaht-i İngiliz*). The powder contained 75 percent of saltpeter and 12.5 percent of charcoal and sulphur, which was the most usual proportion in England and in most of the European countries, even in the first half of the eighteenth century.[101] During the second half of the eighteenth century, Ottoman gunpowder was still manufactured according to this formula; however, by the end of the eighteenth century (1794–1795), the Ottomans produced a better quality of gunpowder mixed in the proportions of 76-14-10, which closely followed the standard European proportions of 75-15-10.[102] It is important to bear in mind that there was no standardized mixture of gunpowder in Europe. In the middle of the sixteenth century, for example, more than 20 different types of powder were produced in Europe, the saltpeter content of which fluctuated between 50 and 85 percent. Besides showing the lack of any standardization, data presented in table 4.4 reveal that the proportions of Ottoman gunpowder closely resembled those of the European mixture.

Again, we need to handle the above data carefully. Maintaining consistent standards was impossible in an empire where gunpowder production was scattered among so many powdermills throughout the empire from Buda to Baghdad. During its "shelf life" and transportation, the powder could have acquired further "post-manufacture defects."[103] This is why contemporary narrative sources present such a contradictory picture of the gunpowder produced in the Ottoman Empire. While some Ottoman chroniclers and Marsigli complained about the quality of Ottoman powder,[104] Raimondo Montecuccoli

Table 4.4
Mixture of gunpowder in selected European countries and in the Ottoman Empire, 1560–1795.

		Saltpeter	Charcoal	Sulfur
1560	Sweden	66.6%	16.6%	16.6%
1595	Germany	52.2%	26.1%	21.7%
1598	France	75.0%	12.5%	12.5%
1608	Denmark	68.3%	23.2%	8.5%
1649	Germany	69.0%	16.5%	14.6%
1650	France	75.6%	10.8%	13.6%
1673	Ottoman Empire	69.0%	15.5%	15.5%
1686	France	76.0%	12.0%	12.0%
1696	France	75.0%	12.5%	12.5%
1696–97	Ottoman Empire	75.0%	12.5%	12.5%
1697	Sweden	73.0%	17.0%	10.0%
1699–1700	Ottoman Empire	75.0%	12.5%	12.5%
1700	Sweden	75.0%	9.0%	16.0%
1742	England and Europe	75.0%	12.5%	12.5%
1793–94	Ottoman Empire	77.1%	12.5%	10.4%
1794–95	Ottoman Empire	75.8%	13.7%	10.5%

For European data, see Arthur Marshall, *Explosives I. History and Manufacture* (London, 1917), p. 27; West, *Gunpowder,* 170; O. F. G. Hogg, *Artillery: Its Origin, Heyday and Decline* (London, 1970).

stated that the Ottomans used "excellent" powder, "as is evident from the noise of discharge, and the velocity and range of the shot."[105] To be sure, this was not only the Ottomans' problem: all those European empires and states that acquired their powder from similarly varying supply sources faced the same difficulties.

Conclusion

We may conclude that the various channels of military acculturation, examined briefly in this essay, helped the Ottomans keep pace with the developments of European military technology. These channels created similarities in artillery hardware and military technology on both sides of the European-Ottoman military frontier. During the seventeenth century and well into the first half of the eighteenth century, when the Ottomans were supposedly inferior to their European rivals, there were hardly any radical differences as far as European and Ottoman gunpowder hardware was concerned. The applicability of this technology and the deployability of weaponry require further analysis, however.[106]

The Ottomans not only hindered the victory of European hegemony but dominated their European adversaries until the late sixteenth century. Montecuccoli, after all, was the first early modern Christian to defeat the Ottomans in a large field engagement at Szentgotthárd in 1664. Unlike many of their rivals, such as the Spanish and Austrian Habsburgs, the Ottomans were capable of manufacturing their weaponry and ammunition within the borders of the empire. Contrary to the received wisdom, they did not depend on foreign sources of supply. In the long run, however, military self-sufficiency is not necessarily an advantage. It can hinder the adoption of new technologies and know-how and can easily promote complacency.[107] The Ottomans might have held on to their traditional military practices until the eighteenth century because their ongoing logistical strengths could compensate for the growing tactical superiority of their Christian adversaries.

Perhaps it was neither the Ottomans' "inferiority" with military technology, as suggested by traditional Eurocentric historiography, nor their supposed shortcomings with ordnance production that brought on their military failures at the end of the seventeenth century. Such factors as two-front engagements and overstrained communications were obviously of greater significance in an empire where weaponry and ammunition manufacturing plants were scattered from Cairo to Buda, often thousands of miles from the theaters of war. It became increasingly difficult to maintain a thriving manufacturing sector in an empire where the economy as a whole experienced the contractions plaguing the entire Mediterranean region. They consequently started to lag behind the Western European economy in fields such as production capacity and productivity. In comparison with the European logistical systems, the Ottoman system proved to be more effective until the introduction of wide-ranging economic and administrative reforms in Europe. After these crucial administrative-bureaucratic reforms had taken place by the late seventeenth century, however, the European adversaries of the Porte were able to supply their ever-growing armies with the necessary weaponry and munitions. Furthermore, such hardware, including the flintlock musket, was of higher quality than that of the Ottomans. The Sultan's adversaries had by then outstripped their mighty rival not only in the field of war industry and military know-how but also in such fields as production capacity, finance, bureaucracy, scientific engineering, and state patronage, to name a few. These factors had been of considerable importance for strengthening the European military machine since the Italian Renaissance. Yet, because of the sudden change in the nature of the Ottoman-European confrontation on land, all these improvements proved to be decisive at the end of the seventeenth century.

Between 1526 and 1683 there were only two major field battles (that of Mezőkeresztes in 1596 and Szentgotthárd in 1664) fought in Hungary, the main theater of the Ottoman-European continental confrontation. Besides minor skirmishes of frontier forces, the Sultan's army and major provincial forces were engaged almost exclusively in siege operations. Consequently, they had to adapt the composition of their artillery and the training of their gunners. At the same time, the Thirty Years' War proved to be a fruitful "laboratory" for Christian armies in major field battles, as Gustavus Adolphus demonstrated. With the exception of the 1663 and 1664 wars of Köprülüzade Fazil Ahmed Pasha, the Ottomans had been given little opportunity to acquaint themselves with the intense firepower, disciplined tactics, and continuous drilling of the reformed European forces. Familiarity with European battle tactics nevertheless became essential, given the changes after the Siege of Vienna in 1683. Siege warfare progressively gave way to field battles: at the Danubian frontier, fifteen major field battles took place between 1683 and 1697. The Ottomans won just two of these battles; one battle ended in stalemate, and all others were won by the allies. By 1699, the Ottomans had lost Hungary, together with Transylvania to the Habsburgs, the Morea to Venice, and Azov to Russia.

One should avoid the temptation to overemphasize the importance of these European victories. Technological developments, such as the adoption of the socket bayonet and flintlock musket, played only a limited role in these victories.[108] It was not better guns that gave the advantage to the Europeans, but better drill, command and control, and bureaucratic administration. Additionally, the Habsburgs were able to defeat their archenemy to the East only in coalition with the other forces of the Holy League, which comprised the German Princes of the Holy Roman Empire, Venice, Poland-Lithuania, and Russia, who in effect represented all the Christian neighbors of the Sultan. The Habsburgs also had to mobilize the economic and human resources of half the continent. As a consequence of this Christian coalition, the Ottomans were forced to fight at four different theaters of war: in Hungary against the imperial forces; in Dalmatia, the Morea, and the Mediterranean against the Venetians; and in Moldavia against the Poles. The Russians, who joined the Liga Sacra in 1686, tied up the Tatars in the Eastern European "steppe frontier."[109] None of the major states in seventeenth-century Europe were capable of waging wars simultaneously at four different frontiers, and the Ottomans were no exception. Neither were the Habsburgs. After the Habsburgs were forced to withdraw their best forces from Hungary to the Rhine frontier to fight the French in the War of the League of Ausburg (1688–1697), the Ottomans quickly recaptured Belgrade in 1690. Like the Ottoman Empire, the Christian archenemy of the Sultan was still too weak to engage successfully in alternative commitments. Because of the Treaty of Karlóca (Karlowitz) in 1699, the expansion of the Ottoman Empire was finally reversed. Yet the Habsburgs subsequently failed to push further south until the late nineteenth century, and plans to "liberate the Balkans" and conquer Istanbul remained a dream.

Acknowledgments

This is a revised version of my paper given at the UCLA/Dibner conference. I would like to thank Jeremy Black, Bert Hall, and Brett Steele for their questions and advice.

Notes

1. Kenneth Meyer Setton, *Venice, Austria, and the Turks in the Seventeenth Century* (American Historical Society, 1991), pp. 6, 100, 450. Setton places too much emphasis on religion. He argues that "the Spanish were caught in an era of religious bigotry, the Turks in a renewal of Islamic fanaticism, and neither people could keep abreast of the technological innovations which had been altering European society from at least the mid-sixteenth century" (p. 6). See also Rhoads Murphey's review in *Archivum Ottomanicum* XIII (1993–94): 371–383.

2. E. L. Jones, *The European Miracle: Environments, Economies, and Geopolitics in the History of Europe and Asia* (Cambridge University Press, 1987), p. 181.

3. Paul Kennedy, *The Rise and Fall of the Great Powers: Economic Change and Military Conflict from 1500 to 2000* (Vintage Books, 1989), p. 12.

4. Jeremy Black, *War and the World: Military Power and the Fate of Continents, 1450–2000* (Yale University Press, 1998).

5. Geoffrey Parker, *The Military Revolution: Military Innovation and the Rise of the West, 1500–1800* (Cambridge University Press, 1988), p. 126. This section is missing from the second edition (1996).

6. Keith Krause, *Arms and the State: Patterns of Military Production and Trade* (Cambridge University Press, 1992), pp. 48–52. Quotation from p. 49.

7. Jonathan Grant, "Rethinking the Ottoman 'Decline': Military Technology Diffusion in the Ottoman Empire, Fifteenth to Eighteenth Centuries," *Journal of World History* 10, no. 1 (1999): 179–201.

8. Jeremy Black, *War and the World*. See also his *Why Wars Happen* (Reaktion Books, 1998).

9. Marshall G. S. Hodgson, *The Venture of Islam: Conscience and History in a World Civilization, volume III, The Gunpowder Empires and Modern Times* (University of Chicago Press, 1974); William H. McNeill, *Pursuit of Power: Technology, Armed Force, and Society Since A.D. 1000* (University of Chicago Press, 1982); McNeill, *The Age of Gunpowder Empires, 1450–1800* (American Historical Association, 1989).

10. See, e.g., Arthur Goldschmidt, *A Concise History of the Middle East* (Westview, 1999), pp. 107–132. Goldschmidt also accepts the Ottomans' technological decline and claims that "by

the end of the sixteenth century, the Ottomans lagged behind the West in weaponry and fighting techniques" (p. 127).

11. See J. Black, *War and the World*; Kelly deVries, "Catapults Are Not Atomic Bombs: Towards a Redefinition of 'Effectiveness' in Premodern Military Technology," *War in History* 4 (1997): 4, 454–470; Bert S. Hall, *Weapons and Warfare in Renaissance Europe: Gunpowder, Technology, and Tactics* (Johns Hopkins University Press, 1997), pp. 212–216; Rhoads Murphey, "The Ottoman Attitude towards the Adoption of Western Technology: The Role of the Efrencî Technicians in Civil and Military Applications," in J.-L. Bacqué-Grammont and P. Dumont, eds., *Contributions à l'histoire économique et sociale de l'Empire ottoman* (Peeters, 1983), pp. 289–298; Murphey, *Ottoman Warfare, 1500–1700* (Rutgers University Press, 1999), pp. 13–15.

12. Murphey, *Ottoman Warfare*, p. 14.

13. The Spanish word *espingarda* originally was used to designate a type of crossbow that came to mean a type of hand cannon (Hall, *Weapons and Warfare*, p. 129). For the etymology of the English word "gun," see Hall, p. 44. For similar problems with regard to terminology in Byzantium, see Mark C. Bartusis, *The Late Byzantine Army: Arms and Society, 1204–1453* (University of Pennsylvania Press, 1992), p. 336.

14. A. M. Belinickii, "O poiavlenii i rasprostraneii ognestrel'nogo oruzhia v srednei Azii i Irane v XIV–XVI. vekah," *Izvestiia Tadzhiskogo Filiala Akademii Nauk SSSR* 15 (1949): 24.

15. İsmail Hami Danişmend, *İzahlı Osmanlı Tarihi Kronolojisi*, volume 1 (Istanbul, n.d.), p. 73.

16. Carlo M. Cipolla, *Gun, Sails and Empires. Technological Innovation and the Early Phase of European Expansion, 1400–1700* (Pantheon, 1965; Barnes and Noble, 1996), p. 90.

17. See İsmail Hakkı Uzunçarşılı, *Osmanlı Devleti Teşkilatından Kapukulu Ocakları*, volume 2, *Cebeci, Topçu, Top Arabacıları, Humbaracı, Lağımcı Ocakları ve Kapukulu Suvarileri* (Türk Tarih Kurumu, 1984), p. 35; Halil İnalcık, "Osmanlılar'da Ateşli Silahlar," *Belleten* XXI/ 83 (1957), p. 509.

18. Mücteba İlgürel, "Osmanlı Topçulugun İlk Devri" *Hakkı Dursun Yıldız Armağanı* (Türk Tarih Kurumu, 1995), pp. 285–293.

19. See Paul Wittek, "The Earliest References to the Use of Firearms by the Ottomans," in D. Ayalon, ed., *Gunpowder and Firearms in the Mamluk Kingdom: A Challenge to a Mediaeval Society*, second edition (Frank Cass, 1978), p. 142. (Ayalon's book was first published in 1956.)

20. Ibid.

21. İnalcık, "Osmanlılar'da Ateşli Silahlar," p. 509.

22. Vernon J. Parry, "Bārūd," in *Encyclopedia of Islam*, second edition (Brill, 1960), p. 1061.

23. Gábor Ágoston, "Ottoman artillery and European military technology in the fifteenth to seventeenth centuries," *Acta Orientalia Academiae Sciencíarum Hungaricae* 47 (1994): 21–22.

24. Djurdjica Petrović, "Fire-arms in the Balkans on the Eve of and After the Ottoman Conquest of the Fourteenth and Fifteenth Centuries," in V. Parry and M. Yapp, eds., *War, Technology and Society in the Middle East* (Oxford University Press, 1975), pp. 175–178.

25. Ibid., p. 173; Ágoston, "Ottoman artillery," p. 22.

26. Petrović, "Fire-arms in the Balkans," p. 175.

27. Bartusis, *The Late Byzantine Army*, pp. 335–336.

28. See Ahmad Y. al-Hassan and Donald R. Hill, *Islamic Technology: An Illustrated History* (Cambridge University Press, 1986), pp. 113–20; Ayalon, *Gunpowder and Firearms in the Mamluk Kingdom*, pp. 2–3, 21–22, 98.

29. Wittek, "Earliest References to the Use of Firearms by the Ottomans," p. 142; Colin Heywood, "Notes on the production of fifteenth-century Ottoman cannon," in *Proceedings of the International Symposium on Islam and Science, Islamabad 1–3 Muharrem, 1404 A.H.* (Government of Pakistan, Ministry of Science and Technology, 1981), pp. 59–61 (reprinted with new notes in Heywood, *Writing Ottoman History*, Ashgate, 2002). Cannon were constructed at the site in 1422 during Murad II's siege of Constantinople; see Bartusis, *The Late Byzantine Army*, p. 337.

30. Ágoston, "Ottoman artillery," pp. 27–28; A. Vasiliev, "Jörg of Nürnberg. A Writer Contemporary with the Fall of Constantinople (1453)," *Byzantion* 10 (1935): 205–209. Doukas claimed that Orban was "a Hungarian by nationality and a very competent technician" (*Decline and Fall of Byzantium to the Ottoman Turks: An Annotated Translation of "Historia Turco-Byzantina" by Harry J. Magoulias*, Wayne State University Press, 1975, p. 200). Since most of the gun founders serving the Hungarian kings were Germans, Orban could have been a German too.

31. Selâhattin Tansel, *Osmanlı Kaynaklarına Göre Fatih Sultan Mehmed'in Siyasi ve Askeri Faaliyeti* (Türk Tarih Kurumu, 1958), p. 52.

32. Nicolo Barbaro, *Diary of the Siege of Constantinople 1453* (Exposition Press, 1969), p. 30; Tansel, *Osmanlı Kaynaklarına Göre Fatih Sultan Mehmed'in Siyasi ve Askeri Faaliyeti*, p. 52; Parker, *The Military Revolution*, p. 225, note 36. On the importance of cannon during the siege, see also Kelly DeVries, "Gunpowder Weapons at the Siege of Constantinople, 1453," in *War and Society in the Eastern Mediterranean, 7th-15th centuries*, ed. Y. Lev (Brill, 1997).

33. Gábor Ágoston, "La strada che conduceva a Nándorfehérvár (Belgrade): L'Ungheria, l'espansione ottomana nei Balcani e la vittoria di Nándorfehérvár," in Z. Visy, ed., *La campana di mezzogiorno. Saggi per il Quinto Centenario della bolla papale* (Edizioni Universitarie Mundus, 2000), pp. 239–246.

34. Cipolla, *Gun, Sails and Empires*, pp. 30–36.

35. See Bernard Lewis, *The Jews of Islam* (Princeton University Press, 1984), pp. 134–135; Stephen Christensen, "European-Ottoman Military Acculturation in the Late Middle Ages," in B. McGuire, ed., *War and Peace in the Middle Ages* (C.A. Reitzels Forlag, 1987), p. 232.

36. See Gábor Ágoston, "Muslim-Christian Acculturation: Ottomans and Hungarians from the fifteenth to the seventeenth centuries," in B. Bennassar and R. Sauzet, eds., *Chrétiens et Musulmans à la Renaissance* (Honoré Champion, 1998), pp. 293–301.

37. Henry and Renée Kahane and Andreas Tietze, *The Lingua Franca in the Levant. Turkish Nautical Terms of Italian and Greek Origin* (University of Illinois Press, 1958), p. 20.

38. Ibid., passim.

39. Péter Sahin-Tóth, "À propos d'un article de C. F. Finkel: Quelques notation supplémentaires concernant les mercenaires de Pápa," *Turcica* 26 (1994): 249–260, especially p. 258.

40. Caroline Finkel, "French mercenaries in the Habsburg-Ottoman War of 1593–1606," *Bulletin of the School of Oriental and African Studies* LV. 3 (1992): 451–471.

41. Sahin-Tóth, "À propos d'un article de C. F. Finkel," p. 258.

42. Finkel, "French mercenaries," p. 465.

43. Quoted in English by Gábor Ágoston, "Habsburgs and Ottomans: Defense, military change and shifts in power," *Turkish Studies Association Bulletin* 22. 1 (1998): 136.

44. Quoted in English by V. J. Parry, "La manière de combattre," in Parry and Yapp, eds., *War, Technology and Society*, p. 228.

45. Quoted in English by Halil İnalcık, "The Socio-Political Effects of the Diffusion of Fire-arms in the Middle East," in Parry and Yapp, eds., *War, Technology and Society*, p. 199.

46. József Kelenik, "A kézi löfegyverek jelentösége a hadügyi forradalom kibontakozásában." *Hadtörténelmi Közlemények* 104. 3. (1991): 80–121.

47. Finkel, "French mercenaries," pp. 463–465.

48. Setton, *Venice, Austria, and the Turks*, 116. See also Murphey, *Archivum Ottomanicum* XIII (1993–94): 374.

49. Rhoads Murphey, "The Ottoman Resurgence in the Seventeenth-Century Mediterranean: The Gamble and Its Results," *Mediterranean Historical Review* 8 (1993): 196.

50. M. Fontenay, "Chiormes turques au XVIIe siècle," in *Le genti del Mare mediterraneo*, ed. R. Ragosta (Lucio Pironti, 1981), p. 890.

51. Ellen G. Friedman, *Spanish Captives in North Africa in the Early Modern Age* (University of Wisconsin Press, 1983), p. 89.

52. Hall, *Weapons and Warfare*, p. 215.

53. Ágoston, "Ottoman artillery," p. 47. Marsigli's comments are based on his observations made in 1683 during his captivity when he was employed for a short time as a laborer in the Ottoman siege works. See also Christopher Duffy, *Siege Warfare: The Fortress in the Early Modern World, 1494–1660* (Routledge, 1996), p. 214. Marsigli remained very critical of the miners and sappers of the allied forces during the unsuccessful siege of Buda in 1684 when,

at an early stage of the siege, he was ordered by Lorraine to inspect their work. See John Stoye, *Marsigli's Europe, 1680–1730* (Yale University Press, 1994), p. 37.

54. He was a typical member of the late-seventeenth-century Italian intelligentsia who managed to acquire considerable knowledge in a wide-range of subjects from history to mathematics, botany or anatomy through private lessons, meetings and conversations with celebrated university teachers and scientists. In 1679–80, during his first stay in Istanbul, and then during his captivity in 1683, he had the opportunity to observe Ottoman troops, and to get first-hand information, both oral and written, concerning the organization, size, weaponry, and siege warfare of the Ottoman armed forces. It was not until the unsuccessful first siege of Buda in 1684 that he had some role in siege work, offering his advice on engineering to Lorraine and Baden. His claim before this siege, that Buda would fall as quickly as ten days, however, cast serious doubt on his "expertise" in siegecraft. In fact, Buda successfully withstood the one-and-half-month siege in 1684 and fell after a two and a half months siege in 1686. See Stoye, *Marsigli's Europe*, pp. 8–12, 36.

55. Ágoston, "Habsburgs and Ottomans," pp. 129–133. Considered "the bastion of the German empire," Győr was recaptured soon thereafter in 1598, as a consequence of a surprise attack after the besiegers had broken down the Fehérvári Gate of the fortress with a petard prepared in Vienna by the French engineer, Jerôme la Marche. Yet Eger and Kanizsa remained in Ottoman hands until the end of the seventeenth century.

56. Paul Rycaut, *The History of the Present State of the Ottoman Empire* (London, 1686), pp. 375–376.

57. Ágoston, "Ottoman artillery," p. 47.

58. Duffy, *Siege Warfare*, p. 213.

59. Fernand Braudel, *The Mediterranean and the Mediterranean World in the Age of Philip II* (Fontana, 1990).

60. See S. Christensen, "European-Ottoman Military Acculturation," pp. 227–251.

61. Gábor Ágoston, "15. Yüzyılda Batı Barut Teknolojisi ve Osmanlılar," *Toplumsal Tarih* 18 (Haziran 1995): 12–13.

62. Halil İnalcık and Mevlud Oğuz, eds., *Gazavât-i Sultân Murâd b. Mehemmed Hân İzladı ve Varna Savaşları (1443–1444) Üzerinde Anonim Gazavâtnâme* (Türk Tarih Kurumu, 1987), pp. 59–60, 68.

63. Parry, "Bārūd," p. 1062; İnalcık, "Socio-Political Effects," p. 204.

64. Parry, "La manière de combattre," p. 224.

65. Max Jähns, *Handbuch einer Geschichte des Kriegswesen von der Urzeit bis zur Renaissance II* (Leipzig, 1880), p. 1214. See also Parry, "La manière de combattre," p. 250.

66. Parry, "Bārūd," p. 1064.

67. József László Kovács, ed., *Ungnád David Konstantinápolyi utazásai* (Szépirodalmi Könyvkiadó, 1986), p. 36.

68. László Szita, *A törökök kiüzése a Körös-Maros közéröl 1686–1695* (Békés Megyei Levéltár, 1995), pp. 205–208. For two slightly different versions of the original inventory, see ibid., pp. 222–24, 242–244.

69. This section is based on Gábor Ágoston, "Merces Prohibitae: The Anglo-Ottoman Trade in War Materials and the Dependence Theory," *Oriente Moderno* XX (LXXXI), n.s. 1 (2001): 177–192.

70. Parker, *The Military Revolution*, p. 127.

71. See R. Murphey, "The Ottoman Attitude," pp. 292–293; İnalcık, "The Socio-Political Effects," pp. 215–216; Kenneth R. Andrews, *Trade, Plunder and Settlement* (Cambridge University Press, 1984), pp. 90–91.

72. S.A. Skilliter, *William Harborne and the Trade with Turkey, 1578–1582* (Oxford University Press, 1977), p. 23. See also V. J. Parry in *The New Cambridge Modern History*, volume III (Cambridge University Press, 1968), p. 368.

73. Braudel, *The Mediterranean*, p. 625.

74. Quoted by Parry in *The New Cambridge Modern History*, p. 368.

75. Murphey, "The Ottoman Attitude," p. 293.

76. Başbakanlık Osmanlı Arşivi (Ottoman Archives of the Prime Ministry, Istanbul; henceforth BOA), Kamil Kepeci Tasnifi (KK) 4738, pp. 1, 6.

77. BOA, Maliyeden Müdevver Defterleri (MAD) 8880, p. 66.

78. BOA, MAD 2730, p. 46.

79. Stanford Shaw, *Between Old and New: The Ottoman Empire under Sultan Selim III, 1789–1807* (Harvard University Press, 1971), p. 143.

80. BOA, Cevdet Askeriye 9594 and 9595.

81. BOA, MAD 10398, p. 102.

82. BOA, MAD 10405, p. 99.

83. See Shaw, *Between Old and New*, pp. 143–144.

84. BOA, Cevdet Askeriye 9756.

85. *Calendar of State Papers and Manuscripts Relating to English Affairs, Existing in the Archives and Collections of Venice, and in other Libraries of Northern Italy* X (London, 1890–1932) (henceforth CSPM, Venice): 525–6.

86. CSPM, Venice, X, p. 318.

87. I.A.A. Thompson, *War and Government in Habsburg Spain, 1560–1620* (Athlone, 1976), pp. 230–241.

88. For more details, see Ágoston, "Merces Prohibitae."

89. Cipolla, *Guns, Sails and Empires*, pp. 95–99; Grant, "Rethinking the Ottoman Decline," pp. 191–192.

90. On the various types of Ottoman cannon, see Ágoston, "Ottoman artillery," pp. 33–45.

91. This type of weapon was also known to the Rumanians and to South-Slavic people, who borrowed the name of the gun from the Hungarian and called it *sacalas* and *sakalus* respectively. See Lajos Tamás, *Etymologish-historisches Wörterbuch der Ungarischen Elemente im Rumanische* (Akadémiai Kiadó, 1966), p. 685; László Hadrovics, *Ungarische Elemente in Serbcroatischen* (Akadémiai Kiadó, 1985), p. 444; Lajos Fekete, "Az oszmán-török nyelv hódoltságkori magyar jövevényszavai," *Magyar Nyelv* XXVI (1930), p. 264.

92. Ágoston, "Ottoman artillery," pp. 43–45.

93. James D. Lavin, *A History of Spanish Firearms* (Herbert Jenkins, 1965), p. 40; Colin Martin and Geoffrey Parker, *The Spanish Armada* (Hamish Hamilton, 1988), p. 215; John A. Lynn, *Giant of the Grand Siècle: The French Army, 1610–1715* (Cambridge University Press, 1997), pp. 501–502.

94. Lynn, *Giant of the Grand Siècle*, pp. 501–502.

95. Suggested by Parker (*The Military Revolution*, p. 128).

96. F. A. Abel, "On the Chemical Composition of the Great Cannon of Muhammed II, recently presented by the Sultan Abdul Aziz Khan to the British Government," *Chemical News* 457 (4 September 1868): 111–112. See also Parry, "Bārūd," p. 1061.

97. Vanoccio Biringuccio, *The Pirotechnia* (MIT Press, 1966). See also Jerzy Piaskowski, "The Technology of Gun Casting in the Army of Muhammad II (Early 15th Century)," in *I. International Congress on the History of Turkish-Islamic Science and Technology, 14–18 September 1981. Proceedings* iii (Istanbul Teknik Üniversitesi, 1981), p. 168.

98. Piaskowski, "The Technology of Gun Casting," pp. 163–168. See also Alan Williams, "Ottoman Military Technology: The Metallurgy of Turkish Armour," in *War and Society in the Eastern Mediterranean*, ed. Y. Lev (Brill, 1997), p. 263.

99. L. Collado, *Practica Manual de Arteglieria* (Venice, 1586).

100. Cf. Parry, "Bārūd," p. 1061; Parker, *The Military Revolution*, p. 206, note 40.

101. Gábor Ágoston, "Gunpowder for the Sultan's Army: New Sources on the Supply of Gunpowder to the Ottoman Army in the Hungarian Campaigns of the Sixteenth and Seventeenth Centuries," *Turcica* XXV (1993): 87–89. In 1742 Benjamin Robins stated that the above proportion was the most usual one in Europe. Cf. Jenny West, *Gunpowder, Government and War in the Mid-Eighteenth Century* (Royal Historical Society/Boydell, 1991), p. 170. See also table 4.4 of the present essay.

102. BOA DBŞM BRİ 18321

103. Murphey, *Ottoman Warfare*, p. 14.

104. Cf. Ágoston, "Gunpowder for the Sultan's Army," p. 87.

105. Cf. Duffy, *Siege Warfare*, p. 213.

106. For such an attempt, see Murphey, *Ottoman Warfare*, pp. 115–122.

107. I owe this point to Brett Steele.

108. J. Black, *War and the World*, pp. 60–95, especially p. 90.

109. Of course, this is William McNeill's term. See his *Europe's Steppe Frontier, 1500–1800* (University of Chicago Press, 1964). One of the best surveys of the war is Ferenc Szakály, *Hungaria Eliberata* (Corvina, 1986).

II

Naval Innovations: Hardware and Software

5

THE *MARY ROSE:* A TALE OF TWO CENTURIES

Alexzandra Hildred

This essay will address the ordnance of early modern naval warfare by interpreting the physical remains on board a vessel that sank on July 19, 1545. The mixed array in size, form, and material of its guns reflects the contrasting techniques used to manufacture them. Their presence together on a mid-sixteenth-century capital ship challenges the simplistic view of a linear evolution in artillery based on the displacement of wrought iron by cast bronze and eventually cast iron. Data gathered from a study of inventories of early modern vessels and fortifications will allow us to examine the armaments within the context of sixteenth-century warfare as a whole. Such evidence will help us ascertain the relative importance of specific guns, as well as the change or stasis in armament design. This essay will then argue that the major developments that revolutionized early modern naval warfare were borne neither in the forge nor in the foundry, but in the shipyards. The revolution was not a qualitative issue of replacing "obsolete" weapons with more reliable ones that allowed gunners to shoot further or reload faster. Instead, it centered on increasing both the size of guns and the quantity each ship could carry. Many of the guns deployed did not represent the state of the art but were the stalwart types that could as easily be assigned to the early fifteenth century as to the sixteenth. This mixed array nevertheless furnished an integrated weapons system, giving long-, medium-, and short-range fire support—not unlike that of a contemporaneous fortification.

When King Henry VIII's warship, the *Mary Rose*, sank at Spithead on July 19, 1545, the practice of gunfounding had been thriving in the West for over two hundred years. Guns are recorded as having been on English Royal Ships as early as 1337–38, when the *All Hallows* cog supported a "small cannon." Ships carrying guns were recorded for the Battle of Sluys in 1340. By 1410–12, the *Marie of Weymouth* carried one gun of brass and two of iron while the *Bernard de la Toure*, weighing 135 tons, carried two iron guns; all were breechloaders. By 1485, the *Grace Dieu* boasted 22 guns and the *Mary,* 58. Ten years later the *Sovereign* carried 141 guns and the *Regent,* 225. Of the 141 guns carried by the *Sovereign,* 110 were serpentines—small breechloaders functioning as anti-personnel weapons. Specific mention of serpentines for shipboard use can be found in the Stowe Manuscript (146, folio 41). It lists the serpentines as weighing 261¼ pounds each and the breech chambers as weighing 41 pounds each. Although these guns varied in size and weight, the quantity of these guns and their placement throughout the ship suggest that their impact in terms of damage to an opponent may have been restricted. During this period, naval tactics remained focused on grappling, boarding, and hand-to-hand combat. Nevertheless, the "fear" value, instigated by both smoke and noise, was an important deterrent.

The evolution of a vessel specifically designed for firepower was a sixteenth-century phenomenon. It relied on refining gun and projectile manufacturing techniques, improving saltpeter and gunpowder processing, and, most importantly, changing the techniques of shipbuilding to permit the placement of lidded gunports close to the water line. These evolutionary trends influenced one another and eventually commingled, having been fueled by severe political and economic pressures.

In April 1509, Henry VIII inherited five vessels from his father. They included the *Regent* and the *Sovereign*—sizeable warships that had been built nearly thirty years earlier. To house these vessels, Henry took over two dockyards: Woolwich dockyard was relatively modest, but the second, at Portsmouth, was well developed. His legacy also included a Clerk of the Ships, Robert Brygandyne, who had filled the post for some fourteen years. Thirty-eight years later—a time span that almost exactly parallels the life of the *Mary Rose*—Henry boasted a fleet of 58 naval vessels, of which 20 were classed as "Great Ships." Also established was a Board of Admiralty with five full-time professional officers and three fully developed dockyards at Woolwich, Deptford, and Portsmouth. Each dockyard employed clerks and craftsmen, in addition to those who served in the central admiralty. Since Henry VIII also fortified the coast of Britain to an extent unprecedented since antiquity, additional armaments were required to arm his dockyards. Undoubtedly, the number of guns produced within this period was "revolutionary" in scale.

The Vessel and Her History

The *Mary Rose* was built between 1510 and 1511 as part of a program of warship building devised by Henry VIII. She was one of only twenty vessels classified as "warships," and, by 1545, she was the second largest and second most

Figure 5.1
The remains of the hull of the *Mary Rose* on display in Portsmouth Historic Dockyard.

heavily armed within the fleet of the king's ships (figure 5.1). The first possible reference to the *Mary Rose*, dated December 1509, describes four new ships being "fitted out" at Southampton. Had the *Mary Rose* belonged to this set, then she probably would have been laid down before April of that year, when Henry came to the throne. A more likely reference is a warrant from January 1510 for two new ships, possibly the *Mary Rose* and the *Peter Pomegranate*. Yet the first explicit reference, dated June 1511, relates to the payments for repairing the *Sovereign* and constructing the *Mary Rose* and the *Peter Pomegranate*. By July there is another reference documenting payment for the conveyance of these same two ships from Portsmouth to the Thames. Between 1511 and 1512, subsequent payments were made for fitting out the *Mary Rose*, which included decking, rigging, ordnance, and the stocking of guns. By April 8 she was ready for deployment in the fleet commanded by Sir Edward Howard.

The *Mary Rose* fought in the First French War (1512–1514). During the latter part of this campaign, she served as the admiral's flagship. She also participated in the preparations for the meeting between Henry VIII and Francis I—the Field of the Cloth of Gold in 1520—and in the Second French War (1522–1525), again acting as flagship of the fleet. On June 4, 1522, Vice-Admiral William Fitzwilliam wrote to Henry VIII praising her excellent sailing qualities. Several references document repairing and re-caulking the vessel, but few refer to her operation at sea. In 1536, she may have been substantially rebuilt, although the costs incurred seem inadequate for such an undertaking. A possible reference to her in 1537 suggested that she may be "unweatherley."

The perceived threat of a Franco-Imperial invasion led to the *Mary Rose*'s mobilization with the rest of the fleet in 1539, as well as in 1543 when the third French war commenced. An "order of battle" for June 1545 locates the *Mary Rose*, at 800 tons, in the powerful central section of the fleet. Within two months she sank while defending Portsmouth from a French invasion fleet (figure 5.2). Although historical sources differ about the sinking, it appears that she reached the 17° angle of heel necessary for destabilization and sank when water entered through her starboard gunports. An entry in the Spanish State Papers dated July 23, 1545, and attributed to Francis van der Delft, Ambassador of the Holy Roman Emperor, included the following description:

> I made inquiries of one of the survivors, a Fleming, how the ship perished, and he told me that the disaster was caused by their not having closed the lowest row of gun ports on one side of the ship. Having fired the guns on that side, the ship was turning in order to fire from the other, when the wind caught her sails so strongly as to heel her over, and plunge her open gun ports beneath the water, which flooded and sank her. They say, however that they can recover the ship and guns.

The entry suggests that the *Mary Rose* fired the guns from her starboard battery and, while maneuvering to bring her port guns to bear, dipped her starboard gunports and sank. This may resolve why all of the port-side guns recovered by the divers in the middle of the nineteenth century were loaded, whereas only four out of five of the starboard in-situ bronze guns were loaded. The cannon amidships were found with neither shot nor powder.

The English sought unsuccessfully to lift their vessel; but items, including ordnance, were raised for some time thereafter. Italian salvors attached ropes to the ship to try to achieve a tidal lift, but this failed, possibly after the foremast and mainmast had been dislodged. The commercial salvors and pioneer divers John Deane and William Edwards worked the site in 1836 and again in 1840. They identified it as the location of the *Mary Rose* because of the discovery of a brass gun dated 1542. All total, they raised four bronze guns, portions of over nineteen wrought-iron guns, and a number of smaller items (figure 5.3). Deane and Edwards abandoned the site after raising all easily accessible objects by placing a number of explosive charges over the site and gathering what emerged. The finding of wrought-iron guns after the explosion may have prompted them to abandon the wreck: they may have incorrectly assumed that these guns were suitable only as ballast, and therefore little of value remained

Figure 5.2
Engraving showing the advance of the French fleet (left) into Portsmouth Harbor. The harbor is being defended by the English fleet. The masts of the *Mary Rose* can be seen just behind Southsea Castle. July 19, 1545. Courtesy the Society of Antiquaries of London.

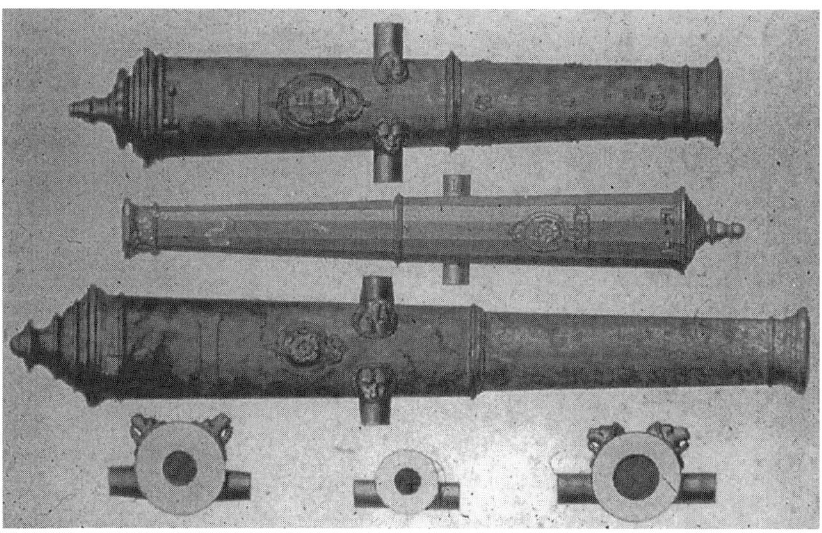

Figure 5.3
(Top to bottom) Bronze cannon, demi-culverin, and demi-cannon recovered from the *Mary Rose* by the divers John and Charles Deane in 1836 and 1840. Courtesy Portsmouth City Museums.

at the site. In actuality, they scarcely touched the wreck, since portions of four decks lay relatively intact and buried beneath the seabed.

Modern searches for the *Mary Rose* commenced under the leadership of Alexander McKee in 1965. After a wrought-iron gun was found in 1970, the search area narrowed. In 1971, the first timbers were sighted, and, between 1971 and 1978, limited excavations were directed to answer specific questions about the integrity of the site. This revealed that a cross-section of a major portion of the starboard hull was present along with the remains of four decks. The *Mary Rose* Trust was formed with the specific aims of raising the ship and associated objects, and housing them in a museum in Portsmouth. Under the archaeological direction of Margaret Rule, a major excavation was carried out between 1979 and 1982. Her work culminated in the recovery of the hull and over 25,000 registered finds. The hull has been on display in Portsmouth Naval Base since 1983, and a museum containing a representative sample of artifacts opened in 1984. The Museums and Galleries Commission judged the collection as "stunning, overwhelming and intensely human," and, in 1997, the museum was one of only 32 museums to receive the Designation Status for an outstanding collection.

Just under half of the hull of the *Mary Rose* survives: the majority of the starboard side from the end of the bowcastle to the rudder. The vessel is made almost entirely of wood, most of which is oak from the South of England. Her construction required not only long timbers (in excess of 12 meters) but also specially shaped timbers that formed key parts of the ship, such as the large wooden knees that support the four decks. The 32-meter keel is of elm, as is the lowermost deck above the hold. Blindages to protect the archers on the upper deck in the ship's waist are present, as are many of the partitions between the cabins. Some of the supporting joists for the upper- and castle-deck superstructure are made from spruce or pine. Examination of the hull shows that she is skeleton built, the keel having been laid first to provide a backbone, followed by the large, curved floor timbers to which the frames were attached. Longitudinal strengtheners called "stringers" were laid against the frames, parallel with the keel. Outer hull planking was added to the outside, and inner planking to the inside between the stringers in some places. Curved knees fixed through the hull provided extra support for the longitudinal deck beams. Half beams were slotted both into the stringers and into longitudinal members called "carlings" that supported the deck planks. Fastenings were by means of oak treenails, iron bolts, and iron nails. Since neither shipwright's plans nor models exist in Britain for vessels of this period, the *Mary Rose* provides the only documentation for shipbuilding at this important time. The likely total length of the *Mary Rose* was approximately 45 meters. She carried low castles

Figure 5.4
Scale model of the *Mary Rose* based on the archaeological remains. The starboard side of the vessel is shown with her seven main gunports shut.

at her bow and stern, not unlike an Elizabethan galleon (figure 5.4). The length of her sterncastle would have been 20 meters; the open waist of the ship, 15 meters; and her bowcastle, 5.5 meters.

The size and number of guns, and thus the firepower, of any ship, were limited by the weight that it could carry without affecting its stability. Until the guns could be positioned low in the hull, their weight and size was limited. The transition from traditional shipbuilding techniques using short overlapping planks to produce a "clinker built" hull to those using the longer butt-jointed planks of the "carvel built" hull occurred towards the end of the fifteenth century with the building of the *Henry Grace à Dieu*. This method of hull manufacture permitted lidded gunports to be placed low in the hull, which in turn increased the size and number of guns that any ship could carry safely. Prior to this, larger guns could have been accommodated only at unlidded ports cut into clinker planking; this is still the case in the upper-deck ports of the *Mary Rose*.

The positioning of guns within the hull appears to have evolved from an initial placement of guns on a lower deck in the stern (and possibly the bow). There the deck was high enough above the water line for the guns to be safely situated without gunport lids. By the time the *Mary Rose* sank, her main deck supported at least seven large gunports on each side (figure 5.5). The dates of

Figure 5.5
Internal elevation of the *Mary Rose* showing the main deck's gunports from the inside. Diagonal braces reinforce the main deck's beams, possibly added during the major refit of 1536. An un-lidded gunport at the upper-deck level can be seen (top right). This is cut through the clinker planking of the sterncastle.

the incorporation of lidded gunports is crucial to understanding the changes in the number, nature, and distribution of guns around any vessel, for they reveal the evolving fighting potential of that vessel.

Gunport lids, located at either end of what appears to be a continuous deck below the weather deck, are depicted in the painting of Henry VIII's 1520 departure for France (figure 5.6). This had been used as an important date for the establishment of the fully fledged gunport. Based on the clothing portrayed, however, the date of the painting has recently been reassessed as the mid-1540s. There would appear to be little pictorial evidence supporting the incorporation of lids all along a lower deck before 1546. Within the series of illustrations in the Anthony Roll inventory, only the two largest vessels, the *Henry Grace à Dieu* and the *Mary Rose*, are shown with lidded gunports on their sides. More vessels, but by no means all, show lidded ports within the stern. The invention of the gunport lid is attributed to a Breton shipwright named Descharges, of whom little is known; this is thought to be an early sixteenth-century advance in shipbuilding technology. The absence of evidence from the Anthony Roll is noteworthy, but it fails to explain the temporal disparity between the introduction of a major innovation and its widespread use.

Figure 5.6
Gunports at either end of a gun deck on vessels depicted in "The Departure from Dover for the Field of the Cloth of Gold" in 1520. This was possibly painted as late as 1540. Reproduced by Gracious Permission of Her Majesty the Queen.

By 1545, the main gun deck on the starboard side of the *Mary Rose* housed a battery of at least seven large-calibre guns that protruded through lidded gunports cut in a carvel-built hull (figure 5.7). There may have been one more in the bow; this area, however, has not been excavated. Earlier documentation suggests a certain number of these ports may be late additions. A list of "ships made new" in 1538 includes the *Mary Rose*, still without her masts. A dendrochronological date around one of the main deck's gunports in the midship area indicates that internal structural work was executed as late as 1541, when a frame now inside the ship was still a tree standing in a forest. This, coupled with a whole group of guns named "port pieces" and listed as "newly made and still at the forge" in 1535, suggests that (some) ports were cut and fitted with lids late in the life of the *Mary Rose* (figure 5.8). Unfortunately, neither the sills of the gunports nor the lids can be dated.

The lids vary in dimensions to suit their individual ports, and their height above the water line varies from 1.40 to 3.00 meters. They are composed of four oak boards: two inner boards with the grain perpendicular to two outer boards. The outer boards are always placed with the grain running parallel with the hull, in keeping with the outer hull's planking. These boards are held

Figure 5.7
Model of the *Mary Rose* showing the lidded gunports on the main deck and the wrought-iron slings positioned on the upper deck above. Looking at the starboard side amidships towards the stern.

together with iron spikes clenched over diamond-shaped roves. The outer face is fitted with large iron straps; these in turn are fitted to hinges stapled into the chamfered edge of the upper thickened wale on the outside of the ship (figure 5.9). The inner face displays up to thirty diamond-shaped heads and a large central ring, allowing the port to be firmly closed within its recessed port opening—the inner boards rebated to fit the sills. The ports are trapezoidal: the width at the top of each port is up to 190 millimeters larger than the width at the bottom, and the diagonal is in the order of one meter. The thickness of the port varies between 100 and 155 millimeters. The upper- and castle-deck gunports remained unlidded. Examination of the hull at the upper-deck level in the sterncastle shows that small ports for swivel guns were blocked or enlarged to accept bigger guns. The inner gunwale was pierced with holes between each frame to support stirrups (miches) for swivel guns. Ten of these were blocked during the life of the *Mary Rose*. One was enlarged to take a culverin; the stirrup-hole is still visible in the bottom sill of the gunport.

A total of fourteen gunport lids were recovered during the excavation. Their location, when studied in relation to their collapsed portside gun carriages, suggests a bilateral symmetry in the positioning of the gun bays. The

Figure 5.8
Photograph of a port piece superimposed onto its correct position on the main gun deck.

orientation and erosion patterns displayed on the lids excavated from the collapsed port side of the hull reveal that these were also open at the time the ship sank. The starboard main-deck gun assemblage consisted of three bronze guns, positioned at the first port in the bow, amidships, and at the second to last port in the stern (figure 5.10). Interspersed with these were the more traditional and supposedly obsolete wrought-iron port pieces. The excavated assemblage from the *Mary Rose* has provided unquestionable evidence of the variety of wrought-iron ordnance. The juxtaposition of these two opposing technologies, one thought to have immediately caused the demise of the other, is one of the most important contributions the *Mary Rose* excavation has made to the study of the evolution of shipboard ordnance.

Although armouries exist containing seemingly large numbers of guns, many weapons listed in inventories lack extant samples. The number of sixteenth-century guns is relatively small. Both bronze and iron were valuable commodities and eagerly recycled. There are no securely dated English examples of existing sixteenth-century guns still mounted on their carriages except those recovered from the sea. Thus, sea-recovered ordnance is virtually the only

Figure 5.9
Inside view of one of the fourteen gunport lids recovered during the excavation.

physical means of adding to the present information about large ordnance and the carriages that supported them (figure 5.11).

Inventory Evidence

We cannot underestimate the importance of the *Mary Rose*. First of all, she is available for close inspection and she is dated precisely. Her hull spans a 35-year period of immense change, as evidenced in the alterations to the hull and the inside of the ship, as well as in the three inventories of guns for her: 1514, 1540, and 1546. Recorded by Anthony Anthony, a clerk in the Ordnance Office, the inventory of 1546 offers the only confirmed illustration we have of the *Mary Rose*. The manuscripts have been elegantly presented together with interpretative essays in The Anthony Roll of Henry VIII's Navy. An extremely detailed inventory was prepared following the death of Henry VIII in 1547. The significance of these inventories within the context of sixteenth-century ordnance and fighting tactics at sea is profound. Vessels must represent a self-contained fighting system: the guns, the powder (type and amount), and the ammunition (form and amount) must function together. The items listed as "spares" encompass those components that will be excessively strained during

Figure 5.10
The main deck armament amidships as viewed from above. This demonstrates the placement of the wrought-iron and cast-bronze guns beside each other.

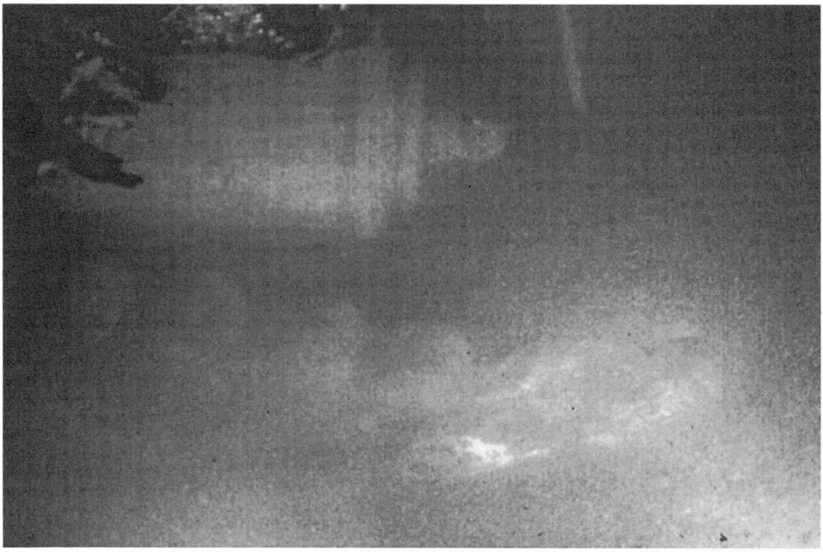

Figure 5.11
Bronze culverin on its elm carriage on the upper deck (MR80A0976).

Figure 5.12
The only illustration of the *Mary Rose* taken from the Anthony Roll inventory of 1546. Courtesy the Master and Fellows, Magdalene College, Cambridge.

the firing operation. It is therefore possible to predict the maximum fighting capability and the sustainability of an engagement. Vessels appear to have used some of the most advanced weapons available; therefore, a study of how their armament changed within a defined period of time (for example, by contrasting the 1540 and 1546 inventories, or the 1546 and 1547 inventories) may reveal which items proved durable and tactically sound. Likewise, such work may also highlight both successful and unsuccessful weapons inventions, and show how long they circulated from inception to combat or scrap heaps. Inventories are crucial for understanding the relative importance of specific guns within the context of ordnance in general. Comparing vessel inventories with those for fortifications may also show which artillery or munitions were used exclusively for naval operations.

The great value of the Anthony Roll of 1546 lies not only in its pictorial representation of the vessels that accompany the inventory (figure 5.12); many of the items listed can be identified within the assemblage excavated from the *Mary Rose*. The position of any named gun within its category reveals the gun's size and weight. How the numbers of named guns fluctuated over time or deployment and what they were replaced with (both numerically and hierarchically) helps establish the evolution of ship-borne ordnance and fighting tactics. Based on inventory and archaeological resources, hitherto enigmatic guns can be identified and given detailed descriptions that include size, weight, and form. Such research depends on the interrelationship of the number and type of guns, on the type of shot listed compared with that found in the bore of a gun, and on notes gathered from additional sixteenth-century sources. The location of ordnance about the ship, provided by the excavation of the *Mary Rose*, together with an analytical appraisal of the number of shot per gun, reveals the functional capability of the ship as a military system. Their potential capability can be predicted since it can be replicated and tested. On a broader scale, the differences in the weapons carried on each type of vessel, along with its crew profile and its engagement sustainability, can be scaled up to assess the operational capability of Henry VIII's navy. The number of listed gunners relative to the number of listed guns invites some interpretation regarding how they serviced the larger guns. *Mary Rose* is listed as carrying 15 carriage-mounted "brass" guns in addition to numerous wrought-iron guns, including 24 carriage-mounted guns, and 52 mounted anti-personnel guns, of which 32 are wrought iron. In addition, she carried 50 handguns, 250 longbows, 300 staff weapons, and 40 dozen darts to be thrown from the fighting tops. When she sank, her complement included 185 soldiers, 200 mariners, and 30 gunners. The *Mary Rose* had also been upgraded from 500 tons to 700.

The *Mary Rose* assemblage of armaments and gun fixtures is unique, thus furnishing invaluable documentation of naval warfare during the period of her life (figures 5.13–5.16). Representing the earliest extant example of a purpose-built warship capable of firing a broadside, the *Mary Rose* featured ninety-one guns deployed over three decks, with fourteen of her heaviest guns located on her main deck. Many of the main deck's guns from the starboard side were found in situ and still on their elm carriages, thus providing vital information about their mounting and operations. The presence of guns on carriages within a specific location on the ship reveals, for example, the constraints on their gunnery performance. Of the fourteen guns on the main deck, six were bronze, with three on each side. These were interspersed with eight large wrought-iron breech-loading guns (four on each side), identified as the hitherto unknown port pieces. Such economical guns are actually listed in many contemporaneous inventories, and represent the majority of the large guns carried by the fleet.

The inventories first identified the guns as either "brass" or iron, and then listed them in descending order of size. A metallurgical analysis of three of the "brass" guns reveal, respectively, 77% copper alloyed with zinc, tin, and lead; 92% copper alloyed with tin, antimony, and lead; and 93% copper primarily alloyed with tin and small amounts of antimony, arsenic, and nickel. Of the ten

Figure 5.13
Array of gun furniture recovered from beside the in situ cannon amidships. Including leather bucket, stone and iron shot, powder ladle and rammer, wooden shot gauges, personal possessions, and clothing.

Figure 5.14
Sequence of differently sized powder ladles to serve individual guns.

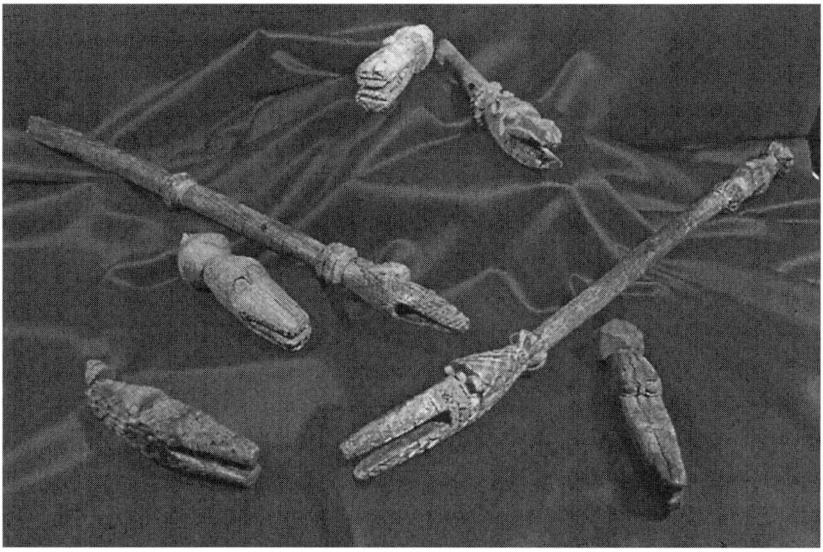

Figure 5.15
Collection of wooden linstocks used to bring a slow match to the touch hole of the guns to ignite the charge and fire the gun.

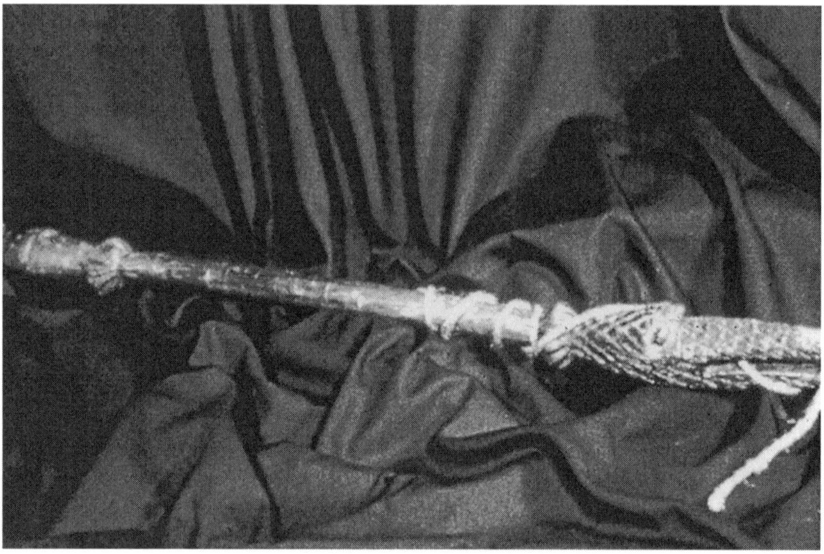

Figure 5.16
Individual linstock.

brass guns recovered, all but two are dated—the earliest and latest bearing the dates 1536 and 1543. This suggests that the ship was at least partially re-armed after her 1535 refit. For the most part, the caliber and weight of the guns vary within prescribed Tudor doctrines for guns of the types listed (figure 5.17). Several are shorter and therefore lighter than expected. Possibly these had been cast expressly for use at sea.

The bronze guns carried on board include two of only nine cannon within the entire fleet (figures 5.18, 5.19). These were mounted amidships one to starboard and the other to port. The other main-deck bronze guns include two demi-cannon in the stern and two culverins in the bow (figures 5.20–5.23). Ten of the fifteen bronze guns listed in the 1546 inventory have been recovered: four during the nineteenth-century salvaging and the other six during recent excavations. Five of these were recovered from their gun stations, still mounted on their carriages. These consist of a bed plank (or planks joined longitudinally), two front cheeks recessed to take the trunnions, and two rear stepped cheeks (figures 5.24–5.27)—all of which were supported on two axles with four solid trucks. The sixth had rolled from the port side and fallen off its carriage. Two additional carriages for bronze guns were found at the very top of the Tudor levels amidships and in the stern. These fit two of the guns recovered during the nineteenth-century salvaging, giving positions on

Figure 5.17
Five of the ten bronze guns recovered from the *Mary Rose*.

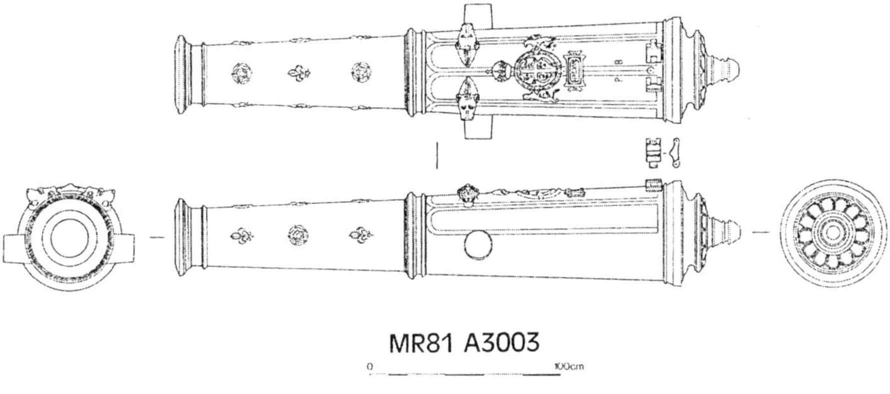

Figure 5.18
Cannon MR 81A3003. Situated on the main deck in the middle of the ship.

Figure 5.19
Detail of device.

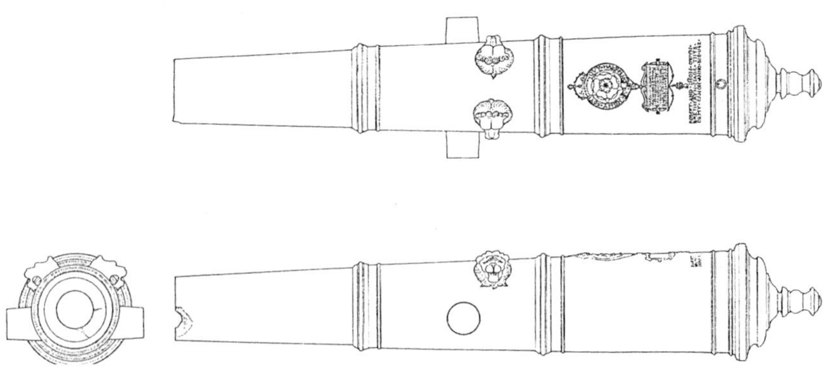

Figure 5.20
Demi-cannon MR 81 A3000. Situated on the main deck in the stern. The muzzle cornice is missing, probably the result of an incomplete casting.

Figure 5.21
Detail of royal device.

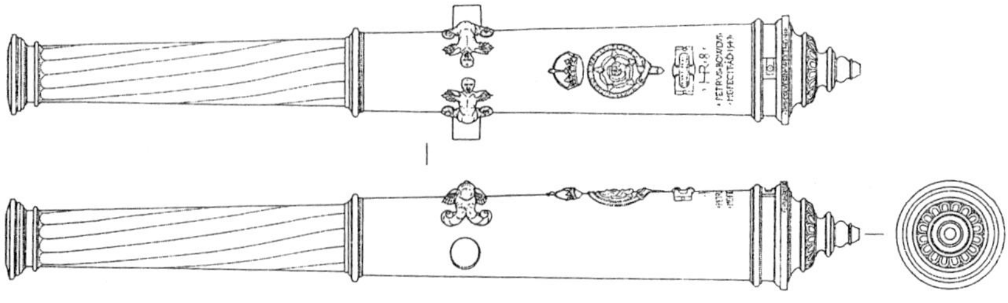

Figure 5.22
Culverin MR81 A1423. Situated on the main deck in the bow. A replica of this gun has been made and fired successfully.

the main deck for a cannon amidships on the port side and a demi-cannon on the port side in the stern. Only one of the six demi-culverins listed was found in position (figures 5.28–5.30). This was recovered from the starboard side of the ship on the castle deck at the front of the sterncastle. In this position at the widest part of the ship, it fired forward, just clearing the bow. A second demi-culverin was recovered by the Deane activity and may be the portside pair to this gun. The four additional demi-culverins could also have been located in four vacant gunports at the upper-deck level in the stern, or in the bowcastle. Either of these locations would have been relatively accessible to Tudor salvors. Isolated carriage elements for small bronze guns found in both the bow and the sterncastles suggest the presence of the falcons and sakers listed.

Twelve wrought-iron "port pieces" are listed in the inventory with 200 stone shot to serve them (16 rounds per gun) (figure 5.31). These guns have been identified solely because of their presence on the *Mary Rose*. The sparse descriptions in inventories that indicate their form, combined with their listing at the top of the wrought-iron gun inventories, suggest that these were the largest wrought-iron guns carried, having bores of between 5 and 8 inches. They are always listed with shot of stone and with two chambers per gun. The largest wrought-iron guns recovered must therefore represent the port pieces. Ten of these could have been positioned on the main deck, four on each side and two from transom ports. Three of these were recovered from the starboard side during recent excavations, with a sledge for the fourth still in position—the gun for this presumably salvaged in the nineteenth century. The last two port pieces listed may have been accommodated on the upper-deck level in the stern or the bow. A portion of the bowcastle lies collapsed, however, and has never been disturbed. Remains of up to eight of the twelve listed port pieces have been recovered, six of these on their elm sledges. These sledges were hollowed out of a single piece of wood that accommodated the barrel, breech

Figure 5.23
Detail of royal device.

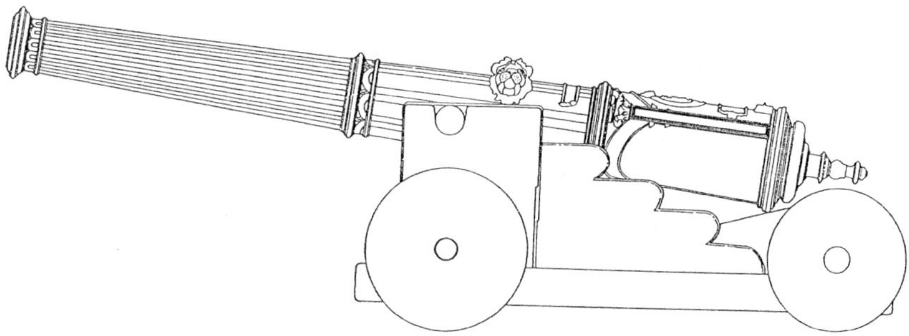

Figure 5.24
Demi-culverin MR79 A1232/1241 on its carriage.

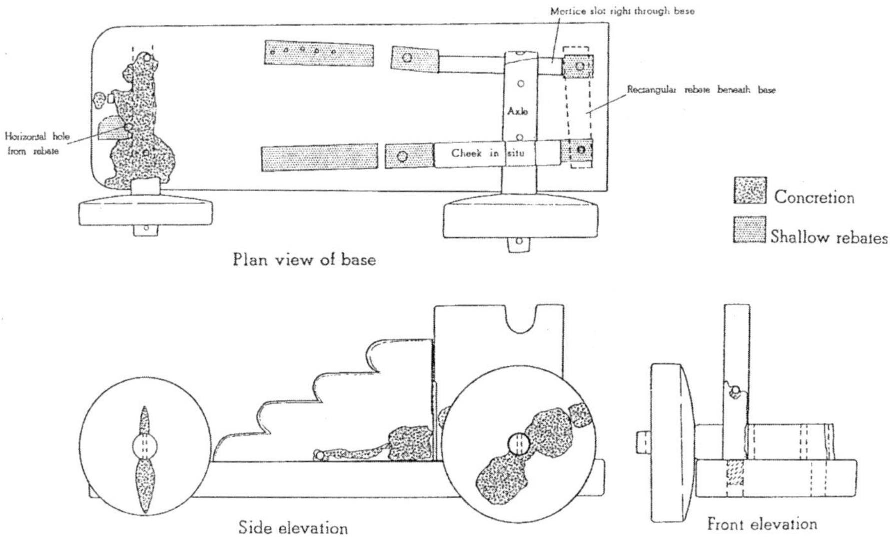

Figure 5.25
Detail of carriage for demi-culverin MR79 A1232.

chamber, breech chamber wedge (forelock), and iron wedges. They were supported at the point of balance by a single ash axle with either a pair of large spoked wheels or a pair of solid smaller trucks. Their relative importance on the *Mary Rose* may be demonstrated by their numerical incidence within the inventory. This is repeated for most vessels listed within the fleet, with a total of 218 carried within the 58 vessels. There are only fifteen cast-bronze guns listed for the *Mary Rose*, ranging from cannon to falcons, but the number of port pieces alone is twelve. Such guns must therefore be considered primarily

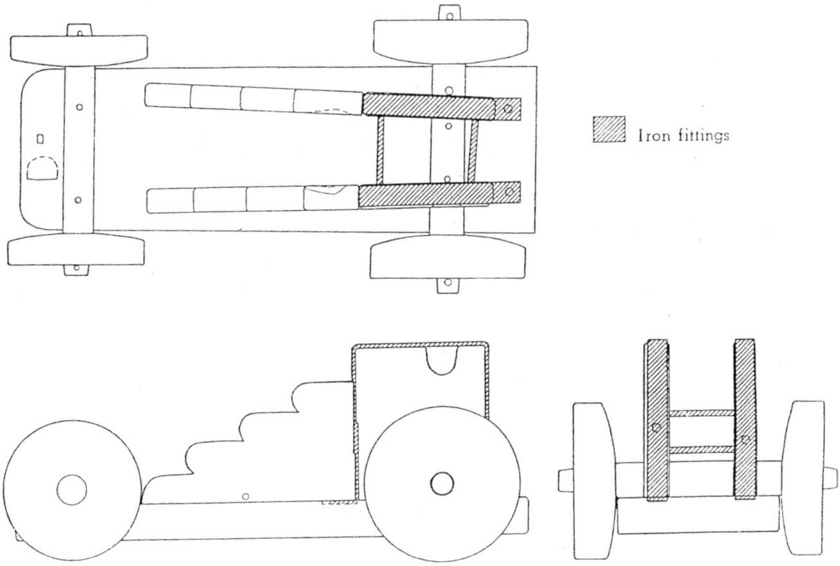

Figure 5.26
Carriage for MR79 A1232, showing iron capsquares.

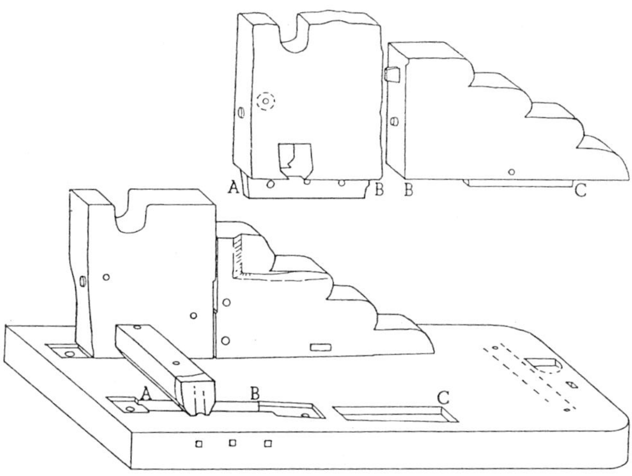

Figure 5.27
Constructional details for carriage MR79 A1232.

Figure 5.28
Demi-culverin MR79 A1232 on reproduction carriage in the *Mary Rose* exhibition.

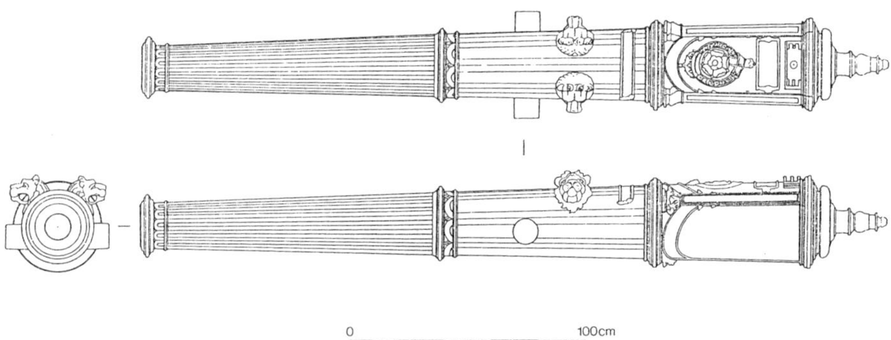

Figure 5.29
Detail on gun MR79 A1232.

Figure 5.30
Royal device on MR79 A1232.

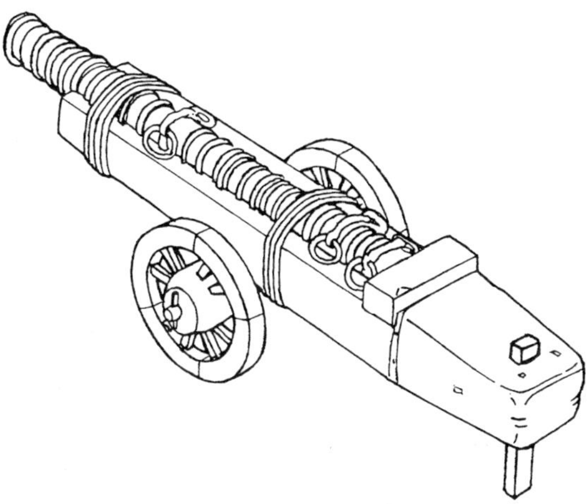

Figure 5.31
Port piece MR81 A2604 on its carriage.

Figure 5.32
Reproduction port piece trained at target.

as anti-ship ordnance, not unlike the newer bronze guns, although they were possibly restricted to closer ranges. Firing trials of a reproduction port piece against a faithful reproduction of hull structure have demonstrated the piercing and splintering capability of stone shot, albeit at a relatively short range (figures 5.32, 5.33).

Similar carriages supported the smaller wrought-iron slings, which were situated on the upper deck, and possibly the enigmatic fowlers (figures 5.34, 5.35). The sling group has only been firmly identified because of their presence on the *Mary Rose*, their listing in the inventory, and the consistent specification

Figure 5.33
Target made up of oak planking and framing of the same dimensions as the hull of the *Mary Rose* at the main-deck level.

of iron shot to serve them. They were positively identified with the discovery of several still loaded with cast-iron shot with bores of between 2 and 4½ inches. Only one complete fowler has been identified. Raised by John Deane in the nineteenth century on its carriage, it is similar to the port pieces but has a bore of 4½ inches. Most of the iron guns recovered from the *Mary Rose* have been radiographed, revealing that the carriage-mounted guns are formed of a stave-built inner tube that is open-ended and covered by a single layer of alternating wide bands and narrow hoops. In some instances, notably with the slings, a double layer of staves has been observed. There were no attempts to weld the barrel or introduce lead between the seams. The chambers were similarly constructed, even though they may have differing widths and layers of staves.

The *Mary Rose* carried the only four known hailshot pieces (figure 5.36), while twenty are listed. These guns have been identified by their size within the iron ordnance listing, and by the presence of iron dice (listed as "iron dice for hailshot" in inventories) in their barrel. All those recovered are cast-iron muzzle-loaders, although wrought-iron hailshot and chambered hailshot also existed. Each gun featured a rectangular bore and a fin-like hook on its underside and may have been placed over a rail (figure 5.37). The guns were found with long wooden stocks, which may have been tucked under

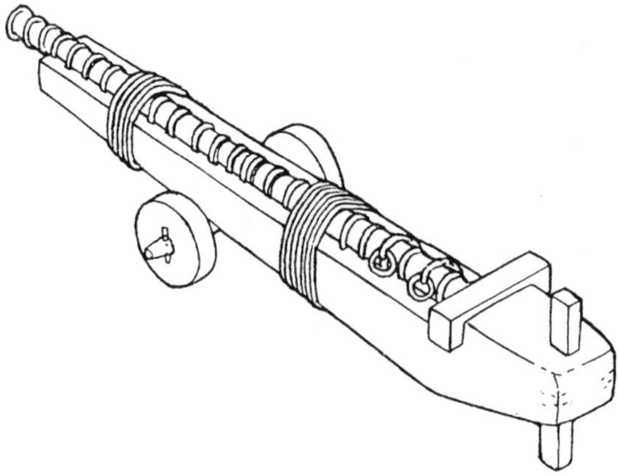

Figure 5.34
Sling 81A0645 on its carriage.

Figure 5.35
Model of the midship section of the *Mary Rose* showing the main gun deck with a port piece and cannon and a sling on the upper deck above. An archer is shooting through removable protective blindages within the upper-deck structure.

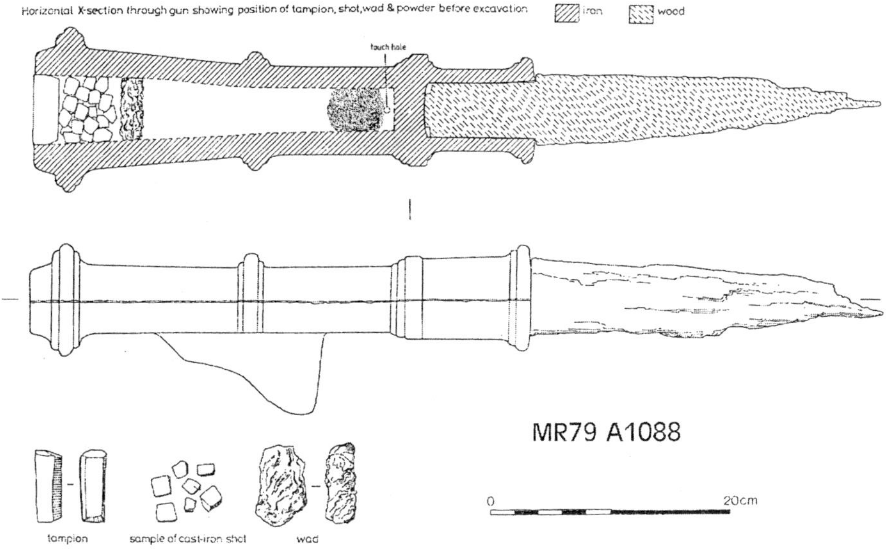

Figure 5.36
Cast-iron hailshot piece.

the gunners' arms. The large number of muzzle-loaders listed over a relatively short period suggests that in 1546 and 1547 they may have been cast rather than wrought. A metallurgical analysis of a sample from the *Mary Rose* indicates that the founders made the gun by pouring the molten metal into a brass mold—the bore mold consisting of sand. Used to cast shot, this technique completely differs from that used to cast large guns, where the outer mold consists of sand and the inner core, of iron. This is the first attempt at the mass production of a cast-iron gun. At this time there are only 27 carriage-mounted cast-iron guns within the fleet; they remain relatively rare, since they do not seem to have the longevity of the wrought-iron guns, such as the bases. Hardly any existed by 1555. Nonetheless, replication and firing trials have confirmed an impressive scatter of cast-iron dice at close ranges (between five and fifteen meters). The hailshot pieces were recovered either from the castle deck in the stern or from positions where they would have fallen from the upper or castle decks. Beyond this unique discovery, their identification is important as they are the second most prevalent type of gun (excluding handguns) listed within the inventory for the *Mary Rose*: they formed 25.30% of the iron-gun assemblage within the fleet in 1546.

The most prevalent single type of ordnance on the *Mary Rose* is the swivel gun, including the double and single bases. *Mary Rose* is listed with 30

Figure 5.37
Hailshot piece showing rectangular bore.

(the parts of about 13 survive), and the total for the fleet comprises 570 single and 221 double bases. Every vessel carried these in quantities of 2 to 60. Because they represent the smallest pieces, swivel guns are always listed as the final guns on the list of wrought-iron ordnance, with bores of between 1 and 3 inches. They may have performed an anti-personnel function as did the serpentines, which were very prevalent in the late fifteenth and early sixteenth centuries. Their identification has been confirmed by those recovered from the *Mary Rose* (figures 5.38, 5.39). In addition, of the 50 handguns listed, five fragmentary examples have been recovered. One is identical in form to one purchased by the Royal Armouries and bears a crest with the initials GARDO, suggesting the Italian town of Gardone in Northern Italy. This also has been replicated and trialed.

The Anthony Roll also lists two different types of powder: serpentine and corned. Two "last" (one "last" weighing 2,688 pounds) of serpentine powder were shipped in barrels carrying 100 pounds of powder each. This equates

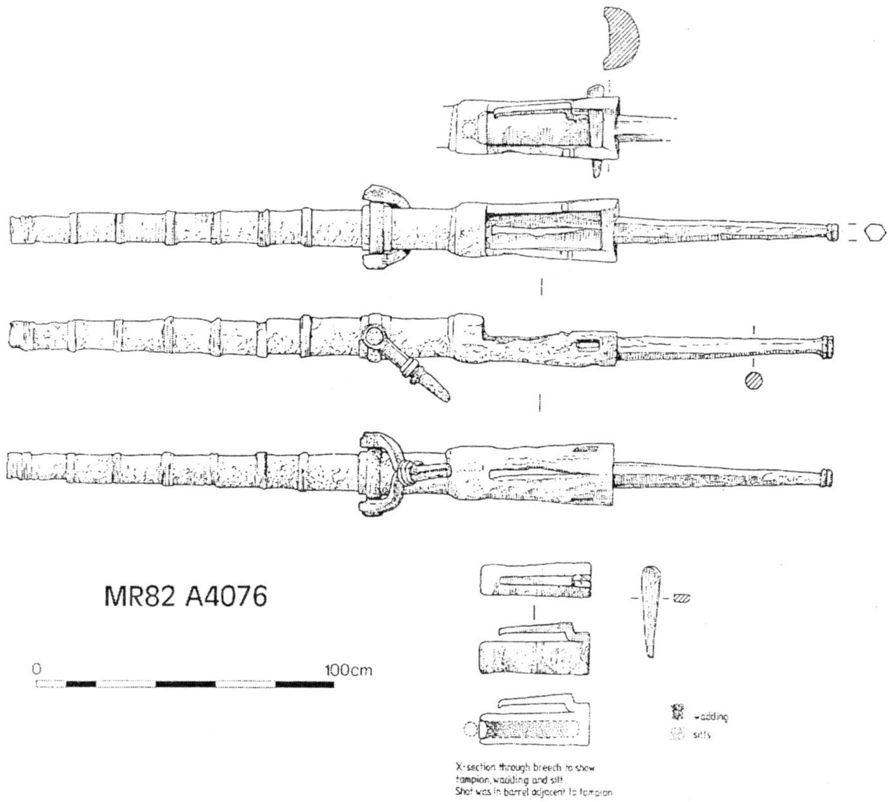

Figure 5.38
Wrought-iron base or swivel gun with removable breech chamber.

to 5,376 pounds of serpentine powder carried in 48 barrels. Only three barrels (300 pounds) of corned powder are listed. Powder listed as "corned" is a relatively recent description in sixteenth-century inventories, and the amounts of each type of powder relative to the ordnance is intriguing, thus offering a new challenge to researching gunpowder technology, use, and storage. It is clear from the amounts listed that the majority of guns utilized "serpentine" powder, and the small amount of corned powder listed raises a question as to what it was used with. There appears to be no single gun listed that coincides with the presence of corned powder exclusively, although the listing of handguns usually guarantees the presence of corned powder. Whatever the powder (unless the amounts used vary greatly from prescribed doctrine), all the vessels are very under-resourced for powder relative to the number of shot carried for each type of gun. Although certain constituents of the powder

Figure 5.39
Collection of bases, chambers, and two hailshot pieces.

have survived within the bores of guns and breech chambers, the potassium nitrate has dissolved; therefore, any potential for identifying the original composition is hopeless. We have to rely on the documentary sources for volume, although the extant barrels recovered may reveal the possible types or sizes used to store gunpowder.

Shot is listed under separate headings of iron, stone, and lead for each named gun, while descending in order of size (figure 5.40). Matching shot with guns based on their position in the inventory and the type of material for the named guns has helped identify certain guns, in particular the bases and slings. Such work has been essential for clarifying the types of wrought-iron guns. Yet the apparently large tolerance for windage (between ½ inch and 1 inch), observed by the excavation of iron shot from within the barrels of bronze guns, makes matching stored shot with guns difficult, especially when some of the iron guns were served with iron shot. It is clear, however, that there are ample quantities of ammunition—in the case of iron shot nearly twice the number listed were recovered. With stone shot, in contrast, 388 were recovered and 390 were listed. The lead shot listed for the bases is actually a composite shot of cast lead containing an iron dice. Nearly 200 of the 400 listed have been recovered, and continue to be recovered during the ongoing monitoring of the site. The 1,000 lead shot for handguns are represented by fewer than 100 recovered.

Figure 5.40
Cast-iron shot laid out in order of size.

These are very small and could have been stored close to the guns on the upper decks and scattered off-site.

The Anthony Roll, combined with the archaeological evidence from the vessel, demonstrates the continued importance of longbows and arrows within the fleet. Every vessel, despite the size and the number of guns, carried longbows. Interestingly, there were no dedicated archers listed within inventories, and on smaller vessels soldiers are not listed. This suggests that either the gunners or the mariners performed this demanding task. The *Mary Rose* archery assemblage is the largest exactly dated assemblage of the later medieval period, with 172 of the 250 listed longbows recovered. Although less than half the listed arrows were found, the presence of a number of partially filled or empty storage chests for arrows totaling seven would accommodate 8,400 of the 9,600 listed arrows, assuming that the documentary claims of a full chest containing 50 sheaves is correct. While there is evidence of their storage barrels, there is none of the bowstrings—they did not survive the particular burial conditions on the *Mary Rose* site.

Staff weapons—including pikes, bills, and halberds—were primarily located on the upper deck in the stern. The remains of 121 bills survive; however, the thinner pikes with much smaller heads are far fewer in number, totaling only 21. These easily could have floated away, as their small leaf-shaped heads are light relative to the long ash staffs. None of the 40 dozen top darts have been identified; if placed on the fighting tops, they probably fell clear of the ship as she sank.

Changes in Armaments

The inventory of 1514 discloses a mixture of wrought-iron and bronze guns. The majority of the wrought-iron guns were small, and, like those on other ships within the same inventory, most were located high up in the castles at the bow and stern. The inventory of 1540 shows a radical change, however (tables 5.1, 5.2). While it demonstrates the growing importance of bronze artillery by listing muzzle-loading, smooth-bore guns that take cast-iron shot, more importantly it reveals that the most dramatic increase is in the number of large breech-loading wrought-iron guns, principally the port pieces (table 5.3). This may reflect the push/drive to rapidly furnish all these newly created gunports with guns, as well as the effort to arm all the newly built or refurbished castles and blockhouses as part of Henry VIII's defence programme of 1539. Alternatively, breech-loading guns may have been desired purely for specific superior qualities—rapid reloading being the most obvious. Our experiments with manufacturing the two gun types have demonstrated that the wrought-iron gun requires negligible resources. One skilled smith with three or four men can build a large wrought-iron gun in a village forge. Because the gun is built up incrementally, and the energy demanded at most stages is limited, the forge can be small. The most taxing requirement (on both skill and temperature) involves installing the plug to seal the back of the breech chamber; both the plug and the chamber have to be hot enough to seal together. During our experiments we had difficulty accomplishing this, undoubtedly due to our inexperience. By contrast, casting the large bronze gun proved utterly different, since the labor force and investment in costly equipment required at all stages was immense. Over three tons of metal must remain molten throughout the pour; therefore, the energy input is of a completely different scale compared to that of wrought-iron guns. So, while village blacksmiths were able to produce wrought-iron guns in small forges all over the country at modest unit costs, bronze-casting masters demanded both heavy capitalization and expensive prices. Consequently, the available facilities were scarce.

At the start of Henry VIII's reign, bronze ordnance was imported from abroad. As monarch, Henry VIII established bronze foundries in and around London and recruited the best practitioners from abroad. One in particular, Peter Baude, is thought to have been one of the key persons in the development of the cast iron industry. Copper, however, still had to be imported from

Table 5.1
Iron guns listed on the *Mary Rose*.

1514		1540		1546	
		Port pieces	9	Port pieces	12 (8 recovered)
Slings	2	Slings	6	Slings	6 (2 complete, parts of 5)
		Fowlers	6	Fowlers	6 (1 complete, parts of 3)
Stone guns	26				
Top guns	3	Top guns	2	Top pieces	2 (none identified)
Cast pieces	2				
Serpentines	28	Bases	60	Bases	30 (4 +9 incomplete)
Murderers	4			Hailshot pieces	20 (4 recovered)
Hackbusshes	9	Handguns	50	Handguns	50

Table 5.2
Brass guns listed on the *Mary Rose*.

1514 inventory		1540 inventory		1546 inventory	
Great curtows	5	Demi-cannon	4	Cannon	2 (2 recovered)
Murderers	2	Culverins	2	Demi-cannon	2 (3 recovered)
Falcons	2	Demi-culverins	2	Culverins	2 (3 recovered)
Falconets	3	Sakers	5	Demi-culverins	6 (2 recovered)
Little brass gun	1	Falcons	2	Sakers	2 (0 recovered)
Total	13	Total	15	Falcons	1 (0 recovered)
				Total	15

Table 5.3
Incidence of port pieces relative to wrought-iron and cast-bronze guns on vessels. (Port pieces are listed in vessel inventories up to 1677.)

	Vessel total	Port piece total	Iron gun total (wrought)	Bronze gun total
1540	10	93	165	85
1546	58	197	396	245
1547	52	218	473	161
1555	18	35	88 wrought, 72 cast	83

Table 5.4
Incidence of hailshot pieces according to Tower of London Inventories (source: Blackmore 1976).

	1547	1559	1568	1589	1595
Number	41	80	173	1	1
Description	iron	forged iron	168 cast, 5 forged, 4 "chambers"	forged iron	forged iron

Table 5.5
Hailshot pieces listed within the fleet, 1546–1555 (source: Anthony Roll Inventory, MS 129, BM Harley MS 1419A, private inventory 1555).

Inventory date	Number of vessels listed	Hailshot pieces total
1546	58	459 in 58 vessels
1547	52	86 in 11 vessels
1555	13	13 in 2 vessels

abroad, and during his reign bronze cost from about seven times as much per ton as iron to about twelve times as much. It is interesting that with the perceived threat of invasion, Henry VIII still sought to arm his ships and forts with wrought-iron guns, heralding an older technology but yielding a cheaper and more rapidly produced weapon (tables 5.1–5.3). The *Mary Rose* reflected this option throughout her life. Her weaponry included the first attempts at the casting of a weapon (rather than a projectile) in iron on an industrial scale: the hailshot pieces (tables 5.4, 5.5). Although she carried only 20, the actual number within the fleet was 459, making them the second most numerous gun within the fleet. This gun is therefore a vital transition from using iron for the projectile to using iron for the gun itself. Its identification is important for comprehending the employment of cast iron for ordnance production. All inventories still demonstrate the inevitable outcome of naval engagements: a vessel would draw close enough to an enemy vessel for sailors and soldiers to use longbows, followed by the 300 staff weapons when boarding was imminent. Longbows, however, ceased to be used in any tactically significant numbers by the end of the century, although small numbers remained during the first quarter of the seventeenth century.

The only near-contemporary drawing of the *Mary Rose* comes from the Anthony Roll. It reveals, albeit with a certain amount of artistic license, a purpose-built ship of war with guns protruding at unusual angles through gunports, both lidded and non-lidded. Based on this illustration, as well as the

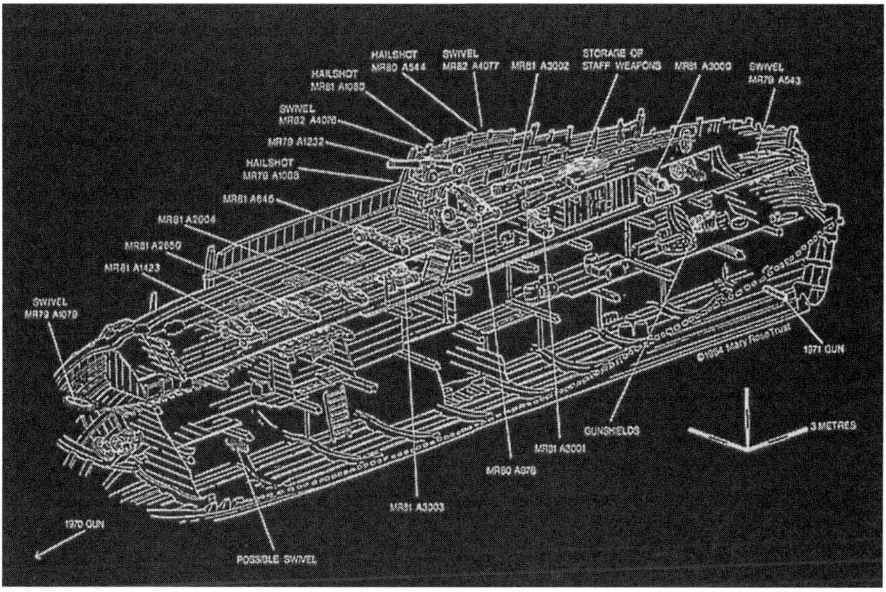

Figure 5.41
Isometric showing the location of armaments.

accompanying list of Ordnance, Artillery, Munitions and Equipment for the War, it is unmistakable that she was a large warship, capable of firing devastating broadsides.

The number and the distribution of specific pieces of ordnance on the decks of the *Mary Rose* suggest that she was a well-adapted fighting machine that achieved a balanced coverage of firepower at least from her broadsides. Not enough of the transom and the bowcastle exist to estimate how guns were located in these positions (figure 5.41). From a distance she could engage the enemy with her bronze weapons armed with cast-iron shot. At closer ranges she could utilize the wrought-iron port pieces and fowlers that fired both stone shot and cracked flints housed in wooden cylinders fabricated like lanterns (hence the term "lanthorn shot"). There is little written about the tactics of sea-borne warfare in the sixteenth century. William Bourne, however, did go into some detail when he advocated placing the long guns out of the stern or bow, and the shorter guns out of the ship's side: "You must fit your ordnance according unto the place that it must be in." He elaborated on the importance of the height and depth of the gunports relative to the carriage wheels, as well as the need to breech the guns and shut the gunports on the "lower orlop"—in our case, the main gun deck. (It is unfortunate that the

Captain of the *Mary Rose* did not feel the same.) Regarding tactics, Bourne specified the following:

> How to make a shot out of one ship unto another that although the sea be wrought, or out of a galley to a ship: If they do mean to enter you, take mark where you do see any scuppers (scottles) for to come up at, as they will stand near thereabouts, to the intent for to be ready, for to come up under the scuppers: there give level with your fowlers, or slings, or bases, for there you shall be sure to do most good, then furthermore, if you do mean for to enter him, then give level with your fowlers and port pieces where you do see his chiefest fight of his ship is, and especially be sure to have them charged and to shoot them off at the first boarding of the ships, for then you shall be sure to speed. And furthermore, mark where his men have most recourse, there discharge your fowlers and bases. And furthermore, for the annoyance of your enemy, therefore you may take away their steeridge with one of your great pieces that is to shoot at his rudder and furthermore at his main mast and so forth.

It is also worth noting Bourne's advice about the choice of shot for a particular engagement: "With their great ordnance, as cannon or culverins, at a great distance, to shoot the whole iron shot (as you do at a battery) as they do approach near, then shoot falcon shot and as they do come nearer, falconet shot or base shot and at hand, all manner of spoiling shot, and dice shot and such other like." Clearly, the guns carried performed different functions, each of which was appreciated if not optimized. At close range the swivel guns and hailshot pieces would be unleashed, and between shots the longbow's arrows would torment the enemy crew. Lastly, the handguns, pikes, bills, and shields pierced with small guns would be used when hand-to-hand combat commenced.

The lifelong inventories for the *Mary Rose* illustrate the change in ordnance carried onboard—most notably, the increase in large guns, possibly during her refitting in 1536 and again after 1540. Interestingly, the major alteration is not the addition of cast-bronze muzzle-loaders, but the dramatic increase in the number of the largest wrought-iron guns (tables 5.1, 5.2). As these large guns were placed on the main gun deck, smaller ports for swivel guns were blocked. This is evidenced by the reduced number of bases (30) in 1546 compared with that of the serpentines (75) in 1514. Listing the names of guns in the Tower between 1514 and 1701, table 5.6 demonstrates the longevity of certain pieces, although their specifications may have changed over time. It also indicates that a number of new "names" appeared in 1540: the port piece, the fowler, and the base. The historical literature reveals that many of these were new additions and their diffusion on board ships within a short period testifies to how quickly such new guns were deployed. It also shows the unmistakable importance of the fleet as part of the nation's defense. This plethora of newly made guns, regardless of the "age" of the technology used to produce them, must reflect both the constrained political economy and the perceived threat of invasion by the European alliance. Before the 1540s, craftsmen were investigating the casting of iron guns. By the time of the inventory of 1555, there were 72 cast-iron, carriage-mounted guns within the fleet of 18 vessels. This figure gradually increased, with both the bronze and wrought-iron guns being largely replaced by iron castings. The motivation was undoubtedly financial, and the response was quite sensible: despite the decreased strength-to-weight ratio of iron as opposed to that of bronze, cast-iron guns appeared worthwhile because they were inexpensive and easy to deploy on robust ships that fought at increased fighting distances. What seems irrationally conservative, however, is the lengthy delay in the cutting of gunports along the lowermost decks of vessels. If gunport lids can be attributed to 1501, why then did it take until the middle of the century to fit them into capital vessels? Perhaps the sinking of the *Mary Rose* reflected the perceived risk that such an invention represented during the early sixteenth century.

This is a tale of two centuries—but perhaps we should extend it to three. The *Mary Rose* carried many wrought-iron pieces that could sit unquestioned beside the serpentines of the fifteenth century. The smooth-bore, cast-bronze muzzle-loaders, by contrast, are hard to distinguish from those of the seventeenth century. Ongoing study of the structural alterations and additions, backed by a dendrochronological study of the timber elements making up the hull structure, should allow us to follow the evolution of the structure and the ordnance that could be accommodated within it. It was this transition that eventually heralded a complete revision in naval warfare tactics. With the *Mary Rose* we have a vessel that carried enough guns to be able to fire formidable broadsides. The difference in the guns' ranges dictated how many guns could have been fired simultaneously to engage their target. The shortest range of any one of the large guns likewise dictated the maximum distance for effective broadside engagement. What is clear from our firing trials is that both the wrought-iron breechloaders and the cast-bronze muzzle-loaders could penetrate the hull of a wooden vessel. We achieved a muzzle velocity of over 384 meters per second for the wrought-iron gun and over 502 meters per second for the cast-bronze gun. Nevertheless, it is also clear that boarding an enemy vessel remained a central battle tactic, as reflected by the presence of longbows and staff weapons, as well as the numerous short-range anti-personnel ordnance. This range of weapons and projectiles carried suggests a formidable

Table 5.6

Gun types listed in the Tower of London inventories. After H. Blackmore, *The Armouries of the Tower of London 1 Ordnance* (HMSO, 1976).

	1405	1495	1514	1523	1540	1547	1559	1568	1589	1595	1603	1620	1635	1665	1677	1683	1688	1698	1701	1720	1726
Base					*	*	*	*	*	*		*	*		*						
Basilisco						*			*			*	*	*							
Bombard			*	*		*	*	*	*	*		*			*						
Bombardel		*	*	*																	
Cannon	*				*	*	*	*	*	*	*	*	*	*	*	*	*	*	*	*	*
Cannon perier						*	*	*	*	*	*	*	*		*						
Culverin			*	*	*	*	*	*	*	*	*	*	*	*	*	*	*	*	*	*	
Curtow		*	*	*																	
Falcon		*	*	*	*	*	*	*	*	*	*	*	*	*	*	*	*	*	*	*	
Falconet		*	*	*	*	*	*	*	*	*	*	*	*	*	*	*	*	*	*	*	
Fowler					*	*	*	*	*	*	*	*	*								
Hailshot Piece						*	*	*	*	*											
Mortar						*	*		*	*	*	*	*	*	*	*	*	*	*	*	*
Murderer		*	*									*	*		*						
Portpiece					*	*	*	*	*	*	*	*	*	*	*						
Pot gun		*	*																		
Robinet						*	*	*	*	*	*	*	*	*	*	*	*	*			
Saker						*	*	*	*	*	*	*	*	*	*	*	*	*	*	*	*
Serpentine		*	*	*																	
Sling				*	*	*	*	*	*	*											
Stone gun				*																	

system of layered defence. Yet we will perhaps always question how much of this was instigated by combat experience and how much by operational reasoning.

Clearly, a simplistic view of the development of artillery based on the simple transition from wrought to cast iron is not borne out by the archaeological evidence. At the same time as bronze guns were being cast for the king's ship in 1535, a whole series of newly made and newly named wrought-iron guns lay in the forge at the Tower of London, the port pieces. During the 36 years between the inception and sinking of the *Mary Rose*, there were undoubtedly milestones that were reached and overturned. Alternations in the platform, the guns, and the constituents and manufacturing of gunpowder resulted in heavier and more powerful guns. Undoubtedly, in the case of some of the weapons, these developments provided an increased range. This period probably represents the only time that so many guns of such different sizes, conflicting traditions, and structural variety were used routinely together. The work on the *Mary Rose* is hardly complete. To ascertain the functional capabilities of her weapons, we anticipate continued study, replication, and testing. Only with such data can we finally grasp the full potential of the vessel and her astonishing ordnance system.

6

Mathematics and Empire: The Military Impulse and the Scientific Revolution

Lesley B. Cormack

> When the seamen of the sixteenth century went to sea, they laid the foundation-stone of the British Empire and, when they returned and made compasses, of modern experimental science.[1]

In London during the 1580s, Thomas Hood, a university-trained mathematician, lectured on mathematical geography and navigation at the home of Sir Thomas Smith in Gracechurch Street. He had earlier attended Trinity College, Cambridge, where he received a B.A. in 1578 and an M.A. in 1581.[2] A merchant and later a governor of the East India Company, Sir Thomas had set up this lectureship to educate those involved with overseas ventures, possibly employees of the Virginia Company. The makeup of the audience is now unknown, although from the tone of Hood's introductory remarks in a lecture titled "A Copie of the Speache made by the Mathematicall Lecturer, unto the Worshipfull Companye present . . . in Gracious Street: the 4 of November 1588," it appeared to consist of mathematical colleagues and mercantile patrons, rather than those mariners he insisted needed training.[3] The contents of Hood's lectures also remain unknown, but the treatises bound with the British Library copy indicate that he stressed navigational techniques, instruments, astronomy, and geometry—all of which he might have learned at Cambridge.[4] A mathematical practitioner himself, Hood belonged to a group of theoretical and practical scholars who were essential to the development of science in the early modern period. Motivated by imperial, mercantile, and intellectual aspirations, these men combined mathematical insights with commercial concerns and laid basic foundations for the Scientific Revolution in England.

In 1941, Edgar Zilsel linked the concepts of navigation, imperialism, and the Scientific Revolution, just as Hood had done in his life and lectures.[5] Since World War II, historians have downplayed this connection, arguing instead for a separation of brain and hand, scholar and craftsman.[6] This separation cannot bear close scrutiny. The development of natural investigation in early modern

Europe was a prerequisite for several ambitious technical, military, and political achievements. Likewise, the basic changes in military organization, technology, and scale, often called the "Military Revolution," inspired new research agendas, patronage, and rhetoric. These in turn contributed to the Scientific Revolution. Mercantilism played its part, as did the growth of the nation-state. A simple causal relationship obviously does not exist between developments as complex as the Military Revolution and the Scientific Revolution. Nevertheless, we can pinpoint how the concerns of expansionist patrons encouraged a particular group of investigators to pursue a research program designed to answer military and imperial problems. These men took up mathematical questions and promoted their solutions as useful to military and imperial endeavours. Mathematical practitioners thus signify a connection between the Military and Scientific Revolutions. This marriage helps explain the massive changes that took place in the scientific domain.

Most historians of the period have downplayed this crucial category of scientifically inclined men: the mathematical practitioners.[7] Mathematics was separate from natural philosophy, and its practitioners usually directed their studies toward commercial and often military applications, including artillery, fortification, navigation, and surveying.[8] These men rose in prominence during the early modern period by promoting measurement, experiment, and utility within the study of nature.[9] Their growing importance resulted from changing economic structures, developing technologies, and new politicized intellectual spaces such as courts. In this way, changes in "science" were related to the development of mercantilism and the nation-state. Crucial conditions for the Scientific Revolution were thus established by mathematical practitioners who bridged the practical applications of their discipline and the universal concerns of natural philosophy. Their success, in turn, was facilitated by the new political, social, and cultural patronage of the princely courts. In other words, key aspects of the Scientific and Military Revolutions were conflated by this interaction between mathematical practitioners and their patrons.

Who Were These Mathematical Practitioners?

In 1941, Zilsel suggested that the emergence of a category of superior artisans was responsible for the origins of modern science.[10] These artisans (artist-engineers, scientific instrument makers, surveyors, and navigators) were distinguished from both university scholars and humanist literati, as well as from the guilds. While participating in the growing capitalist and nationalist enterprise, according to Zilsel, this new group developed a methodology that stressed empiricism, quantification, and cooperation. Although Zilsel's "superior artisans" have inspired my interest in mathematical practitioners, they are not identical categories. According to Zilsel, these superior artisans could not create real scientific knowledge independently; they needed to work in concert with natural philosophers, and it was this critical cooperation that enabled modern science to emerge. In contrast, I see the connection between theory and practice occurring within the minds of individuals. Their careers and connections were consequently fertile grounds for the revolutionary changes emerging in science.

Edward Wright and Thomas Harriot were geographers who pursued both scholarly and practical issues, and could aptly be considered mathematical practitioners. Both were university-educated men who studied the classical foundations of their subject, as well as recent discoveries and theories, but hardly as traditional scholastics. Both went on prolonged voyages of discovery where they learned about oceanic navigation and its shortcomings from rude mechanicals and from skilled navigators. Wright and Herriot went beyond this immediate knowledge, however, and attempted to formalize the structure of the globe, along with their conceptualization of the new world. Connected to important courts and patrons, they also used the cry of utility and imperialism to promote geographic knowledge. The growing interest of patrons and rulers in military and expansionist enterprises thus gave these men a career focus for their research objectives.

Wright, the most famous English geographer of the period, was educated at Gonville and Caius College, Cambridge, receiving a B.A. in 1581 and an M.A. in 1584. He remained at Cambridge until the end of the century, with a brief sojourn to the Azores with the Earl of Cumberland in 1589.[11] Wright's greatest achievement was *Certaine Errors in Navigation* (1599), an appraisal of the problems of modern navigation and the need for a mathematical solution. In this book, Wright explained Mercator's map projection for the first time, providing an elegant Euclidean proof of the geometry involved. In addition to straightforward instructions on map construction, he published a table of meridian parts for each degree to help cartographers construct accurate projections of the meridian network.[12] Likewise, he constructed his own map using this method. Wright's work was the first truly mathematical rendering of Mercator's projection; it also promoted close communication between theoretical mathematicians and practical navigators. This placed English mathematicians, for a time, in the vanguard of European mathematical geography.

Around 1600, Wright moved from Cambridge to London, where he established himself as a teacher of mathematics and geography. Around this time, he contributed to William Gilbert's work on magnetism, thus providing a practical perspective on Gilbert's more natural philosophical outlook.[13] In the early seventeenth century, Wright is said to have become a tutor to Henry,

Prince of Wales (elder son of James). Wright's dedication of his second edition of *Certaine Errors* to Henry in 1610 strengthens this claim.[14] Upon becoming tutor, Wright "caused a large sphere to be made for his Highness, by the help of some German workmen; which sphere by means of spring-work not only represented the motion of the whole celestial sphere, but shewed likewise the particular systems of the Sun and Moon, and their circular motions, together with their places, and possibilities of eclipsing each other. In it was a work by wheel and pinion, for a motion of 171,000 years, if the sphere could be kept so long in motion."[15]

Henry had a decided interest in such devices and rewarded those who could create them.[16] Wright also designed and constructed a number of navigational instruments for the Prince. He even prepared a plan to bring water down from Uxbridge for the royal household.[17] In approximately 1612, Wright was appointed librarian to Prince Henry, who died before Wright could take up the post.[18] In 1614, Sir Thomas Smith, governor of the East India Company, hired Wright to lecture to the Company on mathematics and navigation for a salary of £50 per annum.[19] Whether these lectures were actually delivered is uncertain: Wright died in the following year.

Wright exemplifies a mathematical practitioner who established both intellectual and social connections between geographical theory and practice. He was university-trained and worked as a teacher at various points in his career. Wright addressed theoretical problems, including the mathematically sophisticated construction of map projections, and aided Gilbert in his philosophical enterprise. Nevertheless, he respected practical insights. He experienced first-hand the problems of ocean navigation, constructed instruments, and consulted with sailors and navigators. English competition with other European states, particularly Spain, fueled the demand for improved navigation and geographical understanding. Wright was thus closely connected with the practical problems of the nascent imperial state. His motivations for this balancing of handiwork and brainwork were undoubtedly many, including financial gain, social prestige, and intellectual challenge. Wright certainly stressed the usefulness of his investigations; and, through achieving the patronage of aristocrats, Prince Henry, and the East India Company (for a short time), he demonstrated the utility of geographical knowledge to imperial and mercantile causes.

Thomas Harriot was another preeminent figure in mathematical geography with connections to Prince Henry.[20] Harriot attended Oxford at the same time that Wright studied at Cambridge. He matriculated from St. Mary's Hall in 1577 and received a B.A. there in 1580. By 1582 he was employed by Sir Walter Ralegh, who sent him to Virginia in 1585. Harriot, like Wright, became an academic and theoretical geographer whose sojourn into the practical realm of travel and exploration helped form his conception of the vast globe and the innovations necessary to travel it. Harriot's description of Virginia, seen in his *Briefe and True Report of the New Found Land of Virginia* (1588),[21] represented "the first broad assessment of the potential resources of North America as seen by an educated Englishman who had been there."[22] Harriot was fascinated by the natives he encountered and advocated that English colonists treat them as legitimate occupants of the land. He sought cooperation with the natives and compiled the first word list of any North American Indian language (probably Algonquin).[23] This compilation reveals his desire for practical communication, while simultaneously classifying the world in order to understand and control it. His advice concerning Virginian settlement proved relevant as the Virginia companies of the seventeenth century were established. Such work reflected a manifest awareness of the practical and economic ramifications, if not imperial imperatives, that rational representations of the larger world invariably represented.

Harriot ranked an investigation of the mathematical structure of the globe as more important than exploration or navigation. His mathematics, however, reflected his imperial attitude in general and the experience of his Virginian contacts in particular.[24] He was deeply concerned about astronomical and physical questions, including the imperfection of the moon and the refractive indexes of various materials.[25] Harriot was inspired by Galileo's telescopic observations of the moon, and produced several fine sketches himself after *The Starry Messenger* appeared. He also investigated one of the most pressing problems of seventeenth-century mathematical geography—the problem of determining longitude at sea. Harriot worked long and hard on the longitude question, along with other navigational problems, and he related informally to many mathematical geographers his conviction that compass variation would unravel the longitude knot.[26]

Sir Walter Ralegh recruited Harriot from his old college at Oxford to be his mathematics tutor. Harriot remained in this position for much of the last two decades of the sixteenth century. He advised Ralegh's captains and navigators and pursued research that was of interest to Ralegh.[27] It was under Ralegh's auspices that Harriot voyaged to Virginia and spent time in Ireland. As Richard Hakluyt said of Harriot in 1587, in a dedication to Ralegh:

> By your experience in navigation you saw clearly that our highest glory as an insular kingdom would be built up to its greatest splendor on the firm foundation of the mathematical sciences, and so for a long time you have nourished in your household, with a most liberal salary, a young man well trained in those studies, Thomas Hariot, so that under his guidance you might in spare hours learn those noble sciences.[28]

As Ralegh fell from favor and was eventually condemned to the Tower, Harriot sought the patronage of the ninth Earl of Northumberland. The so-called Wizard Earl was another aristocrat interested in mathematical and geographical pursuits. Although Harriot's relationship with Northumberland is somewhat obscure, he appears to have conducted research within Northumberland's circle and occasionally his household, where he acted as a tutor when needed. Finally, Harriot was also connected with the court of Henry, Prince of Wales, serving as a personal instructor in applied mathematics and geography, just as Wright had.[29] It is likely that Wright and Harriot met at Henry's court. As two university-trained contemporaries, with very similar interests and experiences, they undoubtedly profited much from their mutual geographical and mathematical interests.

Harriot's career displays many of the same characteristics as Wright's. Harriot too drifted in and out of academic pursuits, from the university to Virginia, and again to positions as a researcher and a tutor for Ralegh and Northumberland. Perhaps less connected to practical pursuits than Wright was, Harriot still revealed an interest in voyaging to Virginia and determining longitude. Both appear to have promoted the education of seamen for imperial and mercantile ends. Harriot was also very dependent on patronage, especially that of Ralegh and of Northumberland (poor choices as they turned out to be), and used this patronage to help create an intellectual community where mathematical theory and imperial utility were equally valued.

Wright and Harriot represent only two out of a host of geographers situated in this interconnection between theoretical and practical developments.[30] Their wide-ranging investigations stimulated the development of associations among academic geographers, instrument makers, navigators, and investors. The result was a negotiation between theoretical and practical issues, which encouraged a new intervention with nature. This fruitful association between theory and practice helped generate the kinds of questions these men asked, the kinds of answers they accepted, and the model of the world they developed. Thus, at least in this area of scientific interest, socio-economic and political structures deeply shaped what we now call the Scientific Revolution.

New Intellectual Spaces

One of the changes of early modern Europe that allowed these mathematical practitioners to flourish was an emerging new intellectual space. The sixteenth century was a time of dislocation for natural philosophers. As the Roman Catholic Church lost its professed monopoly on Truth, so too did university scholastics.[31] Although universities continued to be important, their clientele began to change, and gentlemanly training began to vie with theological study for importance.[32] At the same time, a window of opportunity was created through the patronage of the princely courts. To take advantage of this opening, natural philosophers had to espouse the values of their potential patrons. Rather than pursuing syllogistic logic and theological subtleties (the stock-in-trade for university- and church-bound philosophers), princes sought more immediate manifestations of worldly culture, power, and wealth. Therefore, natural philosophers whose work was both prestigious and *practical* (or claimed to be) were more likely to get the coveted patronage positions.[33]

The distinction between the court natural philosopher and the (mathematical) practitioner was therefore blurred. Authors of practical treatises dedicated or presented their works to patrons and princes in order to raise the status of the practitioner and at the same time legitimize empirical knowledge. Alchemists, astrologers, and builders of curious devices all sought the notice of royal and aristocratic patrons for their own material and social betterment. Often they received patronage because of the status they brought to a court and the practical results they might achieve. For example, Portuguese pilots, who in the early fifteenth century had been beneath the notice of king or court, were made royal cosmographers in the early sixteenth century.[34] Their practical mathematical knowledge, combined with a notion of crusading glory, proved extremely attractive in this time of military and imperial expansion.

Most natural philosophers had to compete for these scarce patronage credits by claiming practicality and status for their potential patrons. Those who successfully attached themselves to princely courts gained their reputations both for intellectual acuity and for practical applications. For example, Johannes Kepler and John Dee cast horoscopes for Rudolf and Elizabeth, respectively. Dee advised Elizabeth on the most propitious day for her coronation and consulted with navigators searching for a northwest passage.[35] Likewise, Galileo's activities as a courtier were esoteric as well as applied.[36] His initial attraction to Copernican cosmology, for example, came from its justification for the practical tidal theory that he developed for the Venetian naval establishment. Men such as Dee, Kepler, and Galileo walked a fine line between theory and practice. All three were interested in developing large philosophical systems and desired court patronage for the necessary funding. Yet, as Dee's case makes clear, monarchs sought results more tangible than angelic conversations. All investigators of the natural world with court connections were compelled on occasion to dance for their supper.[37] Even investigators less directly attached to courts, such as geographers William Gilbert or Richard Hakluyt, combined an interest in theoretical cosmographical issues with direct practical results. Hakluyt wished to construct a complete image of the globe but presented his work as an imperial and commercially pragmatic project.[38] Gilbert spoke of mining and navigation, while constructing a new theory of earthly magnetism.[39]

Many English geographers combined a university education with mercantile or commercial experience. A surprising number also participated at either the Elizabethan or the Jacobean courts. In order to be welcomed, they had to combine technical expertise, political savvy, and transcendental knowledge. No sea captains need apply, but a mere university don was inappropriate as well. In other words, the intersection that Zilsel saw taking place between the scholar and the craftsman actually took place within the person of the court geographer.[40] Thus, courtly patronage provided both the locale and the reason for the interconnection of theory and practice.[41]

A claim to utility of knowledge was of prime importance in this new regime, as were direct connections between scholars and craftsmen (or at least scholarly and craft ideas).[42] In this culture, utility was first and foremost a discourse and an ideology, rather than an expected economic payoff.[43] The direct application of ideas was a desire, but the results were often disappointing. This focus on utility was less related to the proto-capitalist economy than to the growth of a new courtly culture.[44] Of course the two are interconnected, since the growth of mercantilism was fundamental to much of the new political organization in Europe. But money was not as important to these new natural investigators as was cultural capital. The mathematical practitioners thus promoted and strove for utility, but usually within the context of the courtly community.

A Research Program

Mathematical practitioners focused on problems that were at least rhetorically useful and had potential applications to early imperial expansionist, military, or mercantile enterprises. In the field of English geography, this translated into an overarching concern for navigation, especially in the northern seas. At the same time, geographical practitioners increasingly valued larger theoretical issues about the configuration of the earth, its continents, and its inhabitants. The professional study of geography therefore necessitated an interaction between such theory and practice.

Sixteenth-century English mathematical geographers developed three main areas of investigation, all centering on navigation. Although rarely applied by navigators and sailors (at least initially), the more complex solutions proposed by these geographers were prompted by questions of utility and service to the empire. The first and most pressing problem was how to determine longitude at sea. This became essential as more transatlantic voyages were made, and it became increasingly profitable to plot the shortest path to the

New World rather than to take the longer course along a fixed latitude. Navigators found it even more desirable to avoid foundering off the Scilly Islands on the return trip, as was the fate of many ships. Second, English geographers faced the difficult question of how to navigate in northern waters. In northern climes, most common charts distorted directions and distances. Because of the extreme magnetic variation, compass needles seemed to behave erratically. The solution lay in the creation of a polar projection map and the charting of the isogonic lines of compass variation. Third, geographers needed to develop a mathematical map projection that would allow sailing and rhumb lines to be drawn in a straight line. This promised to render navigation an easier and more exact art.

The problem of determining longitude at sea plagued geographers from the first Renaissance voyages through the middle of the eighteenth century.[45] Latitude determination was a relatively simple matter in both northern and southern seas, after the Portuguese development of solar declension tables. A navigator required relatively simple angular measurements of either the North Star or the sun.[46] It was necessary to have accurate instruments, of course, and to understand the rudiments of astronomy, but by the middle of the sixteenth century any moderately skilled navigator could find latitude at sea. Longitude was not so simple. All sixteenth-century geographers and navigators were aware that the essential problem was measuring relative time; however, this was extremely difficult aboard ship. Sand glasses, although used to change the watch and establish ship's time for most functions, were simply not accurate enough for calculations of longitude. Likewise, pendulum clocks, as developed by Christian Huygens and others in the seventeenth century, proved too inaccurate. Columbus tried to use lunar eclipses to judge the difference in time between a standard table (probably Regiomontanus') and the observed eclipse.[47] Others also sought to perfect the method of "lunars," but with little success. Galileo successfully developed the method of determining longitude on land by observing the eclipses of Jupiter's moons, but this also proved too difficult to observe from a pitching ship.[48]

One popular idea in England in the late sixteenth century was the possibility of using the variation of the compass to determine longitude. Navigators realized very early that compass needles did not point to true north. Furthermore, sixteenth-century mathematical geographers and navigators soon discovered that this compass variation was not uniform throughout the world, but varied from place to place. Spanish and Portuguese navigators suggested from 1535 on that this variation of compass bearing, if charted, could allow navigators to determine longitude, since they presumed that compass variation varied consistently with geographical location.[49] Some geographers, such as William Gilbert, believed that the compass needle was affected by large land masses, since, for example, it seemed to point to Africa throughout the circumnavigation of that continent.[50] It was consequently maintained that the determination of the degree of variation could be used to establish the degree of longitude, using mathematical calculations or following previously compiled charts.

Neither the theoretician nor the practical craftsman had much luck in cracking the longitude problem. Robert Norman, a self-taught instrument maker, thought the solution might lie in the dip of the compass needle, rather than its variation. Norman felt that variation was too erratic to be useful in finding longitude:

> This Variation is iudged by divers travailers to bee by equall proportion, but herein they are muche deceived, and therefore it appeareth, that norwithstandyng their travaile, they have more followed their bookes then experience in that matter. True it is, that Martin Curtes doeth allowe it to be by proportion, but it is a moste false and erroneous rule. For there is neither proportion nor uniformitie in it, but in some places swift and sudden, and in some places slowe.[51]

He sought the answer in a related phenomenon, the dip of the compass needle (i.e., its pull toward the center of the earth). Norman claimed that this carefully observed dip might be used to determine at least latitudinal position and perhaps longitudinal position as well.[52]

Also in 1581, William Borough, another self-taught mathematician, published *A Discourse of the Variation of the Cumpas*. In it, Borough explained the phenomenon of compass variation, "knowing the variation of the Cumpasse to bee the cause of many errours and imperfections in Navigation."[53] Recognizing that the only way to understand variation was by collecting a mass of inductive data, Borough charged mariners and navigators to make continual, accurate observations.[54] He blamed variation for wreaking havoc with chart construction:

> For, either the partes in them contained, are framed to agree in their latitudes by the skale thereof, and so wrested from the true courses that one place beareth from an other by the Cumpas, or els in setting the parts to agree in their due courses, thei have placed them in false latitudes; or abridged, or overstretched the true distances betweene them.[55]

Borough was credited by Henry Coley, a seventeenth-century English mathematician, with the discovery of compass variation,[56] although Borough himself recognized his predecessors. Borough acknowledged Mercator as the savior of navigators from this variational quagmire, with the publication of his "Universal Map" of 1569 containing his new projection.[57] Borough tentatively proposed variation as a longitude-finding tool, although he temporized that "if there might be had a portable Clocke that would continue true of space of forty or fifty houres together . . . then might the difference of longitude of any two places of knowen Latitudes . . . be also most exactly given."[58] He acknowledged that this was, for the time at least, an impossible dream.

In 1599, Edward Wright translated Simon Stevin's *The Haven-finding Arte* from Dutch.[59] Here the famous Flemish mathematician and engineer claimed that magnetic variation could be used as an aid to navigation in lieu of the calculation of longitude.[60] In his translation, Wright proposed that systematic observations of compass variation be conducted on a world-wide scale, "that at length we may come to the certaintie that they which take charge of ships may know in their navigations to what latitude and to what variation (which shal serve in stead of the longitude not yet found) they ought to bring themselves."[61] Unfortunately, Wright's scheme was not entirely successful. By 1610, in his second edition of *Certaine Errors of Navigation*, Wright had constructed a detailed chart of compass variation—but he had also become more hesitant in his claims concerning the use of variation to determine longitude.[62]

It was not until 1634 that John Pell, John Marr, and Henry Gellibrand discovered that magnetic variation itself varied over time. They discovered that the variation had significantly diminished as two observational sites since Edmund Gunter's measurements in 1622 and 1624. This effectively laid to rest the chimera of variation as a longitude-finding tool, although Edmund Halley would return to variation in the late seventeenth century.[63] While a great deal of time had been fruitlessly spent on the problem of compass variation for longitude determination, these investigations did unite observational, mathematical, and imperial concerns. Compass variation was interesting for a theoretical understanding of the magnetic world, and the practical world of navigation provided the necessary data. This example highlights the difficulty of converting theoretical ideas into profitable techniques.

The difficulty of navigating in northern waters fueled this English attempt to link longitude and variation. Because the English entered the "Age of Discovery" later than the Iberian kingdoms, their share of the New World was a northern one, and their only route to Cathay was through a hypothetical Northwest or Northeast passage. Much ink was spilled, in fact, convincing the various English monarchs that such a passage existed and that the nascent English empire would be aided by its exploration and exploitation.[64] John Dee, for example, explicitly linked the idea of empire and navigational improvement in his 1577 work *General and Rare Memorials*.[65] It soon became apparent to navigators and geographers, however, that the plane charts that sufficed for Mediterranean travel were useless for polar sailing. Some northern charts at least acknowledged this problem by depicting two compass needles, one pointing north, for example, along Labrador and a second pointing along a convergent path following the coast of Greenland.[66] The answer to this problem was a stereographic polar projection showing the pole at the center and the latitudinal lines as concentric circles. John Dee apparently developed such a projection and a companion calculational device, which he called the "Paradoxal Compass" in 1556.[67] Although this technique did not facilitate the use of the compass, it did have the advantage of projecting correct spatial relations and allowed a navigator to use the radiating rhumb lines to plot an approximate great or lesser circle course.

In view of John Dee's secretive nature, it is not clear how many people knew of his innovative map projection. And yet, by 1594, Captain John Davis included this projection and instructions for navigating with it in *The Seaman's Secrets*. Davis may have relied directly on Dee's work in this, since he lived with Dee for a time. Dee taught him cosmography, and he is known to have stolen several books from Dee's library.[68] Likewise, in 1604 George Waymouth offers the following in *The Jewelle of Artes*:

> The demonstration of an Instrument to finde out the degree and minute of the meridian of the paradoxicall chart whose degrees dothe increase from the pole towardes the equinoctiall as you may plaine finde in the demonstration next after this.[69]

Just as the plane chart was ill-suited for polar sailing, it quickly proved inadequate for sailing at most latitudes beyond the equator. The basic problem with a plane chart was that it was drawn as if the lines of latitude and longitude were a constant distance apart. This was approximately true for lines of longitude at only limited distances from the equator; however, in northern or southern waters distortion rapidly became significant. Thus a straight-line course on a plane chart would not have corresponded to a straight course on the globe. Likewise, the optimized great circle course could only have been approximated by a series of steps across the map.

In 1569, Gerard Mercator published his now-famous map projection, which allowed sailing courses and rhumb lines to be drawn as straight lines. He used a geometric progression to increase the distance between lines of latitude in proportion with the increasingly distorted lines of longitude. The result was the map we all recognize, where areas become increasingly enlarged and distorted towards the poles.[70] Distances sailed became difficult mathematical calculations; but proportions between lines of longitude and latitude remained true, and compass bearings corresponded to straight lines, making it possible to plot courses using a ruler and a drawing compass.

Although Mercator drew and published this map in 1569, it was not until the end of the century that the precise technique of using geometric progression for mapmaking was articulated mathematically in print. In 1597 William Barlow, in *The Navigator's Supply*, presented a graphical method for creating a Mercator projection that encouraged navigators to draw their own charts.[71] The second, more significant explanation came in Edward Wright's *Certain Errors in Navigation* (1599). Wright provided an elegant Euclidean proof of the geometry involved in this map projection.[72] Along with straightforward instructions for drawing maps, he published a table of meridian parts for each degree that enabled cartographers to determine accurate projections of the meridian network.[73] He also constructed his own map using this method, published in Hakluyt's *Principal Navigations*.

English mathematical geographers showed an interest in problems that were current, sophisticated, and usually practical. Grounded in the mathematical framework of Ptolemy, they attempted to manipulate their image of the world into a geometrically satisfying design. These mathematical geographers also developed a research program, although implicit, that was driven by the promise of providing useful solutions for navigators and using mathematics to improve England's standing in the world. The questions concerning longitude determination, navigation in northern waters, and map projection were all of paramount importance to England in her drive for new trading routes and new colonies. The problems relating to the determination of longitude and map projection had to be solved before geographers could even begin to understand the magnetic globe in a geometrical manner.[74] This mathematical geography was thus an integrated blend of the theoretical and the practical.

How Did This Cause the Scientific Revolution?

The Scientific Revolution has long been a central concept of causality in the history of science. From the seventeenth century on, analysts have argued that the sixteenth and seventeenth centuries had a fundamental impact on the construction of the modern worldview.[75] Indeed, the discipline of the history of science began in the twentieth century by focusing on the problem of the origin of modern science, and the work of some of its founders concentrated on what this important transformation was and how it came to take place.[76] Edwin Burtt, for example, argued that a metaphysical change in worldview was the foundational moment for modern science.[77] Most traditional accounts would agree that the Scientific Revolution involved the development of a new mechanized and often mathematical model of nature, the acceptance of a new astronomical model, and the rise of experimentation as an important research tool.

In recent years historians of science and sociologists of knowledge have become less convinced that some monolith called "Science" was discovered in this period.[78] They have also questioned the revolutionary nature of the changes to the early modern investigation of the natural world. Medievalists argued for continuity with an earlier period, while others questioned whether changing ideas about the ordering of the universe, which affected only a few hundred people at most and took over 150 years to convince even them, could be called revolutionary.[79] In more recent years, the deeper issue of whether the topics investigated were even "science" has come to the fore. Most historians of this period would now be very reluctant to use this term when dealing with the early modern period. They increasingly employ the term "natural philosophy" but remain concerned with identifying the origins of modern science. Cunningham and Williams have recently challenged the assumption that science owes its beginnings to the changes of the seventeenth century. As they have provocatively pointed out, "natural philosophy" was not simply another word for "science"; instead it referred to an essentially theological and philosophical investigation of the natural world. Those who embarked on this enterprise were not scientists but natural philosophers, a very different intellectual occupation.[80] Cunningham and Williams argue that this was not a revolution *into* science—a "scientific revolution"—but, if anything, a revolution in philosophy—that is, a philosophical revolution. If the Scientific Revolution was neither scientific nor revolutionary, did it happen at all?

There was a profound change that took place in the investigation of science in this period; therefore, some form of the Scientific Revolution is worth saving. While we need to take Cunningham and Williams' point seriously and avoid the present-centred search for modern scientific ancestors, we should not relegate the early modern development of natural investigation to the status of "pre-history." The actors themselves, as Francis Bacon reminds us, were aware of their active participation in the dynamic changes of their era. In the 145 years between Copernicus and Newton, those interested in the Book of Nature developed new methodologies. These included experimentation; new attitudes towards knowledge, God, and nature; a new ideology of utility and progress; and new institutional spaces and practices.[81] The world became quantifiable, investigatable, and controllable. By the end of the period, the exploration of nature, still tied to theological concerns yet increasingly to technological ones as well, was carried out in completely new places for different ends and with quite different results. Perhaps this was not the origin of modern science writ large, but it definitely created the necessary preconditions.

We must look for the key to understanding this transformation in the evolving socio-economic structure of early modern Europe, rather than relying on a metaphysical gestalt switch. Instead of a move "from a closed world to an infinite universe," the Scientific Revolution was a sociological change in the agents, the objects, the locations, and the motivations of investigation.[82]

Edward Wright and Thomas Harriot represent the kind of investigators necessary for the development of this "scientific revolution." These two men, and many other mathematical practitioners, represent a new dialogue between theory and practice, evident in their careers, their ideas, and their interactions with universities, courts, and the workshops of instruments makers, among others. Their careers reveal that new demands were emerging for the pursuit of natural knowledge, especially from the courts and stately homes of aristocratic and noble patrons. Wright's and Harriot's work also betrays a complex mix of theory, inductive fact-gathering, and quantification, which contributed to new approaches to nature and methods of investigation. They were both concerned with prestige and utility, as seen in the rhetoric they used to pursue patronage, and also in the problems they chose and the answers they offered. Although their connections to mercantilism were significant, their science was not dictated by the bottom line of commercial profits, but rather by a more complex interaction among the court, the national and international intellectual communities, and corporate traders.

This changing investigation of nature was also influenced by changing military forces and expansionist desires of European rulers; in other words, it was linked with the Military Revolution. Geographers working for princes captured space intellectually, just as these rulers sought to do so militarily. Henry, Prince of Wales, was inspired to become the Protestant military leader of Europe, both through battle and through his patronage of Wright and Harriot. Their potential improvement of navigation could aid England in military power, commercial prosperity, and imperial control. It was the very same concerns that motivated rulers to expand and better equip their armies that motivated them to pay attention to mathematical practitioners. Those practitioners who looked for research areas also chose issues that were important to military expansion and technological practice. Evolving military concerns, combined with changing political, social, and intellectual perspectives in sixteenth-century Europe, directly shaped the new scientific enterprise.

These mathematical practitioners made important and fundamental contributions to the creation of a new science. Because these men were interested in mathematics, both measurement and quantification became increasingly significant. Their social circumstances ensured that the investigation of nature had to be seen as practical by potential patrons. These practitioners used information from any available source; in response, science acquired a rhetoric of utility and progress, as well as an inductive methodology. Intimately connected to national pride and mercantile profit, the science that developed in this period reflected those concerns, a heritage modern science might like to forget. In essence, in large part because of the work of mathematical practitioners like Wright and Harriot, the investigation of nature began to take place away from the older university venue (although there remained important connections) as science assumed new methodologies, epistemologies, and ideologies of utility and progress. This was a foundation for revolutionary change.[83]

Notes

1. Edgar Zilsel, "The Origins of Gilbert's Scientific Method," *Journal of the History of Ideas* 2 (1941), reprinted in *Roots of Scientific Thought*, ed. P. Wiener and A. Noland (Basic Books, 1957), p. 241.

2. Biographical material on Thomas Hood can be found in E. G. R. Taylor, *Mathematical Practitioners of Tudor and Stuart England* (Cambridge University Press, 1954), 40–41; D. W. Waters, *The Art of Navigation in England in Elizabethan and Early Stuart Times* (Hollis and Carter, 1958), 186–189; *Dictionary of National Biography* 9 (Oxford University Press, 1921; 1964 imprint).

3. Thomas Hood, *Copie of the speache* (London, 1588), sig. A2a ff.

4. Thomas Hood, *The Use of the two Mathematical Instruments the crosse Staffe . . . And the Iacob's Staffe* (London, 1596) and *The Making and use of the Geometrical Instrument, called a Sector* (London, 1598).

5. Zilsel, "Origins of Gilbert's Scientific Method," p. 241.

6. See especially A. Rupert Hall, "The Scholar and the Craftsman in the Scientific Revolution," in *Critical Problems in the History of Science*, ed. M. Clagett (University of Wisconsin Press, 1959).

7. With some modification, I take here Taylor's important classification of the more practical men in *Mathematical Practitioners of Tudor and Stuart England*. For modern treatments of these crucial figures, see James A. Bennett, "The Mechanic's Philosophy and the Mechanical Philosophy," *History of Science* 24 (1986): 1–28; Stephen Johnston, Making Mathematical Practice: Gentlemen, Practitioners, and Artisans in Elizabethan England, Ph.D. thesis, University of Cambridge, 1994. Edgar Zilsel, of course, identifies the important players as the "superior artisans" ("The Sociological Roots of Science," *American Journal of Sociology* 47, 1942: 552–555).

8. Mario Biagioli, "The Social Status of Italian Mathematicians, 1450–1600," *History of Science* 27 (1989): 41–95.

9. James A. Bennett, "The Challenge of practical mathematics," in S. Pumfrey, P. Rossi, and M. Slawinski, eds., *Science, Belief, and Popular Culture in Renaissance Europe* (Manchester University Press, 1991). For an early attempt to claim a different history for mathematics and natural philosophy, see Thomas Kuhn, "Mathematical versus Experimental Traditions in the Development of Physical Science," in *The Essential Tension: Selected Studies in Scientific Tradition and Change* (University of Chicago Press, 1977).

10. Karen Davids, Ivo Schneider, and Wolfgang Krohn discuss the emergence of this group. See Davids, "Zilsel and Stevin: Scholars and Craftsmen in the Early Dutch Republic"; Schneider, "The Relationship between Descartes and Faulhaber in the Light of Zilsel's Craft/Scholar Thesis"; and Krohn, "Leon Battista Alberti: Theory of Beauty." All were delivered at a conference on the Reappraisal of the Zilsel Thesis held in Berlin in 1998 and to appear in a forthcoming volume edited by Diederick Raven and Wolfgang Krohn.

11. As a result of this voyage, Wright wrote *The Voiage of the right honorable George Erl of Cumberland to the Azores* (1589), which was later printed by Richard Hakluyt, "written by the excellent Mathematician and Enginier master Edward Wright," *Principal Navigations, Voiages, Traffiques and Discoveries of the English Nation* 2, part 2 (London, 1598–1600): 155 [misnumbered as 143]–168. M. B. Hall (*The Scientific Renaissance 1450–1630*, London, 1962, p. 204), Waters (*Art of Navigation*, p. 220), and J. W. Shirley ("Science and Navigation in Renaissance England," in *Science and the Arts in the Renaissance*, ed. J. Shirley and F. Hoeniger, Washington, 1985, p. 81) all cite this trip to the Azores as the turning point in Wright's career, his road to Damascus, since it convinced him in graphic terms of the need to revise completely the whole navigational theory and procedure.

12. Wright, *Certaine Errors in Navigation* (London, 1599), sigs. D3a–E4a; Taylor, *Mathematical Practitioners*, #99.

13. Stephen Pumfrey, William Gilbert's Magnetic Philosophy 1580–1684: The Creation and Dissolution of a Discipline, Ph.D. dissertation, University of London, 1987.

14. Edward Wright, *Certaine Errors in Navigation* second edition (1610), sigs. *3a–8b, X1–4. *Dictionary of National Biography*, volume 21, p. 1016; Thomas Birch, *Life of Henry, Prince of Wales, Eldest Son of King James I* (London, 1760), p. 389.

15. "Mr. Sherburne's Appendix to his translation of Manilius, p. 86," in Birch, *Life of Henry, Prince of Wales*, p. 389.

16. R. Malcolm Smuts, *Court Culture and the Origins of a Royalist Tradition in Early Stuart England* (University of Pennsylvania Press, 1987). Smuts especially mentions Salomon de Caus, whose *La perspective avec la raison des ombres et miroirs* (London, 1612), was dedicated "Au Serenissime Prince Henry."

17. Roy Strong, *Henry, Prince of Wales and England's Lost Renaissance* (Thames and Hudson, 1986), p. 218; Edward Wright, *Plat of part of the way whereby a newe River may be brought from Uxbridge to St. James, Whitehall, Westminster, the Strand, St. Giles, Holbourne and London*. MS. 1610. Identified by E. G. R. Taylor, *Late Tudor and Early Stuart Geography 1583–1650* (1934), p. 235.

18. Strong, *Henry, Prince of Wales*, p. 212.

19. Waters, *Art of Navigation*, pp. 320–321.

20. J. W. Shirley, *Thomas Harriot: A Biography* (Oxford University Press, 1983).

21. Harriot, *A Briefe and True Report of the New Found Land of Virginia*. Reproduced verbatim in T. de Bry, *America. Pars I*, published concurrently in English (also Frankfurt, 1590), and in Richard Hakluyt, *Principal Navigations* 3 (1598), pp. 266–280.

22. David Beers Quinn, "Thomas Harriot and the New World," in *Thomas Harriot. Renaissance Scientist*, J. W. Shirley, ed. (Clarendon, 1974), p. 45.

23. J. W. Shirley, *Thomas Harriot. Biography*, p. 133. The manuscript information concerning this expedition is gathered together in D. B. Quinn, *The Roanoke Voyages, 1584–1589: Documents to Illustrate the English Voyages to North America* (Hakluyt Society, 1955).

24. Amir Alexander ("The Imperialist Space of Elizabethan Mathematics," *Studies in the History and Philosophy of Science* 26, 1995: 559–591) argued that Harriot's work on the continuum were influenced by his view of geographical boundaries and the "other": "The geographical space of the foreign coastline and the geometrical space of the continuum were both structured by the Elizabethan narrative of exploration and discovery."

25. J. W. Shirley, *Thomas Harriot. Biography*, pp. 381–416.

26. In 1596, Harriot wrote a manuscript titled "Of the Manner to observe the Variation of the Compasse, or of the wires of the same, by the sonne's rising and setting" (B. L. Add. MS 6788).

27. Harriot's papers contain a large number of notes that seem to be preparations for lecturing novice sailors. B. L. Add. MS 6788. See John W. Shirley, "Sir Walter Ralegh and Thomas Harriot," in *Thomas Harriot. Renaissance Scientist*, p. 21.

28. Richard Hakluyt, introduction to Peter Martyr, as quoted in Shirley, "Science and Navigation," p. 80. See also Shirley, *Thomas Harriot. A Biography*.

29. Shirley, "Science and Navigation," p. 81.

30. For biographies of a number of men who could be classified as mathematical practitioners, see Lesley B. Cormack, *Charting an Empire: Geography at the English Universities 1580–1620* (University of Chicago Press, 1997).

31. For an interesting assessment of this relationship, see Andrew Weeks, *Paracelsus. Speculative Theory and the Crisis of the Early Reformation* (State University of New York Press, 1997).

32. James K. McConica, ed., *The Collegiate University. The History of the University of Oxford*. 3 (Clarendon, 1986), pp. 1–68. A clear majority of English geographers in the late sixteenth and early seventeenth centuries did attend one of the two universities, and Zilsel's and others' dismissal of the universities cannot be accurate. On the university and further careers of English geographers in this period, see Cormack, *Charting an Empire*. For a dismissal of the universities, see Zilsel, "Sociological Roots," p. 548.

33. On patronage of science, see Bruce T. Moran, ed., *Patronage and Institutions. Science, Technology, and Medicine at the European Court, 1500–1750* (Boydell, 1991); Paula Findlen,

Possessing Nature: Museums, Collecting, and Scientific Culture in Early Modern Italy (University of California Press, 1994); Pamela H. Smith, *The Business of Alchemy: Science and Culture in the Holy Roman Empire* (Princeton University Press, 1994).

34. Jerry Brotton, *Trading Territories. Mapping the Early Modern World* (Reaktion Books, 1997), pp. 51–65.

35. Martin Frobisher, Richard Chancellor, Pet, Jackman, Humphrey Gilbert, and Sir Walter Ralegh all took Dee's advice about navigation and strategy. See John Dee, *The Private Diary of Dr. John Dee*, ed. J. Halliwell (Camden Society, 1842), especially pp. 18 and 33.

36. Mario Biagioli, *Galileo Courtier: The Practice of Science in the Culture of Absolutism* (University of Chicago Press, 1993).

37. As Nicholas Clulee makes clear in *John Dee's Natural Philosophy: Between Science and Religion* (Routledge, 1988), Dee's patronage attempts ended in his relative isolation from court, rather than in the glory and international reputation he sought. William Sherman (*John Dee: The Politics of Reading and Writing in the English Renaissance*, University of Massachusetts Press, 1995) disagrees with Clulee, arguing that Dee did receive patronage, just not patronage of the sort he wanted.

38. Hakluyt, *Principal Navigations* (1598–1600). This contrasts with *Divers Voyages touching the discoverie of America, and the Ilands adiacent unto the same, made first of all by our Englishmen* (London, 1582). Zilsel ("Origins of Gilbert's Method," p. 246) names Hakluyt as one of those crossing the divide between scholar and craftsman.

39. Pumfrey, "William Gilbert OR *De Magnete*: Reassessing Zilsel's Thesis," paper delivered at Reappraisal of Zilsel Thesis conference, Berlin, 1998. Bennett, "Practical mathematics."

40. On the presence of geographers at the court of Henry, Prince of Wales, and the resulting research program in imperial geography, see Lesley B. Cormack, "Twisting the Lion's Tail: Practice and Theory at the Court of Henry Prince of Wales," in *Patronage and Institutions*, ed. Moran.

41. Joint stock companies were another locale for this exchange and need to be examined in depth for this contribution. On Hakluyt and the merchants, see Richard Hadden, *On the Shoulders of Merchants. Exchange and the Mathematical Conception of Nature in Early Modern Europe* (State University of New York Press, 1994); Richard Helgerson, *Forms of Nationhood: The Elizabethan Writing of England* (University of Chicago Press, 1992).

42. Utility had been part of "scientific" discourse since at least the early sixteenth century; Francis Bacon articulated rather than invented the concept. In the period after the Glorious Revolution, this discourse became more general and more public, but relied on roots reaching back two centuries. For a good description of this later development, see Larry Stewart, *The Rise of Public Science* (Cambridge University Press, 1992).

43. Katherine Neal ("The Rhetoric of Utility: Avoiding Occult Associations for Mathematics through Profitability and Pleasure," *History of Science* 37, 1999: 151–178) talks about some attempts to make mathematics seem useful. Kathleen Ochs (The Failed Revolution in Applied Science: Studies of Industry by Members of the Royal Society of

London, 16600–1688, Ph.D. dissertation, University of Toronto, 1981) talks about the gap between the ambition of utility and its results.

44. Zilsel, of course, saw the importance of utility as directly connected to the beginnings of capitalism. See "Sociological Roots." R. K. Merton ("Science, Technology and Society in Seventeenth-Century England," *Osiris* 4, 1938) linked it to puritanism.

45. This problem was solved only with the invention of an accurate chronometer, first tested by Captain Cook on his famous voyages. See William J. H. Andrews, ed., *The Quest for Longitude: The Proceedings of the Longitude Symposium, Harvard University, Nov 1993* (Cambridge University Press, Mass., 1996) or, more popularly, Dava Sobel, *Longitude: The True Story of a Lone Genius Who Solved the Greatest Scientific Problem of His Time* (Walker, 1995).

46. Waters, *Art of Navigation*, p. 47.

47. Ibid., p. 58.

48. Ibid., p. 300. The observation of the eclipses of Jupiter's moons was unaffected by the position on the earth, so a table of the eclipses combined with observation would give the difference in time between the standard position and the observer. In practice, too many technical difficulties interfered: the moons proved difficult to observe accurately, and there were differences of opinion as to the exact moment of eclipse. Also, telescopes on board ship that would be accurate enough to spot the Medician planets were cumbersome and often not properly understood by navigators and sailors.

49. See Waters, *Art of Navigation*, pp. 60–71.

50. William Gilbert, *De Magnete* (London, 1600). He used this idea to predict a Northeast passage and a large Terra Australis. Dutch pilots claimed that the line of 0° variation ran through Java, while Gilbert favoured the Peleponese. So reported an anonymous reporter of Guillaume de Nautonier's work, *Metrocomie*, in B. L. Burney MS 368, f. 32a-b. See E. G. R. Taylor, *Late Tudor Geography*, p. 69.

51. Robert Norman, *The New Attractive, containyng a short discourse of the Magnes or Lodestone* (London, 1581), p. 21.

52. Ibid., p. 10.

53. William Borough, *Discourse on the Variation of the Cumpas* (London, 1581), "Preface," sig. ⋆2a. For biographies of Borough, see DNB, volume 2, p. 867; E. G. R. Taylor, *Mathematical Practitioners*, p. 68.

54. Borough, *Discourse on the Variation of the Cumpas*, "Preface," sig. ⋆3a.

55. Ibid., sig. F2a.

56. B. L. Sloane MS 2279, f. 91b.

57. Borough, *Discourse on the Variation of the Cumpas*, sig. F3a. This map, however, remained largely unused by sailors.

58. Ibid., sigs. D1b-2b. E. G. R. Taylor, *Mathematical Practitioners*, #58, claims that this was "a technique for finding the longitude by carrying a number of spring driven watches"—not quite what Borough had in mind.

59. Edward Wright, *The Haven finding art* (London, 1599); E. G. R. Taylor, *Mathematical Practitioners*, #100.

60. Wright, *Haven finding art*, p. 3.

61. Ibid., "Preface," B3a. Waters, *Art of Navigation*, p. 237.

62. Edward Wright, *Certain Errors in Navigation* second edition (London, 1610), sigs. 2P1a–8a; Waters, *Art of Navigation*, p. 316.

63. E. G. R. Taylor, *Mathematical Practitioners*, #158. This finding was announced in Gellibrand, *A Discourse Mathematicall*. For an important interpretation of Gellibrand's discovery, see Stephen Pumfrey, "'O Tempora, O Magnes': A Sociological Analysis of the Discovery of Secular Magnetic Variation in 1634," *British Journal for the History of Science* 22 (1989): 181–214. Edmund Halley charted compass variation in his famous isogonic map of 1701. His declared purpose was "to correct the course of ships at Sea: For if the Variation of the Compass be not allowed, all Reckonings must be so far erroneous." Although the continuing secular variation meant that this chart did not determine longitude for all time, Halley did see that "a further Use is in many Cases to estimate the Longitude at Sea." Halley's map is reproduced on pp. 98–99 of Norman J. W. Thrower, *Maps and Civilization. Cartography in Culture and Society* (University of Chicago Press, 1972, 1996).

64. For example: Anthony Jenkinson, "A proposal for a voyage of discover to Cathay, 1565," B. L. Cotton Galba D. 9, ff. 4–5; "Advise of William Borrowe for the discerning of the sea and coast byyonde. Perhaps whather the way be open to Cathayia, or not," ca. 1568. B. L. Lansdowne 10, ff. 132–133b; Humphrey Gilbert, B. L. Add. 4159, f. 175b; Richard Hakluyt, "The chiefe places where sundry sorts of spices do growe in the East Indies, gathered out of sundry the best and latest authours," Oxf. Bodl. MS Arch. Selden 88, ff. 84–88; William Bourne, *Regiment of the Sea* (London, 1574), f. 76; William Barlow, *Navigator's Supply* (London, 1597), sig. b1a; Sir Dudley Digges, *Of the circumference of the earth: or, a treatise of the northeast Passage* (London, 1612); Henry Briggs, "A Treatise of the NW Passage to the South Sea, to the Continent of Virginia and by Fretum Hudson," (London, 1622), published in Samuel Purchas, *Hakluytus Posthumus or Purchas his Pilgrimes* Part II (London, 1625), 848–454. For a discussion of the imperial implications of these proposals, see Lesley B. Cormack, "The Fashioning of an Empire: Geography and the State in Elizabethan England," in *Geography and Empire*, ed. A. Godlewska and N. Smith (Blackwell, 1994).

65. Sherman, *John Dee*, pp. 152 ff.

66. Borough describes such a map in *Discourse on the Variation of the Compass*, sig. F2b.

67. John Dee announces this compass in *General and Rare Memorials pertayning to the Perfect Arte of Navigation* (London, 1577), claiming that it will appear in a now unknown manuscript, "The Brytish Complement, of the Perfect Arte of Navigation." Waters, *Art of Navigation*, p. 210, claims this paradoxical compass to have been the projection itself, while Sherman, *John Dee*, p. 154, sees this compass as an invention to compute longitudes and

latitudes. Dee did draw a map of the polar regions for Burghley, now contained in Burghley's Ortelius of 1570. See R. A. Skelton, "Maps of a Tudor Statesman," in *Description of the Maps and Architectural Drawings in the Collection . . . at Hatfield House*, ed. R. Skelton and J. Summerson (printed for presentation to members of Roxburghe Club, 1971).

68. John Davis, *The Seaman's Secrets* (London, 1595). See Waters, *Art of Navigation*, p. 201. Concerning Davis' sojourn at Dee's house, see Sherman, *John Dee*, p. 174 and note 86.

69. George Waymouth, *The Jewelle of Artes*. B. L. Add. MS 19,889, f. 78b. Both Waymouth and Davis had been on northern expeditions. Waymouth's travels are recorded in Purchas (1625), Part II, pp. 809–813; Davis' travels, pp. 463–450.

70. Edward Wright, who calculated and explained the mathematical structure of the Mercator Projection, used such a projection to illustrate the title page of the second edition of his *Certaine Errors* (1610). There are astronomical instruments surrounding the title, reminding the reader that proper navigation must be informed by mathematics, proper understanding, and accurate observation. In this, the map was both a tool and a product. See Andrew Ede, "When is a Tool not a Tool? Understanding the role of laboratory equipment in the Early Colloidal Chemistry Laboratory," *Ambix* 40 (1993): 11–24, for a more modern discussion of this question.

71. Barlow, *Navigator's Supply*, sig. A1b, sig. K4b; E. G. R. Taylor, *Mathematical Practitioners*, #95.

72. Edward Wright, *Certain Errors in Navigation* (1599), sig. C4a.

73. Ibid., sigs. D3a-E4a. E. G. R. *Taylor Late Tudor Geography*, p. 76, *Mathematical Practitoners*, #99, *Waters, Art of Navigation*, p. 219.

74. For an interesting discussion of the changing mental construct neccessary to accept Mercator's image of the world, see Samuel Edgerton Jr., "From Mental Matrix to *Mappamundi* to Christian Empire: The Heritage of Ptolemaic Cartography in the Renaissance," in *Art and Cartography: Six Historical Essays*, ed. D. Woodward (University of Chicago Press, 1987).

75. For a historical appraisal of the early use of this term, see David Lindberg's introduction to *Reappraisals of the Scientific Revolution*, ed. D. Lindberg and R. Westman (Cambridge University Press, 1990).

76. See, most particularly, Edwin A. Burtt, *The Metaphysical Foundations of Modern Physical Science* (Routledge, 1924); Herbert Butterfield, *The Origins of Modern Science, 1300–1800* (G. Bell, 1949); Alexandre Koyré, *From a Closed World to an Infinite Universe* (Johns Hopkins University Press, 1957). Merton ("Science, Technology and Society in Seventeenth-Century England") employs a different type of analysis but has a similar definition of the scientific revolution, as does J. Dijksterhuis (*The Mechanization of the World Picture*, Clarendon, 1961).

77. For fuller treatments of Burtt, see Lindberg, p. 16; H. Floris Cohen, *The Scientific Revolution. A Historiographical Inquiry* (University of Chicago Press, 1994), pp. 88–97.

78. Steven Shapin's opening line is instructive: "There is no such thing as the scientific revolution and this is a book about it" (*The Scientific Revolution*, Chicago, 1996, p. 1).

79. See, for example, Pierre Duhem, *Les Origines de la Statique*, 2 vols. (A. Hermann, 1905–06); Lynn Thorndike, *History of Magic and Experimental Science* (Columbia University Press, 1958). See also Lindberg, pp. 13–15. Paul K. Feyerabend (*Against Method*, third edition, Verso, 1993) also argued for a continuity thesis, seeing the revolution as a product of our explanatory model, rather than of the events themselves. Even Thomas Kuhn (*The Copernican Revolution*, Harvard University Press, 1957) had to acknowledge the drawn-out process of this change. R. Hooykaas ("The Rise of Modern Science: When and Why?" *British Journal for the History of Science* 20, 1987: 463–67) problematizes Copernicus' role in the Scientific Revolution.

80. Andrew Cunningham and Perry Williams, "De-centring the 'Big Picture': *The Origins of Modern Science* and the Modern Origins of Science," *British Journal for the History of Science* 26 (1993): 407–432.

81. Shapin (*Scientific Revolution*) does a good job of laying out some of the changes taking place that made up the Scientific Revolution.

82. Koyré, *From Closed World*. Shapin made a case for this new interpretation in "History of Science and Its Sociological Reconstructions" (*History of Science* 20, 1982: 157–211) and then, with Simon Schaffer, provided an extremely influential case study in *The Leviathan and the Air Pump: Hobbes, Boyle, and the Experimental Life* (Princeton University Press, 1985).

83. Geoffrey Parker, *The Military Revolution: Military Innovation and the Rise of the West, 1500–1800* (Cambridge University Press, 1988), p. 84.

7

Harriot and Dee on Exploration and Mathematics: Did Scientific Imagery Make for New Scientific Practice?

Amir Alexander

In a famous passage of the *New Organon*, Sir Francis Bacon challenged the natural philosophers of his time to live up to the example of geographical explorers. "It would be disgraceful," he wrote, "if, while the regions of the material globe—that is, of the earth, of the sea, of the stars—have been in our times laid widely open and revealed, the intellectual globe should remain shut out within the narrow limits of old discoveries."[1] Bacon was not alone in this view: early modern promoters of the new sciences repeatedly cited the great voyages of exploration as a model and an inspiration. The image of the natural philosopher as a Columbus or Magellan pushing forward the frontiers of knowledge became a commonplace of scientific treatises and pamphlets of the period. The newly discovered lands and continents seemed both a proof of the inadequacy of the traditional canon and a promise of great troves of knowledge waiting to be unveiled.

The tales of geographical exploration were a natural stimulus and inspiration for Bacon's experimental method. Taking his cue from the explorers themselves, Bacon developed an experimental philosophy that emphasized direct observation and personal experience in seeking out nature's hidden treasures. It was, after all, only through firsthand experience that the great voyagers discovered new lands and oceans, forever undermining the credibility of medieval scholars who based their geographical speculations on ancient sources.

It is hardly surprising that Bacon and fellow promoters of the experimental and observational approach quickly adopted the imagery of geographical discovery. Following the lead of the voyagers, the new experimental philosophers based their assertions on actual unmediated experience, not on mere speculation. The heroic explorer, searching for hidden riches in an unfamiliar land, was an ideal model for an experimental philosopher seeking his way through trial and error to the hidden secrets of nature. The great voyages of discovery provided an inspiration, a model, and a sense of purpose for the Baconian reform of the sciences.

The close correlation between Bacon's rhetoric of exploration, on the one hand, and his scientific approach, on the other, brings out a more general question. What is the relationship between the imagery of knowledge and that of actual scientific practice? Are metaphors of knowledge, such as voyaging and exploration, merely literary conventions designed to add amusing flourishes to dry scientific texts? Or do they tell us something about scientific content as well? In other words, do different scientific approaches draw on different inspiring tales, and do conflicting metaphors tend to generate opposing scientific practices?

There are, of course, many interrelated factors that contribute to a scientist's methodology, and metaphor is only one of them. In attempting to focus on metaphor as a factor, I will discuss two mathematicians whose careers overlapped to an impressive degree but whose rhetoric and use of imagery differed. John Dee and Thomas Harriot occupied a near-identical position within the Elizabethan world. Both were "Oxbridge"-educated men who were known for their mathematical work and their intense involvement in the Elizabethan imperialist program. Each considered his mathematical work to be directly relevant to the maritime voyages he advocated, and each drew upon his own vision of travel and exploration in determining the meaning and purpose of mathematical practice. These fundamental similarities between the two make them ideal subjects for examining the relationship of rhetoric and imagery to scientific practice. While their respective works suggest a harmony between the meaning of exploration and empire and the purpose and practice of mathematics, they in turn betray profound metaphysical differences.

The Elizabethan polymath John Dee considered Thomas Harriot to be his friend. Dee's copy of Antonio de Espejo's *Viaje* bears the inscription "Johannes Dee: Anno 1590. January 24. Ex Dono Thomae Harriot, Amici mei"[2] ("by the gift of Thomas Harriot, my friend"). Two other notes in his manuscripts, dating from 1592 and 1594, also attest to this relationship, each stating simply, "Mr. Harriot came to see me."[3] But, in spite of his gift of 1590, there is no evidence that Harriot viewed Dee in similar terms. The thousands of surviving manuscript folios bequeathed by Thomas Harriot to his executors contain not a single mention of his supposed friend. Although their careers closely overlapped for decades, the only signs of friendship between them are three laconic notes written in Dee's hand over a span of four years.

In a way, the cool cordiality between the two is hardly surprising if we consider that they belonged to distinctly different generations. Dee was born during the reign of Henry VIII, and was already 63 years old when he received Harriot's book. Harriot, by contrast, was born a year after Elizabeth's accession and, in 1590, was less than half Dee's age. If we examine their fields of intellectual pursuit, however, we might expect to find a greater familiarity between the two, for they occupied a nearly identical space within the Elizabethan intellectual landscape. Both were viewed as the leading English mathematicians of their day, and both were active promoters of and participants in Elizabethan imperialist exploits. Dee set the pattern by connecting high-level mathematical theory with political advocacy and technical support for voyages of exploration. Harriot surpassed his elder in both fields by personally exploring undiscovered lands and pushing the bounds of mathematical theory ever further.

Dee was known as the foremost promoter of exploration and discovery in his generation, and is traditionally credited with coining the term "British Empire." His interest in geographical exploration dates from the 1540s, when he traveled through Europe and studied under the leading geographers of his day: Pedro Nuñez, Gerard Mercator, and Gemma Frisius. In 1553, Dee was recruited as a technical advisor to a voyage sponsored by the Merchant Adventurers of London in search of the Northeast Passage to Cathay. In 1576, Martin Frobisher and Humphrey Gilbert consulted him about their planned voyages to the Arctic in search of a Northwest Passage. Dee provided Frobisher's sailors with geographical instruction, and later drew a map of the Atlantic based on Frobisher's discoveries. Optimistic about the prospects of these voyages, he secured from Gilbert a grant of rights for all discoveries above the 50th parallel. Gilbert's ostensible generosity probably reflected his increasing skepticism about the enterprise, as it would have secured most of what is now Canada for Dee's personal use. In 1577, Dee published a tract titled *General and Rare Memorials pertayning to the Perfect Arte of Navigation*, which argued through historical and legal precedents that Elizabeth possessed title to a vast Atlantic empire. In order to secure those far-flung dominions, Dee advocated the establishment of a "petty navy royall" to patrol the sea routes and protect the queen's interests.[4] Dee, in short, was a scholarly geographer, a technical advisor to the voyages, and a political promoter of expansion and empire.[5]

Harriot was all that and more. In 1585, at the age of 25, he sailed to the New World as a member of Walter Ralegh's first Virginia colony. During his year's stay he explored and charted the North American seaboard, catalogued local plants and animals, observed the customs of the local Indians, and learned the Algonquian language. After his return he published a short treatise titled *A Briefe and True Report on the New Found Land of Virginia*, which also included a series of maps of Virginia based on his exploration.[6] In the 1590s, although no longer a member of Ralegh's household, he served as a geographical and navigational advisor to Ralegh's search for El Dorado in the jungles of Guyana. He lectured Ralegh's captains on navigational matters and eventually drew a map of El Dorado and its lost kingdom. In short, Harriot was not only a technical expert and a publicist for Elizabethan imperialism, he was also a leading New World explorer in his own right.[7]

Although Dee and Harriot were both leading advocates and practitioners of maritime exploration and imperial expansion, their understanding of the meaning and purpose of these projects was radically different. Dee's approach is exemplified in his treatise Britanici Imperii Limites of 1578, in which he argued that Elizabeth was entitled to a vast array of overseas possessions.[8] His arguments were essentially legalistic, composed of a long list of English voyages and "discoveries"—both historical and apocryphal—accompanied by complex dynastic calculations. He argued, for example, that Iceland and Greenland were part of Elizabeth's inheritance because she was the direct successor of King Arthur. Since Arthur fought and defeated the Danes, he was, according to Dee, the undisputed King of Denmark. Centuries later, when the Danes proceeded to colonize the North Atlantic islands, they merely added to Arthur's patrimony; this was then passed through the generations to Queen Elizabeth.[9]

Another example of Dee's extrapolations is his politic treatment of the Spanish claims to dominion in the New World. Concerned that Elizabeth would accept the legitimacy of the 1493 papal division of the new territories between the Iberian powers, he argued that the Pope was in no position to grant what was not his. In case this argument would fail to persuade the Queen, he then added that the Pope did not mean to divide the entire world—only the areas between the northernmost and southernmost latitudes of Spain. In case Elizabeth still felt that the Spanish claims warranted some consideration, Dee presented his ultimate coup de grace: through complex dynastic considerations, he "proved" it was in fact Elizabeth, and not Phillip II, who was the true legal sovereign of the kingdom of Castille.[10] Dee thus concluded:

> Of a greate part of the sea Coastes of Atlantis (otherwise called America) next unto us, and of all the iles nere unto the same from Florida Northerly . . . the Tytle Royall and supreme is due, and appropriate unto your most gratious majestie and that partlie *Iure Gentium*, patlie *Iure Civilis*, and partlie *Iure Divino*, No other Prince or Potentate in the whole world being able to alledge therto any clayme the like.[11]

Dee posited that all those disparate lands belonged to Elizabeth by indisputable universal law. While he supported voyages of exploration, these were not, in the strictest sense, necessary. Dee had proven to his own satisfaction that the English monarch was already master of a grand empire through universal legal consideration. This was true even if no English subject actually set foot in those far-flung domains and no English ships ever set out to explore the New World. Ultimately, for Dee the empire depended not on actual exploration by land and sea but on the correct interpretation and application of natural and divine law.

Dee brought home his claim that disparate parts of the globe lie naturally within Elizabeth's domain by literally mapping them onto her body:

> The single little black circle shown on the left hand side of your majesty's throne, represents Cambalu, the chief city of Cathay. . . . [M]eanwhile, by a wonderful omen the City of Heaven happens to be located at the middle joint of the index finger which encircles the hilt of your sword. . . . Thirdly, at the right side of your majesty, the coast of Atlantis is pleased to have its place. . . .[12]

The Queen's empire and body appear inseparable. Exploration and conquest are irrelevant, as the distant corners of the globe unite in Elizabeth's person.

Harriot's visions of empire were of a very different sort. As a member of Walter Ralegh's household, he could hardly have been concerned about the niceties of universal law. Ralegh was known to his supporters as a privateer and, to his Spanish enemies, simply as a pirate. He dedicated his life to breaking the Iberian stranglehold on the New World and diverting its riches towards England and his own pockets. For Ralegh, America was a wondrous, undiscovered land harboring great riches and awaiting her manly discoverer. If Elizabeth were to earn a stake in this new land, it would be by the noble deeds of her subjects overseas, not by the tedious work of homebound scholars dusting off arcane dynastic documents.

Harriot endorsed such views by participating in Ralegh's various ventures of settlement and exploration. He also voiced his support in a short poem he dedicated to his patron on the eve of one of his voyages. After listing the various navigational aides he had furnished for the expedition, Harriot continued:

> If you use them well on this your journey
> They will be the king of Spaine's atarney
> To Bring you to Silver and Indian Gold
> Which will keep you in age from hunger and cold
> God speed you well and send you fair weather
> And that agayne we may meet together.[13]

Apart from indicating that Harriot's diverse talents did not extend to poetry, this poem suggests a vision of exploration that radically differs from Dee's. For Harriot, as for Ralegh, Drake, and their colleagues, the voyages were a grand adventure for brave men. Empire-building had nothing to do with the universal law of nations, "iure gentium," "iure civilis," or "iure divino," to use Dee's terminology. It also had little to do with complex dynastic calculations, so crucial for the relationships between the Royal houses of Europe. But they had everything to do with courage, the spirit of adventure, and, more often than not, greed.

One way to examine Harriot's and Dee's views is to contrast the visual images they provided to support their views of empire. One is the title page to Dee's *General and Rare Memorials Pertayning to the Perfect Arte of Navigation* of 1577 (figure 7.1).[14] The other is a pictorial map titled "The Arrival of the Englishemen in Virginia," located in the 1590 De Bry edition of Thomas Harriot's *Briefe and True Report of the New Found Land of Virginia* (figure 7.2).[15]

The images obviously belong to two distinct genres: Dee's is a frontispiece designed to convey the general claims which will be discussed in the text; Harriot's is a map illustrating his personal account of the settlement of a new land. In spite of generic differences, their similarity is striking. Both depict fleets of stylized ocean-going ships that serve as the bearers of maritime travel and imperial power. The fleets hover just off shore of rich and promising lands broken by a major river leading to their interior. The similarity ends here, however. Harriot's image is first and foremost a detailed map of an American locale (the North Carolina seaboard) drawn by an eyewitness explorer. It is designed to convey the voyagers' immediate experience to readers at home. Dee's image, on the other hand, is purely symbolic—its ocean a universal sea and its land a place both everywhere and nowhere. It is designed to convey not an immediate experience but an abstract universal idea.[16]

The focus of Dee's frontispiece is Elizabeth, seated at the helm of a grand ship called "Europa" (accompanied by Europa and the bull) that is sailing to seize the castle of "Occasio." On shore is the kneeling figure of "Respublica Britannica" humbly beseeching the Queen to embark on her imperial mission with a "fully equipped expeditionary force." Overhead, lending their support to the enterprise, are the divine light of "Jehova," the archangel Michael, and the cosmic elements of the sun, moon, and stars. The entire universe is waiting for Elizabeth to take her rightful place as ruler of a great empire. All she has to do is take hold of the "occasion" and perfect harmony will reign in the world.

Harriot's map tells a very different story. The Virginian interior is depicted as a rich and desirable land, where Indians fish, hunt, and grow plentiful crops. The road to possession of this wondrous land, however, is extremely hazardous. A long and nearly continuous chain of islets bars the exploring fleet's passage to the Virginia coastline. Attempts to force a passage are extremely risky, as evidenced by the sunken ships that mark each possible passage. Ultimately, a party

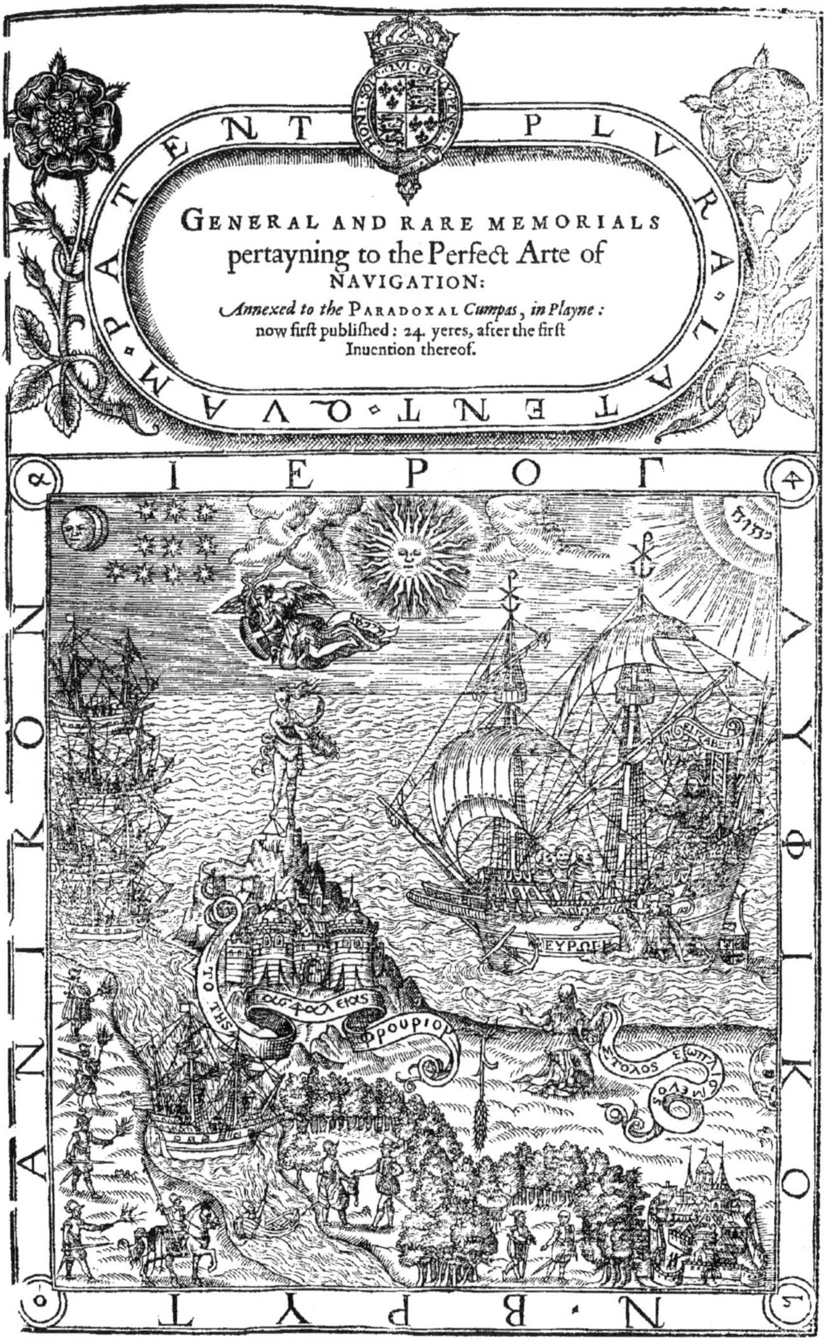

Figure 7.1
Frontispiece to John Dee, General and Rare Memorials pertayning to the Perfect Arte of Navigation (London, 1577).

Figure 7.2
Thomas Harriot and John White, "The Arrival of the Englishemen in Virginia," in Thomas Harriot, A Briefe and True Report on the New Found Land of Virginia, first published by Theodor De Bry, Frankfurt a/M, 1590.

of explorers does manage to break through the obstacles in a small boat and, bearing the sign of the cross, sails toward Roanoke Island.[17]

The contrast between the narratives of the two images is sharp. For Dee, empire belongs to a universal constellation; but for Harriot it belongs to those who actually invest, explore, or settle overseas. For Dee, empire is a divine gift presented to Elizabeth, while for Harriot, it is a product of human efforts. In Dee's scheme, the Queen only had to take advantage of the "occasio" and take charge of her extensive lands. In Harriot's story, the empire is won through financial risks and physical hazards assumed by enterprising explorers who break through formidable obstacles to reach the Promised Land of the interior. In short, whereas for Dee imperial title could be deduced by scholarly cosmic reasoning, for Harriot it could only be won through firsthand voyages of grand adventure.

The other major field in which the careers of Harriot and Dee overlapped was, of course, mathematics. Dee's work on various quantitative problems in cartography and navigation has been amply documented by Eva Taylor and others.[18] Dee drew several maps, built navigational instruments, and offered a solution to the mathematical problem of the Mercator projection. His most famous and lasting contribution to mathematics was, however, of a methodical and promotional nature. His famous "Mathematical Praeface" to Henry Billingsley's first English edition of Euclid's *Elements* (1570) celebrated the unique status of mathematics and its many applications to the sciences and arts, and was reprinted repeatedly throughout the following two centuries.[19]

Harriot's mathematical career was more intense, lasting without interruption until his death in 1621. Unlike Dee, who was quick to publish even controversial works, Harriot published one work only during his lifetime: his treatise on Virginia. After his death, his executors published a book on algebra based on a small portion of his papers and titled *Artis Analyticae Praxis*.[20] The bulk of his mathematical work remains in manuscript form to this day.[21] Yet the little that has been published shows him to be a mathematical atomist and an unmistakable leader in the use of infinitesimal methods. Many of his results were duplicated only decades after his death.[22]

As with their views on exploration and empire, Dee's and Harriot's views on mathematics also diverged. In a few passages in the "Mathematical Praeface," Dee sums up his perspective on the meaning and purpose of mathematics. "All thinges," he quotes Boethius as saying, "do appeare to be formed by the reason of Numbers. For this was the principall example or patterne in the minde of the Creator."[23] Therefore, he argues, "By *Numbers* . . . we may both winde and draw ourselves into the inward and deepe search and vew, of all creatures distinct vertues, natures, properties, and *Formes*: And also farder, arise, clime, ascend, and mount up (with Speculative winges) in spirit, to behold in the Glas of Creation, the *Forme of Formes*, the *Exemplar of Number* of all thinges *Numerable*."[24]

When Dee discusses his vision of mathematics, he adopts the rhetoric of geographical exploration. It was the voyagers who investigated all the distinct virtues and properties of unknown lands and creatures, and it was they who ascended the mountains of the New World. Dee's mathematical voyage is a very special one, however. By pursuing mathematics, he suggests, we are in fact studying the basic patterns according to which God created the universe. We ascend to divine heights, sharing a divine knowledge of the fundamental principles of creation.[25] Armed with this knowledge, we may then descend back down to the physical world and reveal the hidden patterns of number and magnitude that had been placed there by God. It is hardly surprising to find that

Dee considered mathematics to be the purest and most divine of all sciences, second in excellence to theology only.

For Dee, then, the study of mathematics provides a foreknowledge of the physical world, before any actual physical investigations have taken place. The basic patterns are already contained within the familiar body of mathematics, and all we need to do is apply this knowledge to the various fields of human inquiry. If this is done properly, then all problems will inevitably be solved. The "Praeface," in fact, advocates precisely this approach: Euclid's elements are the universal divine pattern by which all earthly things are formed. Once studied, they can be applied to a myriad of fields ranging from surveying to music to navigation.

Harriot's approach was radically different: whereas Dee applied pure and perfect mathematics to his worldly surroundings, Harriot tried to construct an abstract mathematics that would be modeled directly on the physical world. One of his main concerns was with the structure of the geometrical continuum. Since antiquity, it had been known that there were contradictions and paradoxes in the intuitive assumption that the continuum was composed of discrete points. As a result, infinitesimal methods based on this perception had been excluded from mathematics for 2,000 years.[26] Harriot reexamined the problem: he viewed the continuum as deeply perplexing and referred to it as a "labyrinth" harboring great "mysteries of the infinite." He elaborated endlessly on the paradoxes that ensue from an atomistic conception of the continuum, and then he offered his surprising solution: despite all problems, the continuum is in fact composed of indivisible discrete points. An atomistic conception of the continuum, he wrote optimistically, will be the "club of Hercules," which will sweep away the darkness and confusion of the continuum. Based on this faith, Harriot proceeded to develop the infinitesimal approach that presaged the emergence of the calculus.[27]

It is clear that Harriot and Dee offered very different views on the nature of mathematics. Dee attempted to find the perfection of mathematics within the physical world. Harriot, by contrast, examined the inner structure of physical reality and constructed a mathematical approach that closely approximated it. For Dee, mathematics was essentially divine, a form of knowledge we share with God. For Harriot, mathematics was a distinctly earthly construct, modeled on the internal structure of matter and aimed at further exploring its hidden secrets.

As before, it may be helpful to compare the visual images which Dee and Harriot used to present their mathematical approaches. For Dee, geometry is the divine alphabet of nature that structures all else in the world. It is encapsulated in a hieroglyphic Dee calls the "Monad," which is imbued with cosmic significance (figure 7.3).[28] The hieroglyphic is generated from the fundamen-

Figure 7.3
The Monad, from John Dee, Monas Hieroglyphica (Antwerp, 1564), f. 12.

tal geometrical figures—the point, line, circle, and semi-circle, which symbolize the entire universe. Its circular components represent the heavens, and the cross stands for the four sublunar elements and qualities.[29] Dee's Monad is an abstract universal construct that is true everywhere and always. It is the fundamental geometrical sign that generates all else in the universe. Harriot's diagram showcases a very different approach to mathematics. During the course of a general investigation of the nature of the geometrical continuum, it was drawn in an unpublished tract titled *De Infinitis*.[30] The diagram attempts to depict how the continuum looks at a particular point at close range (figure 7.4). There are no general abstractions here, merely a magnified view of a single section of the continuum. Whereas Dee's Monad is an abstract presentation of universal relations, Harriot's sketch is simply a large-scale eyewitness map of the continuum. In principle, it is not very different from the explorer's local map of the Virginia coastline that he drew decades earlier.

Clearly, there is a striking contrast between Harriot's and Dee's respective conceptions of empire and mathematics. For Dee, both empire and mathematics were essentially universal deductive enterprises in which truth is systematically deduced from incontrovertible first principles. Thus, if we scrupulously follow the universal laws of heredity and Royal prerogative, we learn that Elizabeth is the ruler of far-flung territories encompassing several continents. The empire is not the result of maritime military adventures but is simply an integral part of the cosmic order that can be revealed through systematic reasoning. This was also precisely Dee's view on the role of mathematics in the

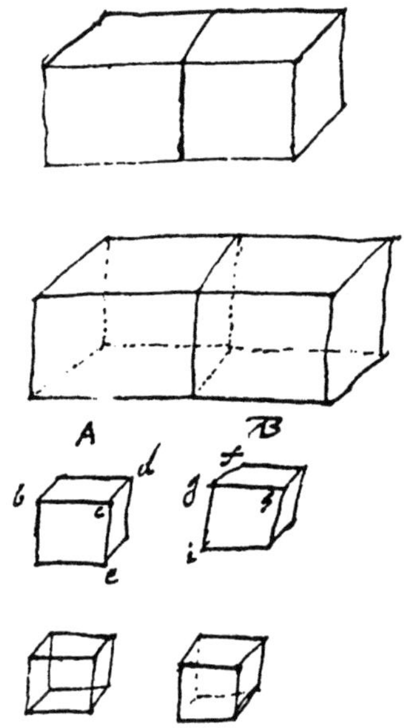

Figure 7.4
Thomas Harriot's depiction of the continuum, B. L. Add. MS 6782, f. 370v (courtesy of the British Library).

universe: the entire universal order—the heavens as well as the earth—could, in principle, be derived by following the rules of mathematical and geometrical reasoning. Dee viewed mathematics as the fundamental pattern of the universe, placed there by God. All of creation follows from it.

In Harriot's case, the parallels between his imperialist and mathematical views are just as close. The empire, for Harriot, could only be won by heroic men voyaging overseas to take possession of distant lands. Dee's attempts to provide Elizabeth with an empire from the confines of his library at Mortlake were bound to seem irrelevant, if not comical, to Harriot, who had confronted the harsh realities of the Virginia coastline firsthand. Only a brave explorer who could prevail over all obstacles and gain access to the hidden riches of the land would win the British Empire. In mathematics, Harriot was similarly skeptical of those universal deductive truths that Dee viewed as the essence of mathematics. Rather than follow abstract principles to their inevitable logical conclusions, he chose to treat mathematical objects as an undiscovered territory to be explored. Just as one could not possess Virginia through legalistic dynastic calculations, one could not comprehend the mathematical continuum on the basis of logical abstract reasoning. What was needed was an actual close survey of the troublesome continuum, and Harriot proceeded to provide just that. When confronted with what he called "the mystery" of the continuum, he examined it closely, charted its outlines, and mapped its inner structure, just as he did on the Virginia seaboard. In the process, he found the seemingly smooth continuum composed of irreducible mathematical "atoms" of definite magnitude. The survey of the continuum thus led him to a mathematical atomism that ran counter to a 2,000-year-old tradition of deductive geometry.[31]

We have already noted the close fit between Bacon's rhetoric and imagery of exploration, and the inductive method he advocated. It will now become evident that this applies to Dee and Harriot as well. For each of them, the metaphors that gave meaning to their enterprise correlated closely with their methodology. In spite of being near contemporaries and occupying overlapping intellectual terrains, Harriot and Dee chose sharply differing courses for both the imagery and the practice of their mathematical pursuits. Harriot's course was similar to Bacon's: not unlike a Baconian experimenter, Harriot's mathematician was a geometrical explorer who directly observed and mapped the hidden realms of the mathematical terrain. Dee, by contrast, sought knowledge about the world—whether mathematical or imperial—through systematic deduction from universal truths. For Harriot, geometrical structures were no different from the ragged and hazardous coastlines he had encountered in his travels. For Dee, geometry represented the perfect and fundamental principles of the universe from which all else was derived.

Both Dee and Harriot followed consistent intellectual strategies when confronted with the diverse challenges of empire, voyaging, and mathematics. For Dee, the British Empire, like a geometrical theorem, could be possessed through proper logical reasoning. For Harriot, even abstract mathematical truths could only be gained, like distant lands, through an actual physical survey.

Notes

1. Francis Bacon, *The New Organon and Related Writings* (Liberal Arts Press, 1960).

2. J. Shirley, *Thomas Harriot: A Biography* (Clarendon, 1983), p. 201.

3. Ibid.

4. John Dee, *General and Rare Memorials pertayning to the Perfect Arte of Navigation* (London, 1577) (reprint: Da Capo, 1968).

5. On Dee, see Nicholas Clulee, *John Dee's Natural Philosophy* (Routledge, 1988); William Sherman, *John Dee* (University of Massachusetts Press, 1995); Peter J. French, *John Dee: The World of an Elizabethan Magus* (Routledge, 1972).

6. Thomas Harriot, *A Briefe and True Report on the New Found Land of Virginia* (Theodor De Bry, Frankfurt a/M, 1590; Dover, 1972).

7. On Harriot, Ralegh, and El Dorado, see Shirley, *Thomas Harriot*; Walter Ralegh, *The Discovery of Guiana by Sir Walter Ralegh* (Argonaut, 1928), first published 1595; Charles Nicoll, *The Creature in the Map* (Morrow, 1996).

8. John Dee, *Britanici Imperii Limites,* British Library Add. MS 59681. See discussion in Sherman, *Dee.*

9. John Dee, *Britanici Imperii Limites*, ff. 30–31.

10. John Dee, *Britanici Imperii Limites*, ff. 71–72.

11. John Dee, *Britanici Imperii Limites*, f. 21. also discussed in Sherman, *Dee*, p. 185.

12. John Dee, *Britanici Imperii Limites*, f. 9.

13. British Library Add. MS 6788, f. 490.

14. Dee, *General and Rare Memorials*, frontispiece.

15. Harriot, *Briefe and True Report*, p. 45 of Dover edition.

16. On Dee's forntispiece see Lesley Cormack, "Britannia Rules the Waves? Images of Empire in Elizabethan England," *Early Modern Literary Studies* (electronic journal), September 1998 (www. humanities.ualberta.ca). For other discussions, see Clulee, *Dee*, pp. 184–185; Frances Yates, *Astraea: The Imperial Theme in the Sixteenth Century* (Routledge, 1975), pp. 49–50; French, *John Dee*, pp. 183–185.

17. On Harriot's map, see Amir Alexander, "The Imperialist Space of Elizabethan Mathematics," *Studies in the History and Philosophy of Science* 26 (1995): 559–591.

18. Eva G. R. Taylor, *Tudor Geography* (Methuen, 1930); Taylor, *The Mathematical Practitioners of Tudor and Stuart England* (Cambridge University Press, 1954); Taylor, "John Dee and the Nautical Triangle," *Journal of the Institute of Navigation* 8 (October 1955): 318–325. See also David W. Waters, *The Art of Navigation in England in Elizabethan and Early Stuart Times* (Yale University Press, 1958). More recently see Lesley B. Cormack, *Charting and Empire: Geography in the English University 1580–1620* (University of Chicago Press, 1997).

19. John Dee, *The Mathematicall Praeface to the Elements of geometry of Euclid of Megara* (London, 1570; Science History Publications, 1975).

20. Thomas Harriot, *Artis Analyticae Praxis* (London, 1631).

21. Most of Harriot's manuscripts are in two main collections: British Library Ad. MSS 6782–6789, and the Harriot Papers in Petworth House, Sussex, HMC 240/i-v, HMC 241/i-x.

22. See for example his work on the logarithmic spiral in J.V. Pepper, "Harriot's Calculation of Meridional Parts as Logarithmic Tangents," *Archive for the History of Exact Sciences* 4 (1967–68): 359–413.

23. Dee, *Mathematical Praeface*, fol. j(r).

24. Ibid.

25. Dee was an ardent practitioner of Hermetic magic, which held that a Magus could access the hidden forms and correspondences which pervade the world, and manipulate them. See French, *John Dee*; Clulee, *Dee*.

26. The paradoxes that follow the assumption that the continuum is composed of discrete points are of two sorts. One is based on the paradoxes of Zeno, which were discussed at length by Aristotle. The other is the "problem of incommensurability." If the continuum is composed of discrete points, this argument goes, then it would serve as a common measure for all magnitudes. Since the time of the Pythagoreans, however, it has been known that not all magnitudes are commensurable. It follows that continuous magnitudes are not composed of discrete points. Knowledge of these problems has strongly discouraged the development of infinitesimal mathematical methods from antiquity to the seventeenth century. For a clear exposition of these issues see Carl B. Boyer, *The History of the Calculus and its Conceptual Development* (Dover, 1959).

27. For Harriot's views on mathematics, see Amir Alexander, "Imperialist Space"; Alexander, *Geometrical Landscapes: The Voyages of Discovery and the Transformation of Mathematical Practice* (Stanford University Press, 2002).

28. The Monad figures prominently in two of Dee's tracts: *Propadeumata Aphoristica* (1558) and *Monas Hieroglyphica* (1564).

29. On the Monad see Clulee, *Dee*, especially p. 86 ff.

30. Harriot's *De Infinitis* is in British Library Add. MS 6782, ff. 362–374v.

31. For more on Harriot's mapping of the continuum and its relationship to his mathematical atomism, see Amir Alexander, "Imperialist Space" and *Geometrical Landscapes*. Harriot's mathematical atomism was later rejected in mainstream mathematics, which relies on internal logical consistency for its coherence and disciplinary integrity. Nevertheless, Harriot's atomism and similar views developed on the continent by his contemporaries were crucial for the development of the infinitesimal methods that ultimately led to Newton's and Leibniz's calculus.

8

Charting the Globe and Tracking the Heavens: Navigation and the Sciences in the Early Modern Era

Michael S. Mahoney

The central problem of navigation and, to some extent, of cartography throughout the period under consideration was the determination of longitude. On it depended both the location of a ship at sea and the accurate mapping of the global world that Europe increasingly viewed as existing for its own benefit. Longitude is a matter of time. There is no fixed point of reference in the sky, but only the uniform rotation of the earth once every sidereal day. If one knows the difference in local time at two points, one knows the longitudinal distance between them. Measuring the time where one is located is relatively straightforward: when the sun is due south, it is noon. Knowing the local time elsewhere at that same moment is another matter. From the middle of the sixteenth century to the early twentieth century, only two methods seemed worth pursuing. One could use the clockwork of the heavens directly or one could build a model of that clock and read it indirectly. Both were perfected in the middle of the eighteenth century, to be replaced only by radio signals a century and a half later.

The story of the determination of longitude has enjoyed considerable publicity over the past several years, in large part owing to a conference on the subject organized at Harvard in November 1993 to commemorate the 250th anniversary of John Harrison's marine chronometer, the clock that did the trick. That conference produced not only its own richly illustrated proceedings, *The Quest for Longitude*, but also provided the basis for Dava Sobel's widely read *Longitude*.[1] I don't want to repeat that story but rather to explore one piece of it in the context of the question at the focus of this conference, namely the relation of war and science in the early modern era, or the "possible links between the Scientific and the Military Revolutions."

In *The Military Revolution: Military Innovation and the Rise of the West, 1500–1800*, Geoffrey Parker cites four points of revolutionary innovation: the introduction of firearms; the sudden growth in the size of armies; increasingly complex and ambitious strategies; and logistical, social, and political demands of warfare on a wholly new scale. On the sea, those innovations took the form of ships designed to carry guns and to bring them to bear effectively on targets at sea and on land, oceanic operations with large fleets, and new financial mechanisms. What differentiates the military and the naval aspects of the revolution are the sites of action. For the most part, the new European armies raised havoc in Europe, while the new fleets carried war to other continents, or rather the seas around and between them.

Parker does not have much to say about the role of improved navigation and the new forms of organization its pursuit required. Yet surely it was an essential element in the strategies and logistics of naval warfare on a global scale. War at sea was aimed at securing the commercial advantages brought by the ability to navigate on the open sea. It put the navies of some countries, including our own, into the business of astronomy and made them their nations' timekeepers.[2]

In the keynote address to the Longitude Conference, David Landes emphasized the contingency of the problem.[3] Clearly, knowing one's longitude precisely is not a prerequisite to travel on the open sea. The Portuguese rounded Africa and crossed the Indian Ocean without such knowledge. Columbus crossed the Atlantic, and Spanish and Portuguese ships soon followed, all without being able to determine their longitude, nor indeed knowing that of their destinations. Once one knew the approximate distance between the old and the new worlds, crossing between them was a matter of "sailing the latitudes," and the prevailing winds and currents conveniently cooperated. As early as Columbus' second voyage, in the middle of which he sent one of his captains back to Spain for more supplies, the voyage had become a matter of routine. Once the path had been opened, others could follow it. Ignorance of longitude could lead to disasters, as in the famous case of La Salle's failed and ultimately fatal attempt to reach New Orleans by sea after first approaching it by coming down the Mississippi. But knowledge of longitude was not a prerequisite to the voyages of exploration.

Exploitation was another matter. Well-traveled paths can be dangerous, or at least expensive, when they mean that pirates and enemy fleets know where to look for you. The trackless ocean becomes safe ground when knowing where you are at any moment means being able to range freely. Coupled with knowing where you are, knowing exactly where you are headed means less time spent circling around, looking for your destination. It reduces the uncertainty of schedules, of provisions, of alternatives along the way. These are clearly advantages to a navy, but they are even greater advantages to commerce, whether private or mercantilist. And where commerce went, the navy would have to follow.

More than war and commerce is involved. One can write much of the history of astronomy and mechanics in the early modern era, technically and institutionally, by reference to the problem of longitude.[4] Through the clock, which it placed at the center of scientific attention, it is deeply embedded in our modern concept of the world. Through the instruments it required, it informed the development of precision tool-making on an industrial basis.[5] It is worth taking a closer look at these connections and then stepping back to consider what light they might shed on the question of early modern science and warfare.

The central figure is Christiaan Huygens, whose interest in the problem was first piqued by the challenge of clocking the pendulum.[6] I put it that way, because it was a question of automating the astronomical pendulum, the accuracy of which was compromised by the need for someone to count its beats and occasionally to keep it going with a slight push. His answer was to attach a mechanical clock, the escapement of which would count the pendulum's swings, while the driving weight provided the push. Viewed from the perspective of the clock, the pendulum provided the tautochronic regulator it had been lacking up to that time.[7] Galileo had first suggested using the pendulum, but Huygens devised how to attach it so as to keep the two mechanisms separate except at the point at which the one could regulate the other while being ever so gently driven by it.

Once Huygens made the connection, the clock became a precision mechanism it had not been before, whatever its fascination as an automaton. Accurate now to within seconds a day, it made seconds count. That had implications both for mechanics and for navigation. It made the clock an interface for theory and practice in mathematical science, an interface at which Huygens worked for his entire career from 1657 until his death in 1695.

Huygens recognized almost immediately that his pendulum clock held promise of a solution to the problem of longitude—but only if he could devise a portable version capable of withstanding the rigors of travel, including first and foremost disturbances of the pendulum itself. Although Galileo had asserted, and perhaps thought he had even demonstrated, the independence of the period of a pendulum from the amplitude of its swing, others quickly determined that a simple pendulum is tautochronous only within a close neighborhood of its center. Long pendulums with heavy bobs swinging over short arcs met that condition for scientific purposes, but that would not work for a portable clock.

Shorter pendulums emphasize the perturbations of wide excursions, so Huygens undertook to examine the nature of tautochrony itself. Or rather, he fell upon the issue in seeking a theoretical derivation of the period of a simple pendulum, which could serve as a basis for experimental determination of the constant of gravity. To make the initial analysis mathematically tractable required an approximation. Subsequent analysis of the approximation provided an exact solution, namely that a pendulum swinging along the arc of an inverted cycloid is tautochronous over its entire trajectory. Moreover, as Huygens then quickly determined, mounting the pendulum by a flexible wire and encasing the wire between two leaves in the shape of a cycloid will constrain the pendulum's bob to follow that same cycloid. As Huygens demonstrated in opening a new area of mathematics, a cycloid is its own evolute.[8]

Tests of the cycloidal clock against the sun confirmed its accuracy to a degree that warranted investigation of the exact calibration of the pendulum. Between 1661 and 1664 Huygens attacked the outstanding problem of the center of oscillation of a compound pendulum, the solution of which led to a small weight adjustable by sliding along a scale inscribed on the rod of the pendulum. The solution itself, based on an application of the principle of conservation of mv^2 (a quantity to which he gave no name), laid the groundwork for the later dynamics of rigid bodies and made extensive use of the new methods of quadrature and cubature that became the foundation of the integral calculus.

At this point, Huygens was ready to send the cycloidal pendulum clock to sea. But before following that story, let us look back to see what had transpired so far, in less than a decade of study: a successful pendulum clock, the derivation of the period of a simple pendulum, the discovery of the cycloid as tautochrone, the new theory of evolutes (later subsumed under the concept of the center of curvature), and the derivation of the center of oscillation of a compound pendulum. At the risk of stating the obvious, we are looking at essential elements of the new mechanics of the Scientific Revolution, all issuing from the clock and its application to the problem of longitude.

There was more to follow. As sea trials brought out the weaknesses of the pendulum clock, Huygens became alert to alternatives. In 1675 he returned to his analysis of motion on a cycloid and determined that what made it tautochronic was the direct proportion between the effective acceleration of the body along the curve and its distance from the center at the bottom. Noting as Hooke did at about the same time that springs act in an analogous way (*ut tensio sic vis*), Huygens determined that a coiled spring would behave exactly like a cycloidal pendulum and hence that it could replace the pendulum as a tautochronic regulator for clocks. Moreover, the spring was only one of a host of mechanisms that obeyed the basic principle of what we now refer to as simple harmonic oscillation and what has since come to be viewed as one of the most fundamental mechanisms of nature. Huygens' search for a "perfect marine balance" based on that principle continued until his death. Pace Otto Mayr, Huygens' work in this regard constitutes a study of regulation by feedback a century before Watt's invention of the steam-engine governor.[9]

But let's go back to the clock at sea. Before it could serve to determine longitude, Huygens had to deal with an old question to which the accuracy of his clock now gave new importance. Ignoring precession, it seems at first a matter of indifference whether one sets the clock to the sun or to the stars. However, owing to the declination of the sun's orbit (as navigators do to this day, we shall assume a Ptolemaic solar system for ease of calculation), the length of the solar day in fact varies over the year. A clock set to noon on February 10 and calibrated to a mean solar day will fall nearly 20 minutes behind the sun by May 14, catching up to within 9 minutes on July 25, but then falling more than 30 minutes behind again by November 1.[10] It will catch up fully only after a year. Those discrepancies add up to serious errors in longitude unless one has a table of daily values and instructions on how to use them. Here the clock as astronomical instrument became the means of making those values precise, while the precise values in turn made the clock accurate as a navigational instrument. In the process a problem of astronomy had been elucidated.

Once underway, Huygens' clocks met with only mixed success under the rigors of travel by sea. A design arrived at in collaboration with Alexander Bruce (later Lord Kincardine) went along with Captain Robert Holmes on voyages in 1663 and 1664 from London to Lisbon and Guinea and then out into the Atlantic. The encouraging data of the first voyage were heightened by the drama of the second, when Holmes, relying on the clocks, overrode the advice of his fellow masters concerning their distance from the Cape Verde Islands and proved to be correct.[11] Trials in the Mediterranean in 1668–69 seemed equally promising. But, eager now to claim suitable reward and recognition from official bodies in England and France, Huygens withheld publication of his treatise on the clock pending long-range tests on an expedition to the West Indies commissioned by the Paris Academy of Sciences in 1670. It was a disaster for the clocks, largely, Huygens believed, owing to the negligence of Jean Richer, the *élève* deputed to care for them. Wherever the fault lay, the clocks remained unproved, and Huygens proceeded with publication of his masterpiece, the *Horologium oscillatorium* (1673), unable to claim a working solution to the problem of longitude.

After a hiatus of some fifteen years, during which he experimented with a variety of new regulating mechanics described in part above, Huygens again sent his pendulum clock to sea in 1686–87, this time to the Cape of Good Hope on board the Dutch East India Company ship *Alkmaar*. The clocks responded badly to heavy seas on the way out, but on the return worked well enough for Huygens to plot a course that could be compared with the ship's logs and with sightings of land. According to the clocks, however, the *Alkmaar* had sailed right through Ireland and Scotland rather than around them.

Far from invalidating the clocks, that result tied the determination of longitude back to fundamental questions of mechanics. On an Academy of Sciences expedition to Cayenne in 1672–73, Jean Richer carried out instructions to determine the length of a 1-second pendulum and found that it was shorter by 1¼ lignes than a 1-second pendulum in Paris. Huygens immediately drew the horological conclusion: a clock set in Paris to mean solar time would run behind by 2 minutes and 10 seconds, owing to the apparent lengthening of its pendulum.[12] The deviation was of the order of the solar inequality and hence affected the calculation of longitude. Subsequently others reported a variation in the length of a second pendulum at various locations on the globe. At issue, of course, was the value of g, as measured through the weight of a body. While some scientists, such as Jean Picard, insisted on its constant value everywhere, the data from the *Alkmaar* voyage decided Huygens in favor of its variation, a conclusion he had reached in the late 1660s in a study of the cause of weight and its application to the period of a pendulum.[13] Then he had reasoned that the rotation of the earth exerted a centrifugal force that decreased the weights of bodies by a factor dependent on the latitude. At the pole, it had no effect; at the equator it caused the greatest decrease, about 1/289. Applied to the data of the voyage, that adjustment brought the course determined by the clocks into line with that of the logs, once allowance was made for the incorrect longitude used by the pilots for the Cape of Good Hope. Reassured by this empirical evidence, which also gave him grounds for rejecting Newton's theory of universal gravity, he now moved to publish the earlier treatise on the subject.[14]

More is involved in this story than Huygens' personal, serendipitous quest. When Colbert decided at the end of 1666 to bring together the French mathematicians and natural philosophers receiving stipends from Louis XIV and incorporate them as a Royal Academy of Sciences, he included Huygens and indeed looked to him for leadership and guidance. In response to Colbert's inquiry about what such an Academy might undertake, preferably to the benefit of its sponsor, Huygens proposed several agendas, one of which included the following projects:

> 10. Observe the motion of the companions of Jupiter and make tables of them.
>
> 11. With the aid of these tables, observe here and in other places in the world, such as in Madagascar, the occultation of each of the said companions behind or in front of Jupiter, to find thereby the true longitude of the said places and to rectify [current] maps.

> 11,1. Observe the declination of the magnet and the change that it undergoes.
>
> 12. Send pendulum clocks to sea with the necessary instructions and a person to take care of them, to carry out [*pratiquer*] the determination of longitudes, which has already succeeded so well in the experiments thus far made.
>
> 13. Measure the times and ratios of the fall of heavy bodies in air.
>
> 14. Measure the size of the earth. Advise on the means of making geographical charts with greater exactitude than hitherto.
>
> 15. Establish once for all the universal measure of sizes [i.e. a universal standard of measure] by means of pendulums, and in consequence [the universal measure] of weight.[15]

Evidently, Huygens was thinking well beyond his clocks and the problem of longitude, or rather he was thinking of the problem of longitude in the context of a much larger program of mechanical and astronomical research. His own experience of the previous decade had shown him the promise of the pendulum as an instrument linking celestial and terrestrial phenomena through its own mechanics

The research agendas that Huygens proposed for the Academy of Sciences reveal an important aspect of the relation of science and the state at the beginning of the latter's institutionalization of the former. Colbert's ministry stands as the historical embodiment of the mercantilist state. To page through his administrative correspondence and the Comptes des bâtiments (which kept accounts of much more than the king's building) is to watch a man dedicated to the organization and expansion of the commercial, industrial, and military infrastructure of Louis XIV's realm. Yet, he himself had no program of research in mind for the Academy of Sciences when he founded it, nor is there much evidence that he brought the resources of the Academy to bear on the projects described in the administrative documents just mentioned. He had a general idea of what he needed, but he could not translate that into an agenda for the mathematicians and natural philosophers he was ostensibly enlisting in the service of the state. He knew he needed their expertise, but he did not know just what that expertise was or how it applied to the problems he was facing. Rather, he left it to the scientists themselves, foremost among them Huygens, to tell him what he needed from them. It is hardly surprising that what he needed is what they happened to be doing, or rather what they wanted to do.[16]

Colbert's request for guidance from Huygens came as Huygens was looking to follow up on the first English test of his clocks. Placing his research agenda at the heart of that of the new Academy placed the resources of the

Academy at his disposal, not only for testing the clocks themselves but also for confirming the astronomical and mechanical theories on which their design was based. Colbert's administrative correspondence reveals the extent of those resources. A letter to Colbert de Terron, Intendant at Rochefort, dated 10 March 1670, informs him that

> M. Richer, having been chosen by the Academy of Sciences to go to the East [*sic*] Indies in order there to carry out astronomical observations that may be related to those being carried out here and to test the clock and pendulums that have been constructed for determining longitude at sea, the King has deemed it proper for him to travel with the squadron of his vessels that is headed to that country. As the said M. Richer is a person of merit who must apply himself to quite abstruse (*curieuses*) matters, I ask you to require the captain of the vessel on board which he shall embark to give him a place at his table and a place at the common table for the man he will be taking with him, seeing to it above all that he [Richer] shall find all the accommodations he will need, both for adjusting the said clocks and for transporting his baggage and instruments, which left six days ago. I am sure you will attend to this most exactly.[17]

The larger research program required a different sort of facility, realized in the Observatory built to the Academy's specifications in 1671 and dedicated to the advancement of astronomy, "that noble science which deals with matters that are obscure but which is thought so useful to public interests, primarily to navigation and geography, as well as to the propagation of the Christian religion."[18] It was on the floor of the west tower that Sedileau and Chazelle, under the direction of Cassini, laid out a planispheric projection of the earth on which researchers placed locations of which the longitude and latitude had been accurately determined by measurement of lunar eclipses.[19]

As several recent studies have described in detail, the Observatory became the center of an extensive effort to map the globe and to measure its surface.[20] That meant establishing and maintaining tables of lunar distances and lunar eclipses (both of our moon and those of Jupiter) used for the determination of longitude of points on land even after Harrison's clock, measurement of the length of the meridian and of the constant of gravity at various places, and the application of those data to the question of the shape of the earth. Along with such theoretical matters went the practical issues of tables, instruments, and instructions for those who had the task of navigating a ship through unknown waters and of mapping the world they were exploring. A similar story may be told of the Greenwich Observatory. It is at such institutions that science became part of the infrastructure of the modern state.

Notes

1. William J. H. Andrewes, *The Quest for Longitude* (Collection of Historical Scientific Instruments, Harvard University, 1996); Dava Sobel, *Longitude* (Walker, 1995).

2. To this day, the United States sets its clocks to the timekeepers of the Naval Observatory in Washington. Initially conveyed to ships in port by means of a ball descending a mast at noon, the time can now be automatically downloaded via the web to any personal computer.

3. David S. Landes, "Finding the Point at Sea," in Andrewes, *The Quest for Longitude.*

4. See Michael S. Mahoney, "Longitude in the Context of the History of Science," in Andrewes, *The Quest for Longitude.*

5. See Richard J. Sorrenson, Scientific Instrument Makers at the Royal Society of London, 1720–1780, Ph.D. dissertation, Princeton University, 1993, esp. Chap. 7 on Jesse Ramsden, and "The state's demand for accurate astronomical and navigational instruments in 18th century Britain," in *The consumption of culture, 1600–1800,* ed. A. Bermingham and J. Brewer (Routledge, 1995).

6. For details, see Michael S. Mahoney, "Christiaan Huygens, The Measurement of Time and Longitude at Sea," in H. Bos et al., eds., *Studies on Christiaan Huygens* (Swets, 1980); Mahoney, "Huygens and the Pendulum: From Device to Mathematical Relation," in H. Breger and E. Grosholz, eds., *The Growth of Mathematical Knowledge* (Oxford University Press, to appear); J. H. Leopold, "The Longitude Timekeepers of Christiaan Huygens," in Andrewes, *The Quest for Longitude.*

7. The central problem of the mechanical clock is converting the accelerated motion of the driving weight or spring into a uniform motion of the hands. That is accomplished through the escapement, which advances a toothed wheel by pallets or pins engaging and releasing it at regular intervals determined by an oscillating mechanism. Prior to the pendulum, none of the mechanisms employed, such as the foliot, had a natural period.

8. The term "evolute" derived from the technique of wrapping a string (or imagining a string wrapped) around a curve and then tracing the locus of its endpoint as the curve is unwrapped (*evoluta*), that is, as the string is pulled away while being held tangent to the curve. The mode of generation and hence the class of curves thus generated was new at the time, and Huygens had to develop much of his mathematical apparatus for handling them. Chapter 4 ("On Evolutes") of his *Horologium oscillatorium* (Paris, 1673) was the first published treatise on the subject.

9. See Otto Mayr, *Authority, Liberty, and Automatic Machinery in the Early Modern Era* (Johns Hopkins University Press, 1986).

10. The dates of the turning points are Huygens' and differ from the modern values, which according to the 1964 English edition of Flammarion's *Astronomy* are February 11, May 15, July 27, and November 4.

11. Sir Robert Moray (Murray) to Christiaan Huygens, January 23, 1665, in Huygens, *Oeuvres complètes,* volume V, p. 205ff. There was some delay in confirming Holmes' report,

because he had been thrown in the Tower for attacking and capturing without declaration of war several Dutch fortresses in the Cape Verde Islands and Guinea (ibid., p. 168, n. 6).

12. Angered by what he considered Richer's negligence on the earlier voyage, Huygens had refused to commit his clocks a second time to the young man's care and so had passed up another opportunity to test them on a long journey.

13. Huygens, "Calcul de la période d'une oscillation (cycloïdale) en un endroit déterminé de la terre en tenant compte de la force centrifuge due à la rotation du globe terrestre," *Oeuvres complètes*, XVII, pp. 285–286.

14. On the role of the voyage in Huygens' critique of Newton's theory, see George E. Smith, "Huygens' 1688 report to the directors of the Dutch East India Company on the measurement of longitude at sea and its implications for the nonuniformity of gravity," *De Zeventiende Eeuw* 12, no. 1 (1996): 198–214.

15. Huygens, *Oeuvres complètes*, volume XIX, pp. 255–256.

16. One can see a parallel in the research agenda that J. C. R. Licklider and other members of the computer community laid out for the Advanced Research Projects Agency in the early 1960s in response to ARPA's expressed need for "command and control systems." See Arthur L. Norberg and Judy E. O'Neill, *Transforming Computer Technology: Information Processing for the Pentagon, 1962–1986* (Johns Hopkins University Press, 1996).

17. Jean Baptiste Colbert, *Lettres, instructions et mémoires de Colbert*, ed. P. Clément, volume 5 (Imprimerie impériale, 1861–1873), pp. 294–295. The following year, Colbert wrote to the French ambassador in Denmark, ordering that he smooth the way there for astronomical observations to be carried out by Picard. Colbert prefaced his order by noting that "entre les grandes choses auxquelles le Roy, nostre maistre, s'applique, celle des sciences n'occupe pas moins son esprit que toutes les autres qui regardent la guerre. . . ." Later, Colbert put the resources of both the navy and the army at the disposal of Cassini to move large lenses from Rome to Paris.

18. Jean-Baptiste du Hamel, *Regiae scientiarum academiae historia*, second edition, book i, section 8, chapter 1 (Paris, 1701), p. 103.

19. Du Hamel, *Historia*, II, 11, 4, v., p. 217. The original was effaced but a copy was published by Jean Baptiste Nolin in 1696 and is reproduced on p. 56 of *The Quest for Longitude*.

20. See most recently Jordan Kellman, Discovery and Enlightenment at Sea: Maritime Exploration and Observation in the 18th-Century French Scientific Community, Ph.D. dissertation, Princeton University, 1998.

III

Gunpowder Production: The Refinement of Waste

9

"The Art and Mystery of Making Gunpowder": The English Experience in the Seventeenth and Eighteenth Centuries

Brenda J. Buchanan

The making of gunpowder was rooted in a world of arcane mysteries in which all medieval crafts operated. The term "Art and Mystery" occurs in documented partnership agreements between gunpowder makers as late as the 1780s. It refers to neither the liberal arts nor the fine arts, but the mechanical arts; and it suggests that both workmen and manufacturers were bound by an agreed-upon body of rules. Today the term may seem archaic, implying secrecy and the workings of the cabal; however, at the time it signified a perfection of workmanship acquired through a process of learning and obeying the rules for craft processes. Yet, in the dangerous industry of gunpowder manufacturing, it was understanding the underlying reasons for these rules, rather than merely following them, that eventually proved most critical.

Gunpowder production in England began on a small but practical scale at royal strongholds, chiefly at and near the Tower of London from the middle of the fifteenth century and at other strategic sites such as Portchester Castle in Hampshire from the early sixteenth century.[1] Until this time there was no incentive to develop a home-based industry. The trading links with the Mediterranean countries, and the cultural and economic links with the coastal regions of northern Europe, were so strong that English merchants were able to purchase saltpeter, sulfur, and gunpowder from these sources. In 1513, for example, letters in cipher to Henry VIII from an English merchant in Burgos chartered the shipment of saltpeter and sulfur from Naples by Florentine factors.[2] Although the eastern Mediterranean continued its trade in sulfur, northern Europe became the central market for saltpeter and gunpowder, since these commodities were available and strategically located for military operations on the Continent. Supplies were stored in the warehouses of ports, especially those of Antwerp; however, when these were insufficient, merchants were able to rely on trading networks developed over the years. Delays occurred nonetheless: in March 1538 an English merchant seeking powder for the King's service learned from Hans Ruckardes that "Three or four months will be required, as it must be bought [in] Do[cheland] and brought to Hamburgh or Antwerp."[3] By contrast, a crisis at Boulogne in September 1544 brought a swifter response. When it was learned that the King had "bestowed upon his sieges so much powder that all he brought is spent," the Tower of London was scoured for further supplies, and a bargain was struck for 200 lasts to be made in Flanders.[4] Gunpowder supplies may also have been available from English sources on the Continent, for in 1540 a lock was ordered for the works at Calais, "where as the gunpowder lieth." A "Powder Mill, known as the Horse Mill," was recorded there in 1556.[5]

Military pressures in Wales, the Scottish Borders, and Ireland further complicated English gunpowder acquisition. In April 1539 only two or three barrels of powder were available at Beaumaris Castle, so that "the King's house [stood in] jeopardy." In August 1544 it was reported from Alnwick that the gunners had been unable to serve "for lack of powder and matches." There was also concern about munitions in Ireland, and a review of the stocks of guns and powder in Dublin was ordered. In Edinburgh Castle the Scots had workmen making guns and operating a mill capable of producing six barrels of powder in three weeks; given their military rivalry with Scotland, the English were further pressured to put their house in order.[6] Their vulnerable position at Antwerp added to this stress. In 1556, when Philip II became King of Spain and Prince of the Netherlands, economic difficulties were compounded by religious and political troubles, culminating in the Dutch Revolt. The sacking of Antwerp in 1576, followed by its capture and blockade in 1585, confirmed the vulnerability of England's gunpowder supply.[7] The Dutch rebellion against Spain also helped stimulate England's preference for firearms, such as the arquebus and the wheel-lock pistol, over the longbow and the crossbow. To meet the growing pressures of this new military demand, the English therefore had to improve and expand their domestic gunpowder production.

Historians have paid little attention to the development of and experimentation with gunpowder manufacturing in England during the early modern centuries.[8] In light of its strategic significance, I will present a thorough analytical history of early English gunpowder acquisition. This essay will examine the process in four phases while providing a novel interpretation of the way in which new scientific insights modified traditional modes of manufacture before the closing decades of the eighteenth century. The first phase was marked by attempts in the sixteenth century to adopt the skills of foreign craftsmen. The English put these into practice from the middle decades by establishing saltpeter beds in the provinces and water-powered mills in the countryside outside London. The second phase was noted for the technological adaptations made by powdermakers in the seventeenth century, especially in response to wartime demands of the 1640s and the 1660s. The third phase concerns the flexibility of provincial partnerships as they addressed the varied demands of the growing economy from the late seventeenth century, especially in relation to mining, trade, and colonial settlement. The fourth phase saw the revival of the Board of Ordnance from the 1740s with the appointment of Charles (later Sir Charles) Frederick. Those who dealt with him recognized his ability to combine traditional and experimental approaches; he knew the rules but was prepared to challenge them.

The Origins of English Gunpowder Manufacture

During the two centuries before the French Revolution, most military operations faced grave gunpowder shortages. The easiest response was to boost the number of staff concerned. The Office of the Ordnance (separated by 1485 from the Royal Privy Wardrobe, in which it had its origins in the early fifteenth century) grew considerably until some fifty persons came under the supervision of the Master of the Ordnance in the middle of the sixteenth century.[9] Its aim was to expand and improve the domestic arms industry, but this was difficult to achieve. The military demand for munitions fluctuated, and the gunpowder (produced under contract, albeit on an irregular basis, by many small-scale manufacturers) was of dubious strength and durability. The problem was compounded by the proximity of continental Europe and the availability of supplies there. Paradoxically, England's strong link with the Continent offered a solution by increasing the potential for convincing foreign craftsmen to come to a technologically backward England to teach their skills.

An early example of this transfer of craftsmanship occurred near the beginning of the reign of Henry VIII. In 1515, Hans Wolf, a foreigner, was appointed one of the King's powdermakers at the Tower of London and elsewhere. He was charged with the task of searching the shires for "stuff" from which to derive saltpeter. Once located, "he and his laborers shall labor, dig or break in any ground," and compensate the owners for any damage.[10] In this tradition, Thomas a Lee, one of the King's gunners, was appointed principal searcher and maker of saltpeter in 1531. In addition to searching and digging, he was authorized to hire workmen, take wood for burning and "trying" saltpeter, arrange transport by land and water, and rent houses for processing the commodity.[11] To boost domestic production further, foreign saltpetermen, including "subjects of the Emperor," were granted the right of denization (that is, of citizenship and residence) in return for their expertise.[12] In 1561 this dependence on foreign "know-how" reached new heights when Queen

Elizabeth granted the Dutchman Gerard Honrick £300 for teaching her subjects the art of saltpeter making. The first instructions for setting up nitre beds were to be found in Conrad Kyeser's *Bellifortis* (1405). More advice followed, yet Honrick's version established the rules. It recommended: black earth, the richer the better; urine, especially from those who drank wine or strong beer; dung, especially from horses fed with oats; and lime made from oyster shells or plaster of Paris. The moistened ingredients were to be layered in beds to which ashes were added (those from oak leaves being recommended to the Queen), and the resuling salts were to be leached out and boiled to form "frozen ykles," probably icicles or crystals.[13]

From the middle of the sixteenth century, the search for technical expertise in saltpeter and gunpowder making also benefited from the government's support of proposals fostering domestic industry. The enterprise of innovators and "projectors" was encouraged by a system of patents that granted the sole right of manufacture to those devising new ways of producing goods. Sulfur provides good examples with the granting of patents in 1565 and 1566 and the acknowledgement in 1572 of proposals from "dealers in brimstone" who hoped to extract this gunpowder ingredient from English copper mines.[14] Some of the innovators were Protestant refugees, perhaps drawn by an invitation such as that from the citizens of Maidstone in Kent, who in 1567 named gunpowder makers among the craftsmen they hoped to attract.[15] Nonetheless, the English gunpowder-making industry could not establish a firm footing by relying on chance contacts. The most significant and far-reaching technological development in these middle decades was the water-powered gunpowder mill. The first mills were set up at Rotherhithe, on the south bank of the Thames beyond London, and subsequently in the well-watered, isolated, yet accessible valleys of the Surrey countryside to the south.

At the time of the Domesday survey, of the late eleventh century, there were over five thousand water mills in England, used largely in the preparation of grain. By the thirteenth century some were providing power for manufacturing purposes, especially the fulling of cloth. Not until the sixteenth century is there any evidence of the use of mills in gunpowder making. The State Papers of February 1555 refer to Henry Reve's erection of a gunpowder mill in Rotherhithe, complaining that the banks had been weakened "by reason of the great abundance of water which came in at the flood gates and sluices made for it."[16] Though brief, this complaint suggests a tide mill, which is powered by trapping the incoming tidal water of the Thames in a reservoir and releasing it as required through a series of gates and sluices. A waterwheel could then be harnessed to provide power for the most essential stage in gunpowder making: the incorporation of the three main ingredients by a prolonged process of hammering with stamps. Albeit a welcome source of energy, this dependence on tides would have limited the operation to six to ten hours per day;[17] at a time when incorporation by stamps generally required more hours, any interruption could have had an adverse effect on quality and safety.

The innovative application of tidal power to gunpowder making may be traced back even earlier, through a lease of 1563 for a second mill at Rotherhithe. It describes a "gonpowder myll," pond, watercourse, and wharf as having been occupied by the Lee family for some twenty years—c. 1543. Additional papers refer to the support of Henry VIII for the building of the mill in c. 1536.[18] This detail recalls the King's appointment in 1531 of Francis a Lee as a saltpeter maker, and suggests that the developments in Rotherhithe may have reflected an early effort by the King to boost the production of gunpowder by supporting this new technology. The Lee (or Leigh) family surface again in the State Papers of Queen Elizabeth's time. In 1575, Francis Leigh, "gunpowder maker to the Council," requested permission to import twenty or thirty lasts of saltpeter to make gunpowder—he and his family having been for fifty years "the greatest dealers therein, and [having] all the implements."[19] This continuity was beneficial for the industry: the apprentice would learn the rules of this "Art and Mystery" from his master and, within the family relationship, the method of teaching by example and catechism would be particularly effective.[20] Leigh supported his case by pointing out the consequences of inadequate domestic production: "provision has to be made in foreign parts, which in the Duke of Alva's time was stayed." This reflected the restriction of trade with the politically troubled Low Countries already mentioned.

The Lee family became a significant supplier of powder because of the advanced technology they had adopted. They nevertheless belonged to an uncoordinated network of producers and merchants who supplied the Ordnance Office on an irregular and inefficient basis. The inadequacies of the system became alarmingly evident in 1588 when it was discovered that the powder supplies were insufficient for fighting the Spanish Armada. This shortage was compounded by the demands of England's military engagement on the Continent, as well as the delays in replenishing the stores.[21] In response to this emergency, George and John Evelyn and Richard Hills received a patent of monopoly for gunpowder making in January 1589, initially for eleven years. Because of subsequent renewals (despite variations in the patentees), the Evelyn family was associated with gunpowder production for almost fifty years.[22] This span of time compares with that of the earlier Lee family. Whereas the Lee family depended on the tidal motion of the Thames, the Evelyns relied on the more constant flowing rivers of the Surrey countryside, which only drought and ice restricted.[23]

The patent of 1589 also addressed the problem of saltpeter production. It granted the Evelyn partnership the sole right to "dig, open and work" for saltpeter in most of the Queen's dominion. Although the patent secured a monopolistic claim for this partnership, the actual collection of saltpeter remained the responsibility of the semi-independent saltpetermen. They worked out in the countryside with a considerable degree of freedom, although under the Letters Patent of the gunpowder makers. After their appointment as Commissioners for Saltpetre and Gunpowder in 1621, the Lords of the Admiralty should have regulated these saltpetermen; however, this was not always the case. In May 1629, a petition claimed that a flax dresser was "making saltpetre in Essex, Suffolk, and Norfolk, and having no experience, knowledge, or discretion, wasted and disabled the grounds, vexed and troubled the King's subjects, and kept back the hire of the petitioners."[24] Clearly, this petition betrays an element of self-interest. Written seventy years after the celebrated purchase of Honrick's recipe, it also reveals that ignorance about making saltpeter persisted. Other complaints in the State Papers show that the saltpetermen searched relentlessly and often destructively for good black earth. They dug out the rich nitrogenous manure from stables and other farm buildings, undermined structures such as dovecots, confiscated carts for the transport of noxious earth to their "boiling houses," and required residents to convey fuel for their boiling operations.[25] They competed with soap makers for ashes and with textile workers for urine supplies.[26]

Here we have a disparate collection of information about saltpeter making, but no evidence about the workshops in which it was done. Fortunately, a chance reference has led to the discovery of the only plan known to date of an English nitre or saltpeter bed. This plan is attached to a grant of land made in 1593 to a powdermaker of Ipswich in the county of Suffolk, northeast of London (figure 9.1).[27] The bed was located in a field called a "dunghill," suggesting that manure may have been piled up until it was sufficiently ripe for processing. The nitre bed was roughly rectangular and covered about 12,000 square feet. In one corner were buildings, perhaps the residence of the saltpeter maker and his team. A covered arcade around three sides of the property shielded the beds so that watering the black earth and ashes with urine could be regulated. The nitrous matter leached from the soil by the process of lixiviation would have been collected and boiled in cauldrons that were housed, presumably, on the western side. In Figure 9.1, steam is seen rising from that side of the property. We also see in the southeast corner what may be a workman raking a vat. After several re-crystallizations and dryings, the saltpeter from Ipswich and from similar nitre beds around the country would have been delivered to the patent holders, probably to the main store at Southwark in London.

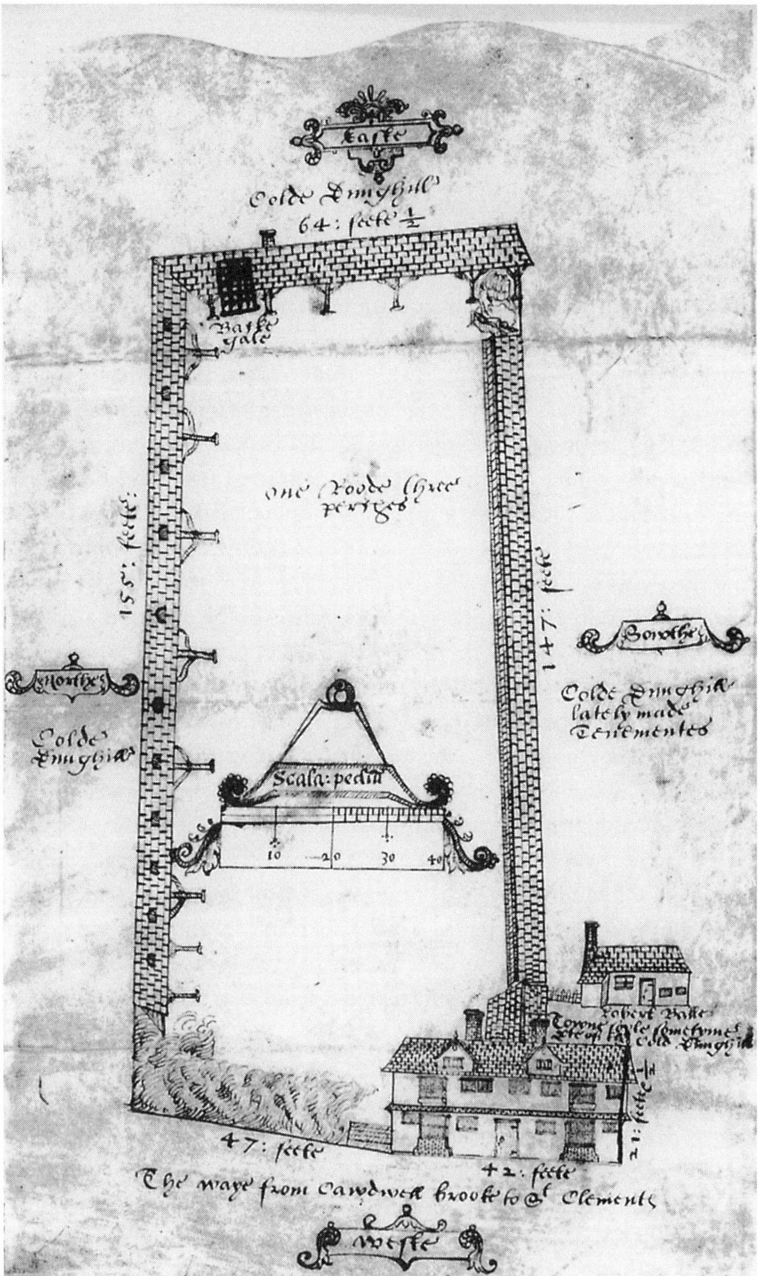

Figure 9.1
Saltpeter works on land granted in 1593 to a Suffolk powdermaker. The only known contemporary plan of such a feature in England, this drawing is held in the Ipswich Branch of the Suffolk Record Office (C/3/8/4/31). Reproduced by courtesy of the Ipswich Borough Council.

Despite attempts to establish a nationwide system for collecting and processing organic waste, the saltpeter industry was unable to keep up with demand, especially in times of war. This led to an ongoing search for overseas supplies, not just from the Low Countries but also from more distant sources. In what must be seen as a great stimulus to shipping and trading, saltpeter was sought in the Baltic ports of the northern waters, on the Barbary coast of northwest Africa, and on the Indian subcontinent. The following monopolistic corporations were authorized to raise capital and regulate trade: the Eastland Company in 1579; the Barbary Company in 1585; and the East India Company in 1600, which experienced the greatest and most enduring success, especially after the saltpeter trade with Bengal was established in the middle of the seventeenth century. The growing volume of imports from the East India Company relieved the pressures on domestic saltpeter making; in contrast to the French, the English consequently were not motivated to improve the national industry, so it was gradually abandoned.[28]

The English saltpeter industry should, however, be counted a success story in one important respect: the spread of nitre beds into the countryside meant that small-scale but enterprising gunpowder makers were able to make illicit purchases of saltpeter beyond the control of the Crown and its Patentees. This inadvertent "opening-up" of the industry was particularly significant for the Bristol region, where gunpowder had been made, usually illegally, since at least the early seventeenth century. Similarly unauthorized activities were discovered in the county of Dorset and at Battle, in Sussex, in 1627; near Chester, in Cheshire, in 1628; in Devon in 1629; and in Bristol again in the 1630s. On the last occasion even the saltpeter itself came from an unauthorized source: the beds at Sherston Magna in Wiltshire.[29] This spread from the metropolis, though repressed in most cases, would become a salient feature in the development of the gunpowder industry.

The Civil War and Its Aftermath

The problems of gunpowder supply became more acute in the middle decades of the seventeenth century. In the Civil War of the 1640s what had not been sufficient for the nation-state could hardly be adequate for its warring parts. During the Dutch Wars of the 1650s, 60s, and 70s, the availability of vital supplies from the Netherlands was severely curtailed. In both cases the manufacturers, especially of gunpowder, had to improvise and adapt. It is these "emergency procedures" that will now be considered.[30]

This logistics challenge was a serious one, for these were intensive gunpowder wars. As late as 1642 the King was petitioned by the bowyers and fletchers, who asked that his campaigns be fought with bows and arrows.[31] These weapons had become obsolete, but to the men of the old order this must have seemed a sensible request as the country moved toward Civil War. Although the division of land and resources between King and Parliament was not clear-cut, Parliament's control of London and the Home Counties gave them considerable advantage in terms of gunpowder supply. They took over the military stores at the Tower, Woolwich, and Greenwich, and profited from the expertise of the workers there. Even though most of the senior officials of the Office of Ordnance joined the King, the craftsmen remained loyal to their workplaces and homes. More significantly, the existing powder mills in Surrey were within Parliament's control. Like those at Chilworth, some of them were damaged by departing Royalist sympathizers, but these were restored by experienced workmen. Parliament also had great influence in the City, whose financiers and merchants facilitated the trade in munitions with the Continent.[32]

Availability in southeastern England nonetheless had to be matched by an ability to deliver gunpowder to outlying Parliamentary strongholds—especially those in the west, where logistic support was uncertain. In these circumstances it became essential for craftsmen to acquire the skills of the saltpeterman and the powdermaker. The cathedral city and riverport of Gloucester, garrisoned since December 1642 by Parliament and besieged for five weeks by the Royalists, provides a good example. Here gunpowder supplies were maintained by artisans who had been successfully initiated into this "mystery." With a goldsmith in charge, a saltpeter "house" and a powder mill were set up on the riverside quay to utilize the waterpower available there. The prepared saltpeter, together with that from another "house" in the town, was stored in outbuildings of the cathedral, in which there was a second powder mill. The crypt of Christ Church served as a magazine. One of the charcoal burners was designated the "ashman" and assigned the task of supplying the saltpeter houses with the finely powdered ashes that were an essential part of saltpeter making. The other versatile craftsmen included the bell-founders, who made "granadoes and other things by the Governor's order"; the bookbinder, who provided the thick, coarse paper required for cartridges; and the "roper," who was in charge of making "match" from hemp. Apothecaries continued to ply their own trade, supplying soldiers and citizens with purges, cordials, and plasters.[33]

Cut off from the arms industry of the southeast, the Royalists had to display an even greater degree of improvisation and versatility. They first drew on the limited supplies accumulated in some of the county magazines; but after the King made Oxford his headquarters, they transformed the university town into a manufacturing center and a military garrison. They achieved the latter more easily. After the army marched into the town on October 29, 1642, carrying the colors captured from the Parliamentary forces defeated at Edgehill, the King established himself and his Court in Christ Church College and billeted his army in the colleges and town. The accompanying train of ordnance and great guns was parked in Magdalen College Grove, where forges and workshops, together with a foundry at Christ Church, were later set up. Cartloads of muskets, powder, and shot (taken largely from the stores in the Guildhall in a gesture that deprived the townsfolk of their arms) were carried into the University Schools Building. The uppermost rooms of its tower thus became a magazine. New College nearby served as the main repository, in the manner of the Tower of London. Horses were grazed at Iffley, outside Oxford but easily accessible; and carts were parked beside the magazine at New College. But there was no ready supply of gunpowder in the city and no team of craftsmen from the old Ordnance Office to produce it; therefore, the Royalists turned to the provinces for help from the "rogue" powdermakers whom they had previously harried. Some stayed on in the county towns producing powder for the Royalist cause, but William Baber of Bristol responded by moving to Oxford with his son, his brother, and another renegade powderman from the port city.[34]

William Baber's life as a powdermaker in Bristol can be traced back to 1619. Unlike those of his contemporaries who were sometimes brought briefly within the law to serve the shipping of the port, he was never licensed. His career nonetheless was marked by occasional entries in the State Papers, as in 1631 when it was recorded that he had received and utilized, for his own purpose, great quantities of saltpeter smuggled into the city at night.[35] It is likely that in this crowded town gunpowder was still produced by hand according to the traditional method shown in an illustration of 1630 (figure 9.2). One workman pounds and incorporates the ingredients with a hand-held pestle, and a second forces the resulting paste through a sieve, corning it to produce separate grains. The scales on the wall reflect the need for an accurate weighing of the saltpeter, sulfur, and charcoal. The hanging sieves may have been used to break up lumpy ingredients or to vary the size of the corned grains. The workmen would have dried the gunpowder on trays set out on a table. Here we also see a small "tryer" for testing the powder's strength. All tasks are manually performed, since the use of animal power could have exposed the illegal operation. It was from this provincial, technologically backward setting that Baber and his colleagues moved to Oxford. Emerging as the leading powdermaker, he delivered nearly 150 barrels in the first six months of 1643. In their first statement about gunpowder, dated January 8, 1643, the Royalist Ordnance Papers recorded that four barrels of powder were received into his Majesty's Stores from William Baber, powdermaker.[36]

Figure 9.2
Powder making by hand, with incorporation by pestle and mortar on the left and corning on the right. As shown in Hanzelet's *Pyrotechnie* of 1630 and reproduced in Guttmann's *Monumenta*.

By what emergency procedures did Baber produce this powder? Based on his provincial background, we can speculate that he did so at first by hand or by using animals within the strong buildings of Oxford. An illustration from a sixteenth-century artillery manual shows how this could have been achieved (figure 9.3). It is documented that powder was received from University College and St. Mary's Church, so both may have played a part in this, as well as from the "Powder Howse," which was strengthened in December 1643 with "Hookes and Hinges."[37] More importantly, several watermills on the outskirts of the city and in nearby villages were adapted for powder making. Anthony Wood, a reliable young observer, noted that "The gunpowder myll was at Osney where the fulling myll stood."[38] Formerly a religious house, Osney Abbey was dissolved in 1539, and its buildings demolished gradually; but by 1546 its fulling mill was restored to working condition.[39] "Fulling" was part of the cloth-making process in which loosely woven material was placed in a trough of water and hammered by wooden mallets until it was thickened and cleaned. The conversion of fulling stocks to gunpowder making may seem

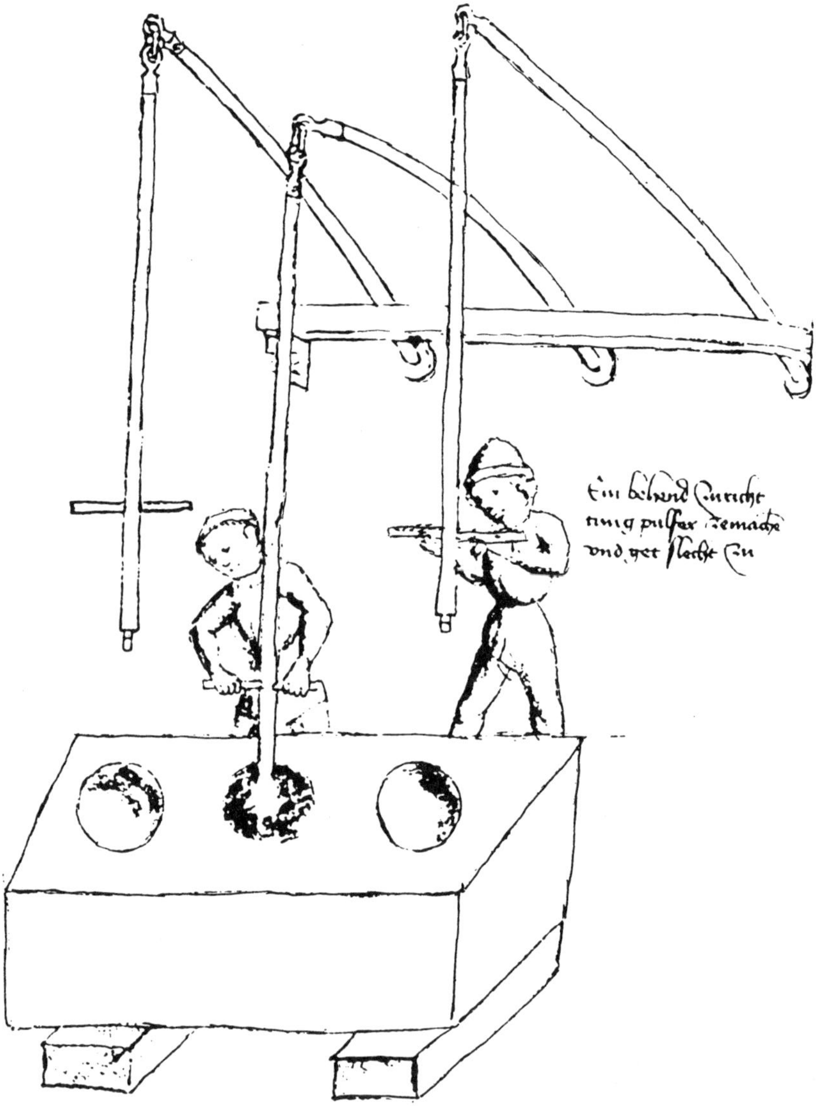

Figure 9.3
Titled "A speedy contrivance for making gunpowder," this illustration from a sixteenth-century artillery manual, inscribed "Hanss Henntz of Nuremberg" and reproduced in Guttmann's *Monumenta*, shows the incorporation of powder by pestle and mortar with the help of flexible wooden spring beams to ease and speed the task of the workman.

unlikely; but, in referring to the methods for mixing the three main ingredients of gunpowder under pressure, Biringuccio described the use of a wooden stamp with a hammer-like handle that descended into a wide-mouthed stone mortar.[40] This suggests the swinging, hammering action of fulling stocks (figure 9.4), rather than the more traditional upright pounding produced by stamps or pestles (though stamps may have been used in early fulling mills). The restoration, and perhaps the extension, of the Osney powder mill began in April 1643 when twelve heavyweight "Mattockes," or pickaxes, were ordered. Baber later claimed that he had invested in these works,[41] the adoption of this emergency procedure indicating how confident he was. Such confidence would later surface in his advice to the Royalists at the strongholds of Worcester and Shrewsbury.[42]

The availability of waterpower at the Osney mill was another novelty that the urban powdermaker had to confront. The need to maintain an adequate flow of water proved to be a challenge here and at other mills. Cannon and other armaments were also being produced at such locations as Holywell and Folly Bridge. Hydraulic control conflicted with plans to extend the lines of fortification and to flood the meadows as a defensive measure. Nonetheless, the importance of maintaining supplies was recognized, and the military engineer Colonel Lloyd was made responsible for "drawing more [wa]ter to ye Powder Mills" by cutting new channels and improving old. Such measures helped ensure a continuous water supply.[43] Oxford Castle may have provided a more traditional location for powder making, but even here waterpower was probably harnessed from a source originally diverted to fill the moat. Near another diverted source of waterpower, a stone edge-runner still survives in a corner of Christ Church Memorial Garden.[44] Its provenance is unknown, although it may be a relic of the Royalist occupation.

Because of William Baber's success in Oxford, the King's Ordnance Commissioners confiscated the powder mills as a neglected source of profit. This action provoked great resentment. Prince Rupert's agent in Oxford reported that the men had refused to cooperate and that, on the morning the patent was eventually sealed in February 1644, "the Drying House and some Six Men Blew upp."[45] It remains unknown whether their anger led to carelessness and a neglect of the rules; it is unlikely, however, to have been sabotage, for Baber received £500 in compensation. He continued to work in Oxford, producing good quality saltpeter such as that delivered to the store on June 3, 1644 and described as "treble refined."[46] Shortly after this he returned to Bristol, a key Royalist stronghold from July 1643 to September 1645. Here Baber's powder-making skills and knowledge of the area would have helped fulfill orders from Oxford and other local garrisons.[47]

Figure 9.4
The recumbent trip hammers of a fulling mill. From Vittoria Zonca, *Novo Teatro di Machine et Edificii* (Padua, 1607).

Despite the adoption of emergency procedures during the Civil War, the demand for continental skills and supplies remained high. Among those particularly welcomed were the military engineers Bernard de Gomme and Bartholomew de la Roche, veterans of the Thirty Years' War in the Netherlands who accompanied Prince Rupert to England in 1642. De Gomme, a Walloon, gave advice on strengthening the fortifications of Royalist strongholds, especially Oxford since it was protected only by obsolete mediaeval walls. De la Roche's specialty lay in the production of explosive ordnance. He was the Master Fireworker of the Train and frequently attended the sieges conducted by the Royalists to supervise the handling of his creations. At Oxford he was based in some "little houses" beside Magdalen College. Here he received supplies of gunpowder and its separate ingredients (including "mealed brimstone") so that he could make the explosives for himself. He was also furnished with oil, spices, beeswax, and glue—as well as brass kettles and pots, tubs, and barrels—for what must have looked like a small laboratory.[48] Yet the production of powder in bulk remained inadequate. On May 8, 1643, only 57 barrels were counted in Oxford, and stocks were low at other garrisons. The situation was saved then, as on other occasions, by the purchase of supplies from the French and Dutch, negotiated by Queen Henrietta Maria with the aid of her jewels. Cargoes were shipped to the northeast coast, especially at Newcastle, and slowly transported south to Oxford, where the arrival of 150 or so barrels of powder could treble the supply in hand. On one occasion the Queen entered Oxford in triumph with a ton of much-needed brimstone in her train.[49] While noting Baber's ingenuity, we must also register the continuing dependence on foreign skills and supplies.

The Civil War ended with the defeat of the Royalists and the execution of the King in 1649. Political uncertainty and commercial rivalry provoked a series of naval wars with the Dutch in the 1650s, the 1660s, and the 1670s. The recurring demand for powder led to repeated shortages. Hence, in the second Dutch War of 1664–1667, the officers of the Board of Ordnance were ordered to "impresse soe many Mills for ye makeng of gunnpowder for his Matie [Majesty's] Service as they shall think fitt." As Baber displayed at Oxford, the requisitioning of mills led to an adaptation of existing techniques. Thus, at Waltham Abbey on the north side of London, there was a conversion of what was "heretofore an Oyle Mill . . . into two Powder Mills . . . with all necessary outhouses for grindinge boylinge corninge & drying of powder."[50] This is probably the first evidence of such a conversion, even though Sir James Hope, a Scottish lead-mine owner, had already remarked on the similarity between oil and gunpowder production. While visiting Zeeland in the Netherlands in 1646, Sir James recorded seeing two horse-powered "oyle milnes . . . lyke unto our ordinarie pouder milnes ... consisting of tuo stones turned upon edge."[51] The conversion would have posed little difficulty, for the stone edge runners used to "bray" the oil seeds (figure 9.5) were to become the standard equipment in powder mills (figure 9.6). Furthermore, the press used to squeeze the brayed oil seeds may have evolved into the gunpowder press that consolidated the incorporated powder.

The adoption of edge runners for this purpose appears innovative. Yet when we consider their earlier adoption in other industries (such as the crushing of dyewoods and grinding of snuff,[52] not to mention the ancient processing of olive oil), their transfer into gunpowder making seems commonplace. Nonetheless, once accepted, their advantages became noteworthy: incorporation by edge runners could be undertaken in fewer hours than that required by the long-established method of using stamps; and the powder produced was thought to be of a better quality because the edge runners not only crushed and compressed the mixture, as did the stamps, but they also ground and mixed the composition more thoroughly. The twisting and shearing motion of the large stones as they revolved in a small circle produced a more homogenized powder. Edge runners were also considered safer than stamps, which were more likely to cause explosions through overheating. The superiority of runners, from an English perspective, was established in 1772 when, by Act of Parliament (12 Geo.III c. 61), the use of stamps was permitted only in special circumstances such as the making of "fine Fowling powder."[53]

Both of the cases examined show the adoption of new procedures, not as the result of a search for better technology but as a means to meet the emergencies of war. The adaptation of the fulling mill did not survive the crisis, nor did most of the new mills set up in Oxford and other strongholds, which went out of service because of the fear of political instability or the choice of unprofitable locations.[54] Technologically advanced edge runners did survive, although the adoption of this innovation was impeded by mills ceasing to produce gunpowder. Thus, at Sewardstone near Waltham Abbey, an inventory, dated 1713, records both an old-fashioned stamping mill with wooden troughs and a newer edge-runner mill with a large bedstone and two large runners. The bed and runners represented a capital investment for the gunpowder mill; but, through a change of ownership, the use of this machinery for powder making was not pursued, and the works were put to other purposes such as snuff making.[55] The problem in general was the peacetime reduction of the military market; therefore, it was in the Bristol region that the first purpose-built edge-runner mills were to continue in operation. From the 1720s, these met the more constant demands of the commercial market as represented by shipping, trade, and mining.

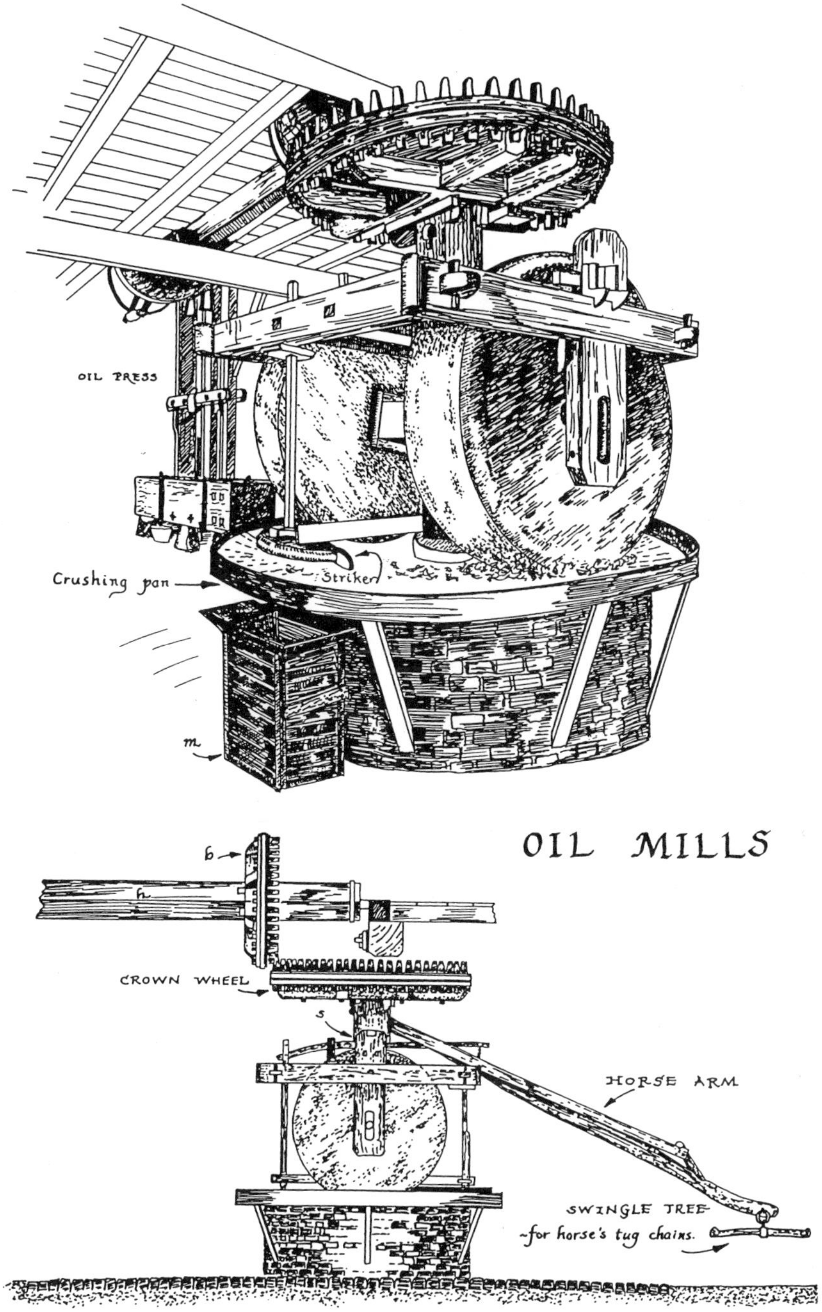

Figure 9.5
A Dutch oil mill. From John Vince, *Power before Steam* (1985). Reproduced by courtesy of the author.

Figure 9.6
The incorporation of gunpowder by water-powered edge runners, from a Notebook of 1798 attributed to John Ticking, master worker at the Royal Faversham Mills. With its possibly unworkable gearing this drawing may be a student copy. From Brayley Hodgetts, ed., *Rise and Progress*.

Provincial Developments after the Restoration

After the Restoration of the Monarchy in 1660, Charles II reestablished the Board of Ordnance, which resumed the practice of contracting with a select band of gunpowder makers who produced in the countryside yet remained accessible to London. The Board's survey in the late 1680s shows that about 60 percent of the productive capacity of these suppliers was absorbed by military commitments, suggesting that 40 percent was available for private trade.[56] There is little evidence of the extent to which this 40 percent was taken up and the influence this had on manufacturing processes. The only well-documented evidence derives from the Bristol region,[57] where records indicate that the mills catered so successfully to the private market that rivals appeared in the provinces only during the second half of the eighteenth century.[58]

When Baber resumed his career in Bristol, he probably returned to work at what became known as "Baber's Tower," which a plan of 1673 locates within the city proper.[59] His facilities were convenient to the saltpeter and brimstone warehouses and the magazine of the port. Yet with the increasing importance of waterpower, such urban locations lost favor. A new series of powder mills were built in the wooded valleys of the nearby countryside, especially at Woolley near Bath in the 1720s and at Littleton, south of Bristol, in the 1740s. Nonetheless, the move did not signify a withdrawal from the busy life of England's second city. Gunpowder making remained a port industry, relying on the import of saltpeter and sulfur as well as the transport—both overseas and along the coast—of the finished product. The partners running the businesses were local merchants and gentry. They had access to the capital funds and credit network of the city's traders, whose prosperity was due largely to the shifting balance of English trade and enterprise as its focus moved from Europe to the New World and Africa. Bristol and the other west coast seaports were well placed to take advantage of this opportunity. When the London-based Royal Africa Company's monopoly of the British slave trade ended in 1698, Bristol ships were quickly dispatched to "purchase" slaves with cargoes that included considerable quantities of gunpowder.[60] They also became a lifeline for the colonists of North America, and gunpowder figured prominently in the cargoes sent there as well. Merchant ships and those fitted out for privateering also required supplies. In addition, there was a growing demand for powder for blasting in lead and coal mining—both locally and in areas accessible to coastal shipping, especially Cornwall.

The association between trading networks and gunpowder manufacturers was close and profitable. Although the main mills at Woolley and Littleton were operated according to established procedures, the mill owners responded to the varied markets in an innovative way. This is shown by a memorandum of 1747 that was prepared for a new partner at the Woolley works. While outlining the standard procedures, it also reveals the powdermakers' willingness to deviate from the norm in order to fulfill the requirements of customers in mining and in various branches of trade and weaponry. The "best powder" was to be made from 64 or 70 pounds of saltpetre, 18 pounds of brimstone (sulfur), and 18 pounds of charcoal, with alder favored. The alternative weights of saltpeter indicate powders tailored to different markets. The 64-pound weight was probably Guinea powder (64% : 18% : 18%), which was suitable for the euphemistic "Africa trade" and mining. The 70-pound weight was likely to have been for Merchant powder (66% : 17% : 17%), supplied to ships plying their trade from Bristol.[61] Even in military circles there was little agreement on the optimal proportions. Based on his muzzle-velocity measurements with the ballistic pendulum, Benjamin Robins acknowledged in *New Principles of Gunnery* (1742) the need for a higher proportion of saltpeter to make a "stronger" mix (75% : 12.5% : 12.5%). His views are especially significant because of his devotion to "Philosophical Researches" and his status as a forerunner of the experimental physicists of the closing decades of the century. Critical of much of the "common powder" made in England, Robins concluded that "worst of all is the powder made for the African Trade, usually styled Guinea Powder: But these weaker Powders are not worth Examination, as there is no established Standard for their Composition."[62] His emphasis on established and testable standards was commendable, but it overlooked the realities of commerce. Different markets had their own distinctive needs; they required a flexibility that accords with Joseph Needham's view that "there cannot be any definitive theoretical set of proportions, only a certain range." If the Guinea powder had been stronger, it would have shattered the African firearms of inferior quality. A higher saltpeter content would also have been inappropriate for blasting powder, where the objective was to shake rather than disintegrate the rock. Indeed, the Board of Ordnance specification from the early 1780s of one formula for all uses (75% saltpeter : 10% sulfur : 15% charcoal) posed a decided disadvantage for military engineers. As late as the end of the Napoleonic Wars it was said that "some of the failures caused in our wars, in attempting to blow up works or demolish bridges, have been produced by the very excellence of the powder,—by its too great strength in short."[63]

Corning represents the second area in which the provincial powdermakers improved upon standard practice in order to meet their customers' needs. In the important corollary to the process of incorporation, the powdermaker forced the dampened powder through a parchment sieve to produce consolidated grains. This would prevent their separation into the original ingredients during transport. Such consolidated powder would also burn more rapidly and hence generate greater power than the uncorned Serpentine Powder. At the Woolley Mills, from at least the 1740s, corning evolved to the stage where grains of a specified size could be produced. For stock-taking, these were listed in an increasing order of fineness and value as "F," "FF," or "FFF," with a cheaper form designated as "F&M," which was possibly blasting and/or Guinea powder. This technique of graded corning did not engage the attention of the Board of Ordnance until the mid 1780s, perhaps because their expectations were more limited than those of private customers. Only one grain size was required for small arms and ordnance until Major William Congreve of the Royal Laboratory at Woolwich—possibly influenced by experiments such as those of John Ingenhousz published in 1779—introduced the distinction between large grains for cannon and small grains for muskets.[64]

By the middle of the eighteenth century, the provincial gunpowder makers' embrace of innovations extended to the use of cast iron in incorporating beds and runners, as well as in drying stoves. In the case of beds and runners, this may have been a response to the shortage of traditional stones: the fissure-free limestone from Namur in France, for example, or the basalt from Eifel in Germany, used also in oil mills.[65] The innovation may have owed more to the local merchant network, for all the cast-iron ware came from the famous works at Coalbrookdale, where close links were maintained. The founder, Abraham Darby, had previously worked in Bristol, and the purchases were made through his agent in the port, the Bristol merchant and banker Thomas Goldney.[66] At the longer-established Woolley mills, the beds and runners were purchased from 1759 to supplement the existing stones. By contrast, at Littleton they were purchased from 1749 as part of the initial capital investment of a new partnership on a new site. Comparative evidence is hard to come by, but the mills of the London area seemed reluctant to adopt this innovation. It was more than a decade after their introduction at Littleton that the Board of Ordnance equipped two new horse mills at the recently purchased site at Faversham with cast-iron beds and runners. They justified the decision because cast iron is "less expensive and not so liable to accidents as stones."[67] Beds and runners were also introduced at this time at Ewell in Surrey, where the partners were gunfounders[68] and therefore well placed to furnish the new equipment. Their recommendation as late as 1787 by George Napier suggests that the innovation diffused slowly.[69]

The pattern of deliveries from Coalbrookdale to Littleton in 1749 is revealing. Three cast-iron beds, each just over one ton in weight, were followed by two cast-iron runners, each over two tons, and two "Cast Iron Rings for Grinding Runners," each weighing around one ton. This suggests an initial

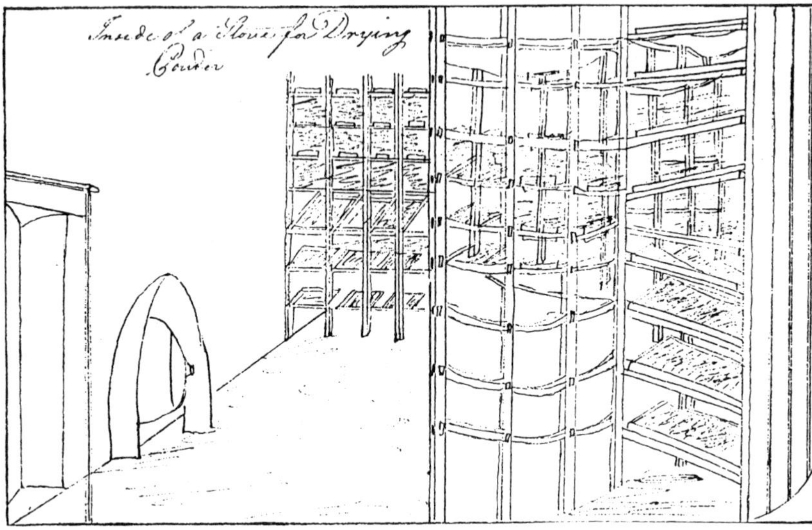

Figure 9.7
The interior of a gunpowder drying house. Powder was laid on racks in a room into which heat was conveyed through a cast-iron dome or cockle seen here on the left. Ticking's Notebook 1798. From Brayley Hodgetts, ed., *Rise and Progress*.

phase, in which each of the cast-iron beds may have been served differently—one by traditional stone edge runners, one by stones with cast-iron rings or tyres, and one by cast-iron runners. A year later, after the trial period, Coalbrookdale delivered the first of six complete sets of cast-iron units. The parts steadily increased in weight until by 1764 those purchased for Woolley included a bed of four tons and runners of five tons. At this time, Coalbrookdale also delivered cast-iron "cockles" to both the Woolley and the Littleton mills. These stoves were attached to but separated from the gunpowder-drying house and designed to convey heat through a cast-iron dome. Their main advantage was the superior control of smoke and sparks (figure 9.7). Until the present research, cast-iron "cockle" stoves were known only as a simple form of central heating, dating from the early nineteenth century.[70] Their use a half-century earlier highlights the innovative orientation of the provincial gunpowder makers.

The Board of Ordnance and the Royal Society: Art *and* Science?

The re-establishment of the Board of Ordnance in 1660 was mirrored by the founding of the Royal Society in the same year.[71] The former was concerned—

through its contractors—with arms procurement, and the latter—through its active members—with natural philosophy in general and both chemistry and mechanics in particular. It therefore seemed likely that the Board would embrace the art of military technology, while the Society would embody the science, especially of gunpowder. This is too simple a generalization, however. The Royal Society and the Board of Ordnance incorporated elements of both approaches, albeit for different ends.

In the early years of the Royal Society there was a considerable interest in explosives, perhaps reflecting past experiences and the concerns of the first president, Viscount Brouncker, who presented a paper on the recoil of guns. Robert Hooke wrote on schemes for determining the force of gunpowder while Robert Boyle discussed its expansion when fired.[72] The chemists likewise pursued the issue of saltpeter. A paper by Thomas Henshaw, which placed the subject in perspective, was followed by reports on how to increase its domestic production, along with reports on how it was generated in the "Mogol's Dominions." Some members even took the practical step of consulting associates of the East India Company on the subject.[73] The Fellows of the Royal Society had less interest in the technology of gunpowder making, although Henshaw and Prince Rupert made significant contributions. Henshaw's paper on "The History of Making Gunpowder" of 1662 showed the same regard for those rules epitomizing the "Art and Mystery" of a trade. In particular, Henshaw's understanding of the whole "Secret of the Art" was evident in his advice: not only must the ingredients be assembled in their correct proportions, but there must be an "exact mixture of them, that in every the least part of *Powder* may be found all the Materials in their just proportion."[74] This advice celebrates the high standard of flawless workmanship required to achieve this aim as it downplays the experimental approach.

The paper Prince Rupert sent to the Society in January 1662 (translated and made available some eighteen months later) made a significant claim about a new kind of powder whose strength exceeded the best in England. The instructions nevertheless called for the same careful workmanship as that described by Henshaw. The details of manufacture suggest it was written by an experienced powdermaker. The recommendation, for example, that a clean brush dipped in pure water should be used to sprinkle the mix to avoid making it too wet, indicates a hard lesson learned. Similarly, practical experience is implicit in the advice that when moist powder sticks in the holes of the sieve, a "little twig" may loosen it. Given that Prince Rupert's paper was written in "High Dutch," its contributor must have come from the Continent, thus revealing how much the Prince was maintaining his contacts there.[75] The gunpowder did well in a trial reported in November 1663. A charge of "common

English powder" passed through three boards and stuck in the fourth, while the Prince's powder passed through four and stuck in the fifth.[76] The Board of Ordnance faced more practical problems, however, and may not have always welcomed the advice of Royal Society luminaries. Only two months earlier, in September 1663, the Privy Council had called for a survey of war stores to be made to resolve the post-Restoration confusion. They sought to establish "what quantities . . . are in our stores and what further is yet wanting for the full completing those magazines royal."[77]

Prince Rupert's concern with the manufacture and performance of powder surfaced again in 1681 during an episode that linked him to the Royal Society, the Board of Ordnance, and the major powder producer in England, Sir Polycarpus Wharton. The Board, anxious to send an expert to Germany to study the techniques of gunpowder making practiced there, sought the advice of Sir Polycarpus, who would later claim that the Prince discussed the matter with him. It is not known if the visit to Germany was made, however. Sir Polycarpus later claimed to have erected mills near Windsor that were "much differing from the common Sort."[78] These may have been edge-runner mills built under continental influence. The expertise of the Germans and especially the Dutch in this matter is well attested,[79] although the influence of the nearby oil mills converted in the 1660s (mentioned above) should not be discounted.

A note in the Royal Society papers of October 1681 suggests a different tack. It shows that Prince Rupert was then seeking a patent for a person who had discovered a means to produce gunpowder ten times stronger than the common powder. The "strength" of powder was traditionally associated with its saltpeter content. Even after making an allowance for hyperbole, this claim suggests an increase in the saltpeter ratio rather than a change in the manufacturing process.[80] The latter continued to interest Sir Polycarpus Wharton, but his experiments in gunpowder manufacturing were cut short in 1708 by the financial problems that forced him to sell his powder mills at Sewardstone in Essex. An inventory of that year records that in the "great mill" there was then "a large pair of stones bedstonesweep all fixed and in order." A second inventory, of 1713 (mentioned above), provides even clearer evidence of an edge-runner mill, listing "One large bedstone, two large runners, shaft, spindle, wheeles etc" in the "new mill."[81] With a further change of ownership in that year, and the conclusion of the War of the Spanish Succession, the powder works were put to other uses, including the working of copper and snuff.

Apart from this confusing example of technological initiative, the long period of warfare and political and commercial rivalry from 1689 to 1713 was distinguished more by Great Britain's struggle to maintain supplies of gunpowder than by experimental improvements of its performance. These pressures were revealed in April 1702 by the storekeeper who described arriving at the Tower at 8 o'clock and working sometimes until the evening, with the prospect of working on Sundays "till this business is over."[82] The "business" was over in 1713 with the Peace of Utrecht, and the tempo of work on supplies then flagged for more than a quarter of a century. It was revived by the resurgence of military activity in the middle decades of the eighteenth century, when the war between Britain and Spain from 1739 broadened into the conflict over the Austrian Succession in the 1740s and culminated in the struggle between Britain and France during the Seven Years' War from 1756 to 1763. Once more a great demand for military supplies existed. The challenge of procuring gunpowder of both quality and quantity was not addressed seriously, however, until Charles Frederick (1709–1785) began his rise to positions of influence. He has been a neglected figure in the history of gunpowder procurement. His contribution will now be reviewed.

Charles Frederick (figure 9.8) was a Board of Ordnance man par excellence. His early life seems incongruous with this eventual role. He was a scholarly man with an abiding interest in Roman antiquities, coins, and medals; however, it was these concerns that brought him into contact with those who recognized his talents and advanced his career. He was elected a Fellow of the Society of Antiquaries in 1732 and its Director in 1736, taking charge of its drawings, engravings, and books. One of his contemporaries in the Society was William Bogdani, Clerk of the Ordnance Office in the Tower. Martin Folkes was another, who was appointed the first Chief Master of the Royal Military Academy at Woolwich when it was set up, by a Royal Warrant dated April 30, 1741, as a "School for Practitioner Engineers etc." Folkes then became President of the Royal Society from November 30, 1741 until 1752, and President of the Society of Antiquaries from 1750 until his death in June 1754. An even more significant connection for Charles Frederick may have been the Duke of Montagu, also a Fellow of the Society of Antiquaries. The Duke served as Master General of the Board of Ordnance for most of the 1740s until his death in 1749, and it may have been to him that Charles Frederick owed his preferment. A Member of Parliament since 1741, Frederick was appointed Comptroller of the Laboratory at Woolwich in 1746 and Surveyor General of the Ordnance in 1750. These offices dovetailed to give him responsibility for the testing and quality control of all war goods. He held both offices until 1782 and undoubtedly stayed too long. But this does not detract from his noteworthy responsibility for the materiel of war during thirty-six of the most significant military years of the eighteenth century.[83]

The Laboratory at Woolwich, to which Frederick was appointed, had its origins in Henry VIII's tilt-yard at Greenwich. When that was demolished in

Figure 9.8
Portrait of Charles (later Sir Charles) Frederick (1709–1785), by Andrea Casali (1700–1784). Charles Frederick went on the Grand Tour with his brother, 1737–1739, and this portrait was painted in 1738 while he was in Rome. Reproduced by courtesy of the Ashmolean Museum, Oxford.

1650, marking the obsolescence of armor, its materials were used to repair the Great Barn where "fireworks" were made. Because of further changes at Greenwich, the materials were again transferred to Woolwich, where they were erected in 1696 as "a new Laboratory for fixing shells and Carcasses."[84] This remained active until 1716. During the generally peaceful years of Sir Robert Walpole's subsequent administration, its staff was greatly reduced and its stocks utilized only for exercise and proof. The war with Spain in 1739 brought about a token revival until February 1746 when steps were taken to establish the Laboratory on a proper footing. By an Order in Council the "defective state of the Laboratory at Woolwich" was to be redressed and a Comptroller and staff appointed so that "the Art of making Fireworks for real use as well as for Triumph may be again recovered."[85] From this low point Frederick built up a staff consisting of a Chief Firemaster, a firemaster's mate, a clerk, and workmen, especially *matrosses,* who produced fireworks and cartridges and charged bombs, carcasses, and grenadoes for the tests conducted in the shot-yard at Woolwich.[86]

Since 1683 the Board of Ordnance had been operating under a new Royal Warrant. Its Master General and Lieutenant General held military responsibilities, but its Surveyor General was a civil officer in charge of all stock and the proof of all war stores. If serviceable, such materiel was to carry the government mark. Like the Clerk of the Ordnance, who was responsible to Parliament for expenditure, the Surveyor General was responsible for quality. The duties were onerous, and about 1762 two committees were set up to help test new equipment and to evaluate new proposals. Sir Charles (knighted in 1761) nevertheless remained "hands on" in his approach.[87] His expertise is undeniable: surviving accounts by those who knew him and correspondence by those who sought his advice together reveal his expertise with gunpowder making.

Frederick's commitment to powder making was disclosed in his first public test: the firework display of great splendor to celebrate the conclusion of the War of Austrian Succession in 1748. Having been appointed to make fireworks for use in war rather than triumph, he was now challenged to revert to the latter. He was assisted by Captain Thomas Desaguliers, his Chief Firemaster and son of the famous natural philosopher, but engaged himself fully in the preparations. The display was to be held in Green Park, and Frederick's sister-in-law reported to her husband, Admiral Boscawen, that he was busy in the office built for him there from eight in the morning until four in the afternoon. Frederick may have tried to discourage visitors, for Horace Walpole, prolific writer and member of Parliament, found him "talking to himself . . . on the superiority that his fireworks will have." Walpole's description of him as "bronzed over with a patina of gunpowder" confirms somewhat alarmingly his intense involvement in the business of powder making.[88] In the event, the display over-reached itself, and an unplanned explosion along a trail of "corned powder" came close to causing disaster.[89] Nevertheless, the bravura performance confirmed the revival in the fortunes of the Royal Laboratory, although these were to be severely tested in the mid-century struggle for European, colonial, and naval power.

The second public test came with the pressing need for supplies in the Seven Years' War of 1756 to 1763. Frederick's commitment to powder making had already been demonstrated in a lively fashion, but what was now to be seen was his ability to give advice and encouragement. The procurement of gunpowder was plagued by the continuing dependence on private contractors. Although the system allowed the Board of Ordnance to ride out periods of peace unhampered by a costly war machine, it left Britain gravely disadvantaged in time of war. The problem produced some somber comments from Sir John (later Lord) Ligonier, the distinguished old soldier who was Lieutenant General of the Board of Ordnance from 1748 to 1757 and its Master General from 1759 to 1763. With the French and Indian conflict in North America already underway, he wrote to Charles Frederick in October 1755: "The Powder Runs in my head and [I] think all the Contractors ought to be summon'd and talked to, Pressed to work for the king, for I fear our Powder will go abroad if a Better Price is offered." Shortly after, noting the reduction of the barrels in store from 11,000 to 7,000, he worried "what a fatal thing this might prove to be." Various measures, such as penalty clauses and restrictions on the sale of powder overseas, were introduced, but Ligonier remained fearful that "all . . . is Gone for nothing Without this material."[90]

From the perspective of the private provincial powdermakers, the restrictions on sales had a particularly unwelcome effect. Licenses had to be obtained from the Board if they wished to send their powder coastwise or overseas, and so, for the first time since before the Civil War, they were brought under the scrutiny of the metropolis. However, through these negotiations in the early 1760s, the Bristol partners came to realize that with their commercial outlets blocked by war, the Board of Ordnance might prove to be a useful alternative market. They offered to produce 500 barrels in three months, but this was not taken up because the samples sent to London failed to meet the standards of proof set by the Board. I have discussed this episode more fully elsewhere,[91] and have concluded that the reasons for rejection were social rather than technical. Bristol was far from the Board's sphere of influence and its coterie of established suppliers. The provincial powdermakers were disappointed and puzzled, but their commercial loss is our historical gain: their dealings with Sir Charles drew from him advice on the technology of gunpowder making, which is available from no other source. This account of best practice in the middle decades of the eighteenth century does not come from the institutional records of the Board of Ordnance. It has survived in personal correspondence, among the business papers of a partner in a provincial powder works. The rare survival of this well-informed advice is important, both for the information it conveys and the evidence it provides of the professional approach of the Surveyor General, Sir Charles Frederick.[92]

In December 1761 the Bristol partners sent samples of gunpowder to the Board. They feared these did not match the strength of those sent from London, despite the trials undertaken and the fact that these were the strongest they had ever made. They claimed that the powder had gone through "the regular process of manufacturing, it have been dried in the Stove its proper time, it have been ground the full number of hours, and it has its full quantity of Petre . . . we can make it no better." Their only doubts lay in the weight of their edge runners, which at two and a half to three tons were thought to be lighter than those of five to six tons used by the powdermakers of London.[93] They learned in ten days that the trials had gone badly but that Sir Charles was willing to discuss the matter further. In the meantime, his advice was conveyed in a detailed letter that combined observations on best practice with an encouragement to experiment. He wanted to know the composition of the failed powder, including the exact quantity of each ingredient. He emphasized the importance of the quality and purity of each ingredient. In the case of the saltpeter, the "Defect is most likely owing to want of refining the petre well of its Sea Salts [perhaps acquired in the hold of an East Indiaman?] and other Dross," but it must not then be ground too wet or too long, or dried too long in the stove. The sulfur too must be well refined. The best charcoal was of hazel, then alder. It must not be made from recently cut green wood and must not be kept too long or it is "good for nothing." There must be no "undue proportion of the Ingredients"; in particular, "Too much Petre spoils the powder." He paid special attention to the process of incorporation, advising that a "want of weight" in their runners was unlikely because those at Faversham scarcely exceeded two tons. These had, however, been "*turned Smooth* (in a kind of Turning loom)" after their delivery by the makers. He thought the smoothness of runners and bed to be important. The grinding of the charge should be without interruption for "the least stoppage of the runners spoils the powder." The Board approved of the "Grain and Cleanness of the Samples" sent by the Bristol partners, an indication that the powder had been corned efficiently to produce discrete grains and then sieved thoroughly to remove the dust. It is notable that in his survey of good practice, Sir Charles makes no reference to the use of a press. This may have been because the need for extra pressure was obviated by the weight of the runners. The cast-iron ones of five tons were the same weight as those at Augusta, Georgia, 100 years later. There the compression was so effective the press was used only in making fine ground powder.[94]

Sir Charles' understanding of current practice, as shown by this advice, must have been enhanced by his experience at the Faversham gunpowder mills. In an attempt to meet wartime shortages, these mills were purchased by the Crown in 1759 and administered by the Board. The venture was not a resounding success because much of the effort went into the difficult task of recycling the damp and damaged powder returned in ships' holds after a time at sea. At least it provided firsthand experience of the problems encountered. This can be seen in the reference by Sir Charles to the looms (or lathes) for turning the iron runners. Some of these had recently been delivered at Faversham, as the government suppliers began to catch up with the private provincial partnerships.[95] Having established the rules and procedures, Sir Charles went on to encourage the powdermakers to experiment. They should "keep *one mill* at work trying Experiments (for a little while) on different Compositions different Times of grinding etcr," by which they might achieve "the right method as others have done." The partners responded by sending samples of the ingredients used, revealing the composition of the materials put under the runners, and experimenting with the time spent on incorporation. Identical batches were ground for five and six hours, respectively. These were sent to London distinguished only by symbols so that the Board could not automatically opt for the longer incorporation. But success eluded them. They returned to such lucrative private trade as wartime conditions allowed, for as one partner said, "our Privateers like it and so do the Merchants."[96]

This episode has a significance that goes beyond the discouraging experience of the provincial powdermakers. By setting his advice on good practice within the context of an experimental approach, Sir Charles demonstrated how far gunpowder manufacture in England had moved towards embracing both art and science. By the early 1780s, however, various influences were coming together that required changes to be made by the Board of Ordnance and the Royal Laboratory. There was dissatisfaction with military supplies in the War of American Independence and a growing desire for financial and administrative reforms at home. For example, Edmund Burke's Bill for Economical Reform, passed by Parliament in 1782, aimed at controlling "subordinate treasuries" such as that operated by the Board. Plans for a new Royal Warrant were put in hand, and Sir Charles retired from both his offices in 1782.[97] His effective successor was William Congreve. As Comptroller of the Royal Laboratory from 1789 he had responsibility for the Faversham and Waltham Abbey mills,

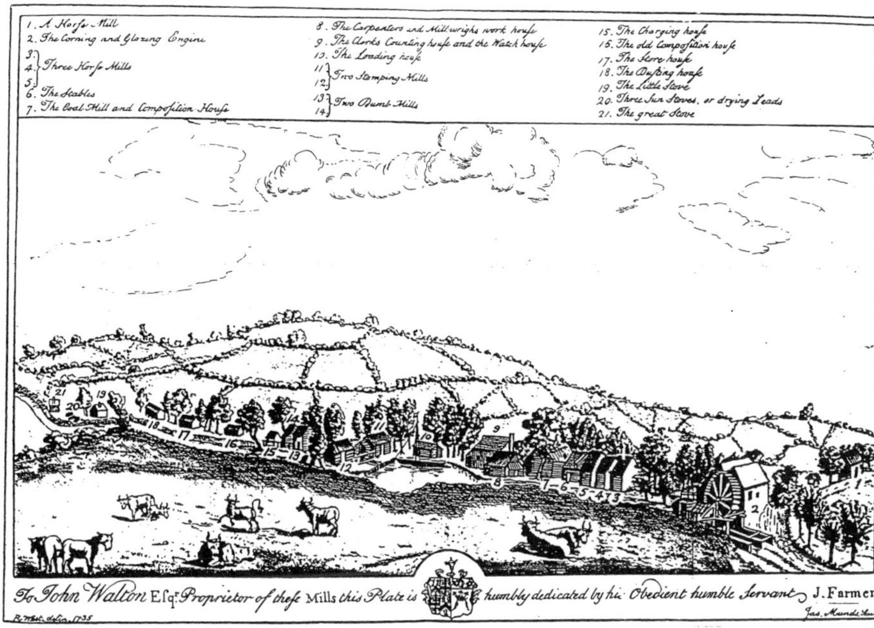

Figure 9.9
Engraving of the Waltham Abbey Powder Works by John Farmer, *History of Essex* (1735). See Buchanan (1998–99) for a study of "The Old Establishment" shown here. The purchase of this rural powder mill by the Crown in 1787, and its development by the Board of Ordnance, marked the beginning of a new phase in the manufacture of explosives in Britain.

the latter (figure 9.9) having been purchased by the Crown in 1787. He was made directly answerable to the Master General for the manufacture and testing of supplies.[98] These changes show the Board recognized that the expertise which had sustained the gunpowder industry thus far, even though based on skills and experimentation, was no longer adequate, and that the insights being developed by scientific talents such as Congreve himself, along with Charles Hutton and John Ingenhousz, must inform the processes of production.[99]

But gunpowder is a complex subject, and as this fourth phase was concluding, it began to assume new significance at political, intellectual, and technological levels. It had long since lost its early position of esteem as an agent of progress (with the printing press and the compass); yet it was still respected as being, in the grand words of the political economist Adam Smith in 1776, "favourable both to the permanency and to the extension of civilization."[100] Yet gunpowder was by then already falling from grace through its association with revolution in America and France. In 1787 Joseph Priestley, the religious and political nonconformist and scientific experimenter, confirmed this identification in what became known as his "Gunpowder Sermon"; here he referred to "laying gunpowder, grain by grain under the old building of error and superstition, which a single spark may hereafter inflame." In 1794, he sought refuge in the United States from the political reaction to his views, yet the myth still had some time to run. Fears were not calmed when, in 1796, Edmund Burke made inflammatory references to the likely destruction of buildings of the old regime by French chemists attempting to extract an "insurrectionary nitre" from their mortar.[101] Implicitly Burke criticized science as much as gunpowder's role in revolution. In both respects they return us to our subject: gunpowder and the vexed question of its vital ingredient, saltpeter. Although a sweeping generalization, the greater certainty of supplies from India allowed the British to apply their growing scientific understanding to matters of technique in both the purification of ingredients and the processes of manufacture. In turn, the need for the French to develop a "home-grown" supply of saltpeter stimulated a scientific understanding of the subject. This related especially to its chemistry.[102]

Conclusion

With its emphasis on rules and a perfection of workmanship, the concept of "Art and Mystery" mattered more in the dangerous and unpredictable industry of gunpowder making than in most, and it prevailed long into the eighteenth century. But this study has also shown that there can be traced from at least the 1640s a growth of observation and experimentation that heralded a systematic if not scientific approach. Under the stimulus of national need and private trade, certain innovations were adopted based on expediency or the observation of procedures in similar industries at home or abroad. Examples include the conversion of fulling and oil mills into gunpowder mills, the introduction of beds of stone and then iron to edge-runner mills; and the development of different formulae and grain sizes to satisfy different markets. The members of the Royal Society tended to be less interested in the technology of gunpowder making than in the chemistry of its constituents and the nature of explosions. Nonetheless, Henshaw's understanding of the importance of incorporation and Prince Rupert's search for a stronger powder were both revealing. So were the links made in the early 1680s between the Royal Society, the Board of Ordnance, and government powder suppliers, since these may have led to the introduction of Dutch-style edge-runner mills into the country. Official interest in gunpowder had flagged in the peaceful decades of the 1720s and 1730s, making the work of the private powdermakers all the more important; yet, during the 1740s, it revived with the resumption of military activity and the appointment of Charles Frederick to positions of importance. Contemporaries respected him because he had "studied his Art more than any Man in England, and made more Experiments."[103] It was the combination of knowing the rules and being willing to experiment that made Sir Charles' tenure at the Board of Ordnance and Royal Laboratory so significant. This marked an important but overlooked stage in the growing trend towards establishing a consciously scientific approach to gunpowder making. The systematic empirical approach pioneered by Sir Charles was to be developed by his successors, especially William (later Sir William) Congreve senior. In the years before the last quarter of the eighteenth century, the "Art and Mystery of Making Gunpowder" thus evolved from a craft-based practice to a process based more securely on scientific methodology.

Notes

1. State Papers (SP), referred to in this and following notes, have been calendared in the volume edited by Brayley Hodgetts (BH, 1909). In 1457 John Judd was made Master of the Ordnance for life, and in 1461 a "Powderhous" in the Tower of London was first mentioned. Patent Roll, 1 Edward IV, pt. i, mem. 5. BH, pp. 182–183. At Portchester there had been defenses against invasion from the sea since Roman times. SP Henry VIII, Exchequer T. R. Misc. Books, 215, pp. 7–577 (under date). BH, p. 184.

2. SP Henry VIII, Vesp. C. i. 69. BH, p. 84.

3. SP Henry VIII, Galba B. x. 74. BH, p. 187.

4. SP Henry VIII, sec. 192, ff. 18, 99. The following were dispatched by ship from London— "2½ lasts of fine corne powder, 4 lasts of coarse corne powder and 23½ lasts of serpentyn powder." BH, pp. 198–199. Note: 1 last = 24 cwt. or 24 barrels of gunpowder, each barrel normally holding 100 lb. (Zupko 1968, p. 96; Roy 1964/1975, p. 62). A "rundlett" was a cask of varying capacity.

5. SP Henry VIII, Accounts etc., Exchequer, Q. R., 206/10. BH, p. 189; Ffoulkes 1937, p. 58.

6. SP Henry VIII, sec. 150, pp. 111–112; Add. MSS. 32,655, f. 129. Hamilton Papers II, No. 298; Ireland, vol. ii, No. 19; SP Henry VIII, Add. MSS. 32,646, f. 167. Hamilton Papers, No. 70. BH, pp. 188, 193, 190, 189.

7. Rich and Wilson 1967, pp. 167–181, 293.

8. For a recent survey, see Buchanan 1995–96, pp. 126–127.

9. Stewart 1996, pp. 6–19. This volume provides important information about the Ordnance Office in the years 1585–1625. It includes a chapter on "Gunpowder and saltpeter," but as the subtitle of the book indicates, the focus is on the bureaucracy rather than the technology of the subject.

10. SP Henry VIII, sec. 10, p. 154 (1515). BH, p. 185.

11. SP Henry VIII, Privy Seal (1531). BH, p. 185.

12. Rhys Jenkins Collection, Science Museum London, Box 46, Folder 3, item 105, Saltpetre 1544, Letters of Denization and Acts of Naturalization for Aliens in England. Martyn Gruffill and Nicholas Lamay were Robert Spencer's workers; Peter Grynty, saltpeter maker, was from the dominion of the Emperor; Nicholas Lyon, saltpeter maker, was born in Normandy. Sir John Bowyer had five foreigners in his saltpeter works, with whose assistance in 1544 "forty thousand [cwt?] weight of saltpetre was made."

13. See Williams 1975 for Honrick's recipe and his own experiments.

14. Thirsk 1977, pp. 43–59; Lansd. MSS. No. 14, Art. 13. BH, p. 213.

15. Thirsk 1977, p. 47.

16. Court of Requests, Proc. Phil. and Mary, Bundle 24, No. 119. BH, p. 208.

17. Reynolds 1983, pp. 66–67.

18. Augmentation Office, Particulars for Leases (Surrey), Roll 139, No. 23. BH, pp. 211–212.

19. S. P. Dom. Elizabeth, Addenda, vol. xxiv, No. 50. BH, pp. 213–214.

20. This method of teaching was portrayed in *Das Feuerwerkbuch* of the early fifteenth century. See Kramer and Barter Bailey in Buchanan 1996.

21. See Stewart 1996, p. 8, especially details in the footnotes.

22. Patent Roll, 31 Elizabeth, pt. 8, mem. 10 (25). BH, pp. 218–219; Stewart 1996, pp. 86–87.

23. For a survey of the Surrey mills, see Crocker and Crocker 1990.

24. SP Dom. Charles I, vol. cxlii, No. 32. BH, p. 256.

25. SP Dom. Charles I, vol. ccxcii, f. 222; vol. cclxxviii, No. 4. BH, pp. 272–273. John Giffard complained in 1634 that people carried off "the earth from their pigeon-houses to manure the land," and also refused to carry coal to his boiling-house.

26. SP Dom. Charles I, vol. cclxiii, No. 1; Rymer's "Foedera," xviii, p. 813. BH, pp. 269, 242.

27. For more details and comparable archaeological evidence, see Buchanan 1995–96, pp. 130–132. In 1577, Cornelius (de Vos?) offered to make saltpetre in the New Forest, Hampshire, Acts of Privy Council, 1577, p. 142. BH, p. 214.

28. Zins 1972; Willan 1968; Chaudhuri 1965. The reliance on supplies from India was such that premiums proposed by the Society of Arts during the Seven Years' War (1756–1763) to revive the domestic saltpetre industry, were never claimed (Partington 1960, p. 319).

29. See in particular, SP Dom. Charles I, vol. ccxxviii, f. 94a (1633), for the making of saltpetre "without authority." BH, p. 268.

30. Some of the ideas in this section were developed for presentation at the 26th Symposium of the International Committee for the History of Technology (ICOHTEC) in Belfort, France (1999), in the session organized by Patrice Bret.

31. The Royalist Ordnance Papers (1642–1646), referred to in this and following notes, were edited by Ian Roy and will henceforth be cited as "Roy." The papers are divided into Receipts; Issues and Inventories; and Correspondence. They form a valuable insight into the Royalist organization during the Civil War. See the reference to the Petition to the King, 15 September 1642, p. 7.

32. Documents C: Correspondence, Letter from the King to Sir John Heydon, Lieutenant General of the Ordnance, November 17, 1642. With reference to the powder mills at Chilworth in Surrey, "Our will and Command therefore is, That you forthwth take effectuall Order for ye absolute demolishing and destroying of ye said Milne, by letting out ye Water out of ye Ponds or otherwise; for as there may bee noe more use made of them." Roy, p. 359. On Parliamentary influence, see Lewis 1976, pp. 94–95.

33. This account is based largely on Evans 1992.

34. For general surveys, see Roy 1964/1975; Edwards 1995.

35. SP Dom. Charles I, vol. cc, No. 26; vol. cciv, No. 9. BH pp. 266–267. See Buchanan 1995–96, pp. 128–130.

36. Documents A: Receipts, January 8, 1642/3. Roy, p. 64.

37. Documents A: Receipts. Powder was, for example, received from University College on April 21, 1643, and from St. Mary's Church on February 16, 1644 (Roy, pp. 76, 123). For repairs to the powder house, see Roy, p. 121.

38. Wood (volume I, 1632–63, 1891), p. 74. Roy (1964/1975, Introduction, p. 29) notes that only two powder mills were reported in Parliamentary records at the surrender of Oxford, but suggests that others may have been put out of action before that point was reached.

39. Foreman 1993, p. 64.

40. Biringuccio 1540, 1966 issue, pp. 414–415.

41. SP Dom. Charles II, volume xxix, No. 76; volume ccxxxii, No. 193. BH, pp. 297, 298.

42. Roy 1964/1975, Introduction, p. 37.

43. Documents B: Issues and Inventories, Roy, pp. 187, 469. Charles Lloyd (c. 1602–1661) was a military engineer in the service of the Dutch before becoming Engineer in Chief and Quartermaster General to Charles I.

44. For a general survey of the Oxford mills, see Foreman 1993, especially 76–79 and illustrations pp. 8–9. For the castle, see Hassall 1976.

45. Roy 1964/1975, Introduction, p. 30. Trevor to Prince Rupert, 29 February 1643–44. The incident was also reported by Parliamentarian spies.

46. Documents A: Receipts, Roy, p. 137.

47. Roy 1964/1975, Introduction, pp. 37–46. Bristol was also of great importance as a Royalist port, trading in the means of war. See Edwards 1995, pp. 123–124.

48. Documents B: Issues and Inventories, Roy, pp. 198, 427. Bernard de Gomme (1620–1685) was a military engineer who served in the Low Countries until 1642. For the delivery of "all things necessary for his fierworks" to "Monsr: La Roche," see for example the issue dated 20 March 1643, Documents B: Issues and Inventories, Roy, pp. 207–208. Although his work was held in high regard, it was expensive with regard to materials, and in early 1644 even Prince Rupert was kept waiting for the arrival of explosives from Oxford. After the fall of Oxford he left England in the Prince's entourage. See also Roy's Introduction, pp. 32–33.

49. Documents B: Issues and Inventories, Review of Stores, 8 May 1643, Roy, p. 226; and Documents A: Receipts, 19 July 1643, "Brimstone—1 tonne p estimacon in 6 barrells 2 rundletts," Roy, pp. 107–108.

50. For the document quoted, see Fairclough 1985, p. 14. For an account of the Waltham Abbey Powder Mills before their purchase by the Crown in 1787, see Buchanan 1998–1999.

51. Hope 1646/1958, pp. 156–158. Sir James recorded the industrial processes observed on his journey—gunfounding and charcoal making in Kent, sulfur extraction near Liege. In Amsterdam he renewed his acquaintance with the Dutch chemist Frans Rooy, who had made saltpetre in Edinburgh.

52. Reynolds 1983, pp. 70–75.

53. For a fuller review of the introduction of edge runners, see Buchanan 1995–96, pp. 144–146. It is suggested that they may have been introduced into the industry by stages, first for a preliminary mixing of the ingredients, and later for their full incorporation. For comments on the technical advantages of edge runners, see E. F. Clark, "Written Contribution," in Buchanan 1995–96, pp. 157–158. For the continuing use of stamps in France and the United States, see Patrice Bret and Robert Howard in Buchanan ed. 1996.

54. For example, Rimington (1966) describes the chance discovery of a rural and probably short-lived Civil War powder mill, revealed during the excavation of a country manor house.

55. For the inventory of 1713, see Crocker and Fairclough 1998, pp. 33–34. For the conversion of the powder mill to other uses by at least 1715, see Fairclough 1996, pp. 132–133. It was probably used for copper working, and by 1740 there were mills on the site producing smalt (pigments for glass), rice, and snuff.

56. This calculation is based on several documents, especially PRO, WO49/220, Survey of 1687. See Buchanan 2002.

57. Brenda J. Buchanan, Capital Investment in a Regional Economy: Some Aspects of the Sources and Employment of Capital in North Somerset, 1750–1830, Ph.D. thesis, University of London, 1992. For a study of the relationship between the documentary evidence (Buchanan) and physical remains (Tucker), relating to the Woolley Gunpowder Mills in particular, see Buchanan and Tucker 1981.

58. It seems that no other provincial mills were established until that at Thelwall in Cheshire (1758), some 100 miles distant from Liverpool but within its hinterland; and those in the mining areas of south Cumbria (from 1764), Derbyshire (1801), and Cornwall (from 1809).

59. See Jacob Millerd's Plan of Bristol (1671, updated 1673), reproduced in Buchanan 1995–96, pp. 132–133.

60. The author's paper on "The Africa Trade and the Bristol Gunpowder Industry" has been published in the *Transactions of the Bristol and Gloucestershire Archaeological Society* (2001). It was presented at a conference on "The Atlantic Slave Trade and Provincial Britain" held in Bristol, April 1999.

61. Somerset Record Office (SRO), Strachey Papers, DD/SH, Box 27, Memorandum of 1747. See Buchanan and Tucker 1981; Buchanan 1995–96; Buchanan 1996. The proportions given here are in the order advised by Needham (1986, p. 109) as being "the normal usage of explosives chemists." The order is saltpetre, sulfur, carbon.

62. Robins 1742, p. 60.

63. Needham 1986, p. 111; *Encyclopaedia* (c. 1812–1814), p. 349, in the author's possession.

64. For evidence on the Bristol region, see Buchanan 1995–96, pp. 151–52. For Congreve, see West 1991, p. 184 and an unpublished paper by Cocroft.

65. See Buchanan 1995–96, pp. 144–146.

66. The Goldneys were prominent Bristol Quakers, merchants and bankers, with shares in the Coalbrookdale works. Thomas Goldney (1694–1768) kept a record of his business and private dealings from 1742–1768, now in the Wiltshire Record Office (WRO), Goldney Account Book, ref. 473/295.

67. West 1991, p. 156.

68. Tomlinson 1976, p. 398; West 1991, pp. 202–203.

69. Napier (1787, p. 108) recommended that the iron cylinders should be grooved to allow a better penetration of the paste for a greater intermixture of the parts. Napier was briefly Sir Charles Frederick's successor at the Royal Laboratory.

70. See the "Written Contribution" by Richard Hills to the paper by Buchanan 1995–96, p. 159.

71. Sprat 1667; Birch 1756–57; Hall 1992.

72. Brouncker, "Experiments on the Recoiling of Guns," 2 Jan. 1661 and 15 Jan. 1662, Birch 1756–57, volume I, pp. 8, 20, 74; published in Sprat 1667, pp. 233–239. Hooke, "Scheme for Determining the Force of Gunpowder by Weight," 9 Sept. 1663, Birch 1756–57, volume I, pp. 302–303, at which meeting Hooke also produced "a microscopical observation of the several parts of a fly." Boyle, "What is really the expansion of gunpowder when fired," 27 July 1664, a subject returned to in 1667 with experiments on "Bending a Spring by Force of Gunpowder," Birch 1756–57, volume I, p. 455; volume II, p. 142.

73. Henshaw, "The Making of Salt-peter," published in Sprat 1667, pp. 260–276. Schroter, 1663, in Birch 1756–57, volume I, pp. 173–174, in which sheep dung and urine was preferred. "Of the way, used in the Mogul's Dominions, to make Saltpetre," *Philosophical Transactions* (1665–66), pp. 103–104. MSS "Journal of the Royal Society," volume II, 26 Oct. 1664. If answers could not be given by East India Company men in London, their factors overseas would be consulted.

74. Henshaw, "Making Gunpowder," published in Sprat 1667, p. 277.

75. The translation of the paper sent by Prince Rupert was presented by Sir Robert Moray on 22 July 1663, with the observation that the powder was some ten times stronger than the best English powder, Birch 1756–57, volume I, pp. 281–285.

76. Report by Sir Paul Neile on the trials, 16 Nov. 1663, Birch 1756–57, volume I, p. 335.

77. Tomlinson 1979, pp. 136–137. This study of the administration of the Board of Ordnance under the later Stuarts offers little information on gunpowder making, but it does provide a valuable survey of the storage and supply of Ordnance stores, and examines in particular the efficiency of the Office.

78. Crocker and Fairclough 1998, p. 27.

79. Mussmann in Buchanan 1996, pp. 329–350.

80. Report of Prince Rupert seeking patent for stronger gunpowder, 26 Oct. 1681, Birch 1756–57, volume IV, p. 99. The repetition of the stylized phrase "ten times stronger" may indicate aspiration rather than achievement, but it also suggests that the question of the "strength" of gunpowder deserves further study, especially in relation to the ingredients used and the testing methodology of the product.

81. Crocker and Fairclough 1998, p. 32–33.

82. Tomlinson 1979, pp. 68–71.

83. See the library and archives of The Society of Antiquaries of London for information on its Fellows. On Folkes and the Royal Military Academy, see also Hogg 1953, pp. 1–5; Hogg 1963, p. 286. In June 1999, I presented a paper on Sir Charles Frederick, FSA 1732, FRS 1733, MP 1741, KB 1761, to the Society. It is now being prepared for publication. Details of the career of Sir Charles have been drawn from the library and archives of the Society, and from Brabrook 1910 and Fellowes 1932. I am also grateful for advice from the present Sir Charles Frederick.

84. For references to what was to become known as the Royal Laboratory, see Hogg 1963, especially pp. 105–106, 111, 222, 257, 260.

85. Warrants (King's and others), Woolwich, 1744–1748, and PRO, WO/55/408, p. 113. Hogg 1963, pp. 295–297. Charles Frederick was appointed Comptroller of the Laboratory on 12 Feb. 1746.

86. Hogg 1963, p. 492.

87. For the Royal Warrant of Charles II, 25 July 1683, see Hogg 1963, p. 868. Charles Frederick was appointed Surveyor General on April 10, 1750. For the duties of the officers, see Appendix IV, especially pp. 1052–1054. For the two new committees see Appendix XVI, especially p. 1432.

88. Aspinall-Oglander 1940, p. 123. Letter from Fanny Boscawen to her husband, November 15, 1748. The Treaty of Aix-la-Chapelle had been signed on 7 Oct. 1748. Walpole 1738–1758/1937, volume 37, p. 297.

89. Brock 1949, pp. 48–52. Handel composed a "grand overture on warlike instruments" for the occasion.

90. BM Add MSS 57318. Letters from Sir John Ligonier, 1754–1757. This unpublished correspondence was quoted in a paper presented by the author at Fort Ligonier, Western Pennsylvania, USA (October 1995), named in 1758 in honor of the Commander-in-Chief of HM Forces. The paper has now been published under the title "Sir John (later Lord) Ligonier (1680–1770), Military Commander and Member of Parliament for Bath," *Bath History* 2000.

91. Buchanan, in Buchanan 1996, pp. 237–252.

92. The Strachey Papers, with their information on the gunpowder industry and on Sir Charles Frederick, have been deposited at the Somerset Record Office (SRO, DD/SH, Box 27). I am preparing them for publication by the Somerset Record Society.

93. SRO, DD/SH, Box 27. Letter from Isaac Baugh in Bristol to Henry Strachey in London, 5 December 1761.

94. SRO, DD/SH, Box 27. Letter from Henry Strachey in London to Isaac Baugh in Bristol, 15 December 1761. On the Augusta Powder Works, see Rains 1882, p. 24.

95. West 1991, Faversham Mills, pp. 149–166. West describes the production and supply of gunpowder to the government in the middle of the eighteenth century, but the Surveyor General to the Board of Ordnance of the time, Sir Charles Frederick, does not appear in the index and figures only briefly in the text. However, an item on p. 52 refers to a request in April 1757 that he should provide information about the import of powder from the Dutch in 1740. This gives an intriguing indication of the continuing importance of continental supplies, noted earlier in the present study.

96. On the role of the partners in these negotiations, see Buchanan, in Buchanan 1996; see also SRO, DD/SH, Box 27, correspondence dated 26 October 1761, 5 December 1761, 15 December 1761, 9 January 1762, 11 January 1762, 6 February 1762, and 22 February 1762. The lobbying on this and other matters continued, for Henry Strachey the Younger wrote to his father on August 1, 1762 that it had cost him "a great deal of tongue and foot fatigue."

97. Hogg 1963, pp. 1042, 1080–86. By Royal Warrant of 1783 the responsibility for the inspection of armaments was placed with named artillery officers.

98. Hogg 1963, pp. 465, 481. Captain William Congreve had been active at Woolwich since 1778 when he was appointed Superintendent of Military Machines, with an establishment suitable for training artillerymen.

99. Hutton 1778; Ingen-Housz [*sic*] 1779. On William Congreve senior see the unpublished paper by Cocroft.

100. Smith 1776/1786, volume 3, book 5, pp. 70–71.

101. See Crossland 1987; Golinski 1992.

102. On the significance of saltpeter supplies from India, see Buchanan (in press).

103. SRO, DD/SH, Box 27. Letter of 15 Dec. 1761.

Bibliography

Aspinall-Oglander, C. 1940. *Admiral's Wife.*

Bailey, S. 1966. "The Royal Armouries Firework Book." In Buchanan, ed., *Gunpowder.*

Birch, T. 1756–1757. *The History of the Royal Society of London.*

Biringuccio, V. 1540/1966. *The Pirotechnia of Vannoccio Biringuccio.*

Brabrook, E. 1910. *On the Fellows of the Society of Antiquaries Who Have Held the Office of Director.*

Brayley Hodgetts, E., ed. 1909. *The Rise and Progress of the British Explosives Industry.*

Bret, P. 1996. "The Organization of Gunpowder Production in France, 1775–1830." In Buchanan, *Gunpowder.*

Brock, A. 1949. *A History of Fireworks.*

Buchanan, B. 1992. Capital Investment in a Regional Economy. Ph. D. thesis, University of London.

Buchanan, B. 1995–96. "The Technology of Gunpowder Making in the Eighteenth Century." *Transactions of the Newcomen Society* 67: 125–159.

Buchanan, B, ed. 1996. *Gunpowder.*

Buchanan, B. 1996. "Meeting Standards." In Buchanan, *Gunpowder.*

Buchanan, B. 1998–99. "Waltham Abbey Royal Gunpowder Mills." *Transactions of the Newcomen Society* 70, no. 2: 221–250.

Buchanan, B. 2000. "Sir John (later Lord) Ligonier (1680–1770), Military Commander and Member of Parliament for Bath." *Bath History* 8: 80–105.

Buchanan, B. 2001. "The Africa Trade and the Bristol Gunpowder Industry." *Transactions of the Bristol and Gloucestershire Archaeological Society* 118: 133–156.

Buchanan, B., and M. Tucker. 1981. "The Manufacture of Gunpowder." *Industrial Archaelogy Review* 5: 185–202.

Chaudhuri, K. 1965. *The English East India Company.*

Cocroft, W. 1996. "William Congreve (1743–1814), Experimenter and Manufacturer." Presented to 24th Symposium of ICOHTEC, Budapest.

Crocker, A., and G. Crocker. 1990. "Gunpowder Mills of Surrey." *Surrey History* 4, part 3: 134–158.

Crocker, G., and K. Fairclough. 1998. "The Introduction of Edge Runner Incorporating Mills in the British Gunpowder Industry." *Industrial Archaeology Review* 20: 23–36.

Crosland, M. 1987. "The Image of Science as a Threat." *British Journal for the History of Science* 20: 277–307.

Edwards, P. 1995. "Gunpowder and the English Civil War." *Journal of the Arms and Armour Society* 15: 109–131.

Evans, D. 1992. "Gloucester's Civil War Trades and Industries." *Transactions of the Bristol and Gloucestershire Archaeological Society* 110: 137–147.

Fairclough, K. 1985. "Early Gunpowder Production at Waltham." *Essex Journal* 20: 11–16.

Fairclough, K. 1996. "The Hard Case of Sir Polycarpus Wharton." *Surrey Archaeological Collections* 83: 125–135.

Fellowes, E. 1932. *The Family of Frederick.*

Ffoulkes, C. 1937. *The Gunfounders of England.*

Foreman, W. 1993. *Oxfordshire Mills.*

Golinski, J. 1992. *Science as Public Culture.*

Hassall, T. 1976. "Excavations at Oxford Castle, 1965–73." *Oxoniensia* 45: 232–308.

Henshaw, T. 1667/1959. "The History of the Making of Salt-peter" and "The History of Making Gunpowder." In Sprat, *History of the Royal Society of London.*

Hogg, O. 1954. "The Royal Military Academy in the 18th Century." *Royal Artillery Journal* 81, part 1: 1–16.

Hogg, O. 1963. *The Royal Arsenal.*

Hope, J. 1646/1958. "The Diary of Sir James Hope, 24 January-1 October 1646." *Miscellany of the Scottish History Society* 9 (1958).

Howard, R. "The Evolution of the Process of Powder Making from an American Perspective." In Buchanan, *Gunpowder.*

Hutton, C. 1778. "The force of fired-gunpowder. . . ." *Philosophical Transactions of the Royal Society of London* 68.

Ingen-Housz, J. 1779. "Account of a new dind of inflammable air or gass. . . ." *Philosophical Transactions of the Royal Society of London* 69.

Kramer, G. 1996. "*Das Feuerwerkbuch*: Its Importance in the Early History of Black Powder." In Buchanan, *Gunpowder.*

Lewis, D. 1976. The Office of Ordnance and the Parliamentarian Land Forces 1642–1648. Ph. D. thesis, Loughborough University of Technology.

Mussmann, O. 1996. "Gunpowder Production in the Electorate and the Kingdom of Hanover." In Buchanan, *Gunpowder.*

Napier, G. 1787. "Observations on Gun-Powder." *Transactions of the Royal Irish Academy* 2: 97–117.

Needham, J., et al. 1986. *Science and Civilisation in China*, volume 5, part 7.

Partington, J. 1960. *A History of Greek Fire and Gunpowder.*

Rains, G. 1882. *History of the Confederate Powder Works.*

Reynolds, T. 1983. *Stronger Than a Hundred Men.*

Rich, E., and C. Wilson. 1967. *The Cambridge Economic History of Europe*, volume 4.

Rimington, F. 1960. "Excavations at the Allerston Manor Site." *Transactions of the Scarborough and District Archaeological Society* 2: 19–28.

Robins, B. 1742. *New Principles of Gunnery.*

Roy, I., ed. 1975. *The Royalist Ordnance Papers 1642–1646.*

Sprat, T. 1667. The *History of the Royal Society of London.*

Stewart, R. 1996. *The English Ordnance Office.*

Thirsk, J. 1977. *Economic Policy and Projects.*

Tomlinson, H. 1976. "Wealdon gunfounding." *Economic History Review* 29: 382–400.

Tomlinson, H. 1979. *Guns and Government.*

Walpole, H. 1937. *Horace Walpole's Correspondence*, ed. W. Lewis, volume 37.

West, J. 1991. *Gunpowder, Government and War in the Mid-Eighteenth Century.*

Willan, T. 1959/1968. *Studies in Elizabethan Foreign Trade.*

Williams, A. 1975. "The Production of Saltpetre in the Middle Ages." *Ambix* 22: 125–133.

Wood, A. 1891. *Anthony Wood's Life and Times*, ed. A. Clark, volume 19.

Zins, H. 1972. *England and the Baltic in the Elizabethan Era.*

Zupko, R. 1968. *A Dictionary of English Weights and Measures.*

10

Chemistry in the War Machine: Saltpeter Production in Eighteenth-Century Sweden

Thomas Kaiserfeld

Since the Middle Ages, saltpeter has been an important resource for the manufacture of gunpowder, constituting approximately three-quarters of its weight. In Sweden, gunpowder's other constituents—ten to fifteen percent each of sulfur and charcoal—were in abundance in pyrite deposits and in deciduous forests near gunpowder works. Saltpeter, on the other hand, was scarce, and the manure used for its manufacture was coveted by peasants. As such, agricultural interests opposed those of the military, who had used gunpowder more extensively since the seventeenth century. In addition, the eighteenth century saw gunpowder-blasting slowly replacing the older use of fire-setting in mines, the traditional method of cracking the ore.[1] In short, saltpeter in eighteenth-century Sweden was a site of contestation for agrarian, military, and mining interests.

The importance of saltpeter, and the competing interests surrounding it, also meant that its production entered the discourse of those interested in natural philosophy and economics. Dissertations on the topic were published and defended at universities in Uppsala and Lund, as well as at Åbo Academy in Finland, then a part of Sweden. Other publications came out of the Royal Swedish Academy of Science in Stockholm and the War College. Thus, Swedish saltpeter production represented a rich topic for eighteenth-century intellectuals.

This chapter deals with the organization of saltpeter production in Sweden and its transformations during the eighteenth and early nineteenth centuries. More specifically, it addresses how science participated in organizing production. How did the various parties refer to scientific theories in their debates on optimizing saltpeter production? Who employed which theories and with what agenda? In the end, did arguments based on scientific findings have any impact on the organization of saltpeter production? As we will see, technical change based on scientific findings does not necessarily imply new institutional conditions. Instead, it can revive long-abandoned rules and regulations. Moreover, when studying the militarization and demilitarization of different sectors of society, it is not enough to highlight scientific, industrial, or military interests alone. Other interest groups may have played an equal or even more important part in these transformations.

The Methods of Saltpeter Production and Their Institutional Contexts in Sweden

In the sixteenth and early seventeenth centuries, saltpeter production had been organized by forcing farmers to deliver raw materials to saltpeter works run by the Crown. This enforcement was called the saltpeter aid (*salpeterhjälpen*).[2] In 1616, a decree stated that each homestead (*hemman*) in the vicinity of saltpeter works had to deliver four barrels of soil, one barrel of sheep's dung, half a barrel of ashes, three loads of wood, and two sheaves of straw.[3] In effect, the 26 Swedish saltpeter works operating in 1624 used such locally supplied raw materials to extract saltpeter from a nitrogenous mix of soil, dung, and ashes through a wood-consuming process of leaching and boiling.[4] The mix of soil, dung, and ashes was soaked with water that was then filtered. The filtered liquid was used once or twice more for leaching the mix and was finally heated in large copper cauldrons holding 800 liters or more (somewhat more than 200 gallons). Saltpeter workers then skimmed off impurities and, through a very complicated and tedious step-by-step process that took weeks of continuous heating, extracted raw saltpeter from the lye.

This system of duty deliveries to saltpeter works did not provide enough to meet the national need. Therefore, during the seventeenth century, saltpeter had to be imported in large quantities.[5] In 1642, the saltpeter aid was transformed into a monetary tax that soon had to be paid by all homesteads, not simply by those who earlier supplied the saltpeter aid.[6] Towards the end of the seventeenth century, the monetary saltpeter aid was complemented by mobile saltpeter "boiler teams" (*pannelag*). Organized in an allotment system, they traveled from homestead to homestead during the summer season in order to collect the soil and extract the saltpeter on site.

In the saltpeter boilers' allotment system (*salpetersjuderiindelningsverket*), Sweden was divided geographically into seven inspection areas (*inspektionsområden*). With the aid of the county governor (*landshövdingen*), each inspector took inventory of his respective area, approved the saltpeter delivered, paid out remunerations, and collected fines. In each inspection area, 20 to 30 "boiler teams" were responsible for extracting saltpeter from soil by leaching and boiling. Within each team, a foreman and two to five boilers were enjoined to produce at least 10 lispounds (85 kilos, somewhat less than 200 pounds) of saltpeter;

otherwise, they were fined.[7] As compensation, these saltpeter men were exempt from conscription to the armed forces. In addition, the seven inspection areas comprised districts (*sjuderidistrikt*), each corresponding to a boiler team consisting of a number of military parishes (*rotar*), with eight to ten homesteads in each. One boiler team then circulated among the homesteads within a specific military parish for one year so that soil for the boiling of saltpeter would not be taken from the same homestead more than once every six years, or in the southernmost district not more than once every four years.[8] Instead of having the farmers deliver the material to saltpeter works run by the Crown, the saltpeter boiler teams now brought their pots and pans to the farms to extract the saltpeter from the soil on site.

This organization of saltpeter production mobilized every farmer, cow, and sheep in Sweden, or at least in the southern and eastern parts where the soil was rich enough in saltpeter to be exploited and the wood abundant enough to keep the lye boiling. Obviously, this required advanced monitoring of both the population and the land. Such administration reflected the military importance of saltpeter, hence the transfer of its supervision to the Armed Forces in 1612. In the 1620s, the Swedes developed a system of parish registration (*kyrkobokföring*) to monitor citizens. Organized by the Church on the command of the Crown, the registration was launched to levy the Swedish soldiers and sailors of the Thirty Years' War.[9] This bureaucratic innovation involved a fairly thorough control of the Swedish population: their civil status, their income, and their domicile, criminal, and clerical records were documented. In 1680, the system functioned as an efficient control apparatus that gave the military and tax administration the means to exploit the human and material resources of the country. In fact, this parish registration is among the oldest still running and renders Sweden a genealogist's heaven.

The introduction of the saltpeter boilers' allotment system depended even more so on the existing taxation system (*indelningsverket*), which originated in the Middle Ages, when taxes were collected in goods and not in money. By allowing the Crown's officials themselves to collect their subsistence from the copyholders and tenants in a specific area, this system admitted taxation without involving monies. Moreover, to secure the supply of troops, the War College administered a military allotment system in 1682 (*det ständiga knektehållet*) that augmented the more general system established during the Thirty Years' War. This addition entailed the division of each county into military parishes where the taxpayers were required to recruit a soldier, who then had to be lodged and given a piece of land to farm during peacetime. From this perspective, the introduction of the saltpeter boilers' allotment system signified a new use of a familiar and reliable institutional framework.

Nevertheless, the introduction of a saltpeter boilers' allotment system proved ineffectual until 1723, when the War College made a new inventory of the country's saltpeter-rich soils. Not until a decree in 1746, however, did the organization of saltpeter production succeed on a national level.[10] By devising a finer-meshed division of homesteads and individuals, the allotment system became an effective tool for collecting and exploiting materials such as soil, dung, and wood that were needed for producing saltpeter. The decree of 1746 remained valid until 1805, when the parliament, the *Riksdag*, revoked the King's right to the saltpeter-rich soils. In the meantime, the use of the saltpeter boilers' allotment system facilitated the War College's collection of the monetary saltpeter aid and licensed experienced saltpeter men with adequate equipment to exploit the richest soils. The system thereby enhanced the opportunity to raise domestic production.

In the middle of the eighteenth century, another novelty was introduced to further raise the saltpeter productivity of Swedish soils. This novelty, however, did not fit the production organization of the existing allotment system of saltpeter boilers. The building of saltpeter barns nevertheless became a more common method for treating soil before its actual leaching and boiling. These barns were rather simple wooden sheds where soil—mixed with manure, stale urine, composted wastes, or even carcasses—was stored in oblong pyramidal heaps. Their roofs kept the rain from leaching the soil, which was turned every fortnight. The idea behind the saltpeter barns was that saltpeter could be generated or grown more efficiently in certain favorable environments such as well-tended heaps of soil in shelters. In general, the soil from barns could also be used for processing every third or fourth year, at most every second. This should be compared to the rate of production in the allotment system, where the soils were collected at most every sixth year, except in the southernmost parts where the rate was every fourth year.[11] Since the barns were immobile and needed continuous tending for three to four years before their content could be used for the evaporation of saltpeter, this alternative production method did not fit the organization of the mobile saltpeter men.

The German historian Ottomar Thiele has described how the building of saltpeter barns had spread to Germany from France, although it never became well established there.[12] In France, however, the method of preparing the soil in saltpeter barns was tried as one of many ways to enhance saltpeter production when it was centrally reorganized under the well-known French chemist Antoine Lavoisier in 1775. The French historian Patrice Bret has documented how this was done through the privately financed authority La Régie des Poudres et Salpêtres.[13] Established in 1775, its financiers were given the exclusive right to produce saltpeter, on the assumption that this set-up also safeguarded the interests of the Crown, namely, continuous access to usable saltpeter at a reasonable price.

In Sweden, saltpeter barns had been referred to from the end of the seventeenth century, but there are no indications that they had come into more general use before the 1740s. One of the central documents encouraging the foundation of saltpeter barns was a report submitted by the War College in 1746, the same year a royal decree was issued that seemed finally to organize the traveling boiler teams into an effective national system of production. The War College had published this report after the hereditary prince Adolf Fredrik described the use of the method in foreign countries.[14]

In the report, the War College described two different kinds of saltpeter barns. (See figure 10.1.) The simpler kind was cheaper, easier to build, and "could serve the more simple-minded," while the more elaborate kind was more expensive but also would "richly reward the trouble."[15] The reasons for the publication of the report were not only to "avoid the load the Kingdom, in days of yore, must have suffered, to let considerable Capitals go out for the procurement of Saltpeter for the defence of the Country and other necessary uses, but one has likewise been conscientious, to provide the Mines with enough gunpowder, especially since it has been clear to each and everyone, that those with a lot of saved Forest and relieve of working load, now almost everywhere have begun to use gunpowder for blasting Ore in the Mines."[16] The efforts of the War College were not limited to words alone. The Crown also promised to pay a reward for every lispound of saltpeter delivered that could be proven to have been produced from soil treated in a saltpeter barn, a reward significantly higher than the remunerations given to the saltpeter men.

During the following decades, recommendations and discoveries that promoted saltpeter barns were printed in Stockholm almost annually. Soon, the Crown also approved of loans to encourage the construction of barns. To start with, only nobility and persons of rank were eligible, but eventually freeholders could also apply for the loans. In 1748, the Crown gave a loan of 3,000 daler copper coins free of interest to the town clerk of Linköping, Carl Fredrich Lund, for expanding his saltpeter works with barns. Soon, more loans followed and, between 1752 and 1754, loans totaling 74,000 daler copper coins were given to nobility for the erection of saltpeter barns. Later, the system was expanded to include copyholders as well.[17] In 1756, a letter from the King stated that every barn erected, larger than 60 × 16 ells (around 40 × 10 yards) with soil heaps higher than 3 ells (2 yards), was to be rewarded with a bounty of 300 daler copper coins.[18]

These loans and bounties resembled those supplied to owners of manufacturing operations (mainly for the production of textiles) by the Estates' Office of Manufacture (*Riksens Ständers Manufakturkontor*), which was established in

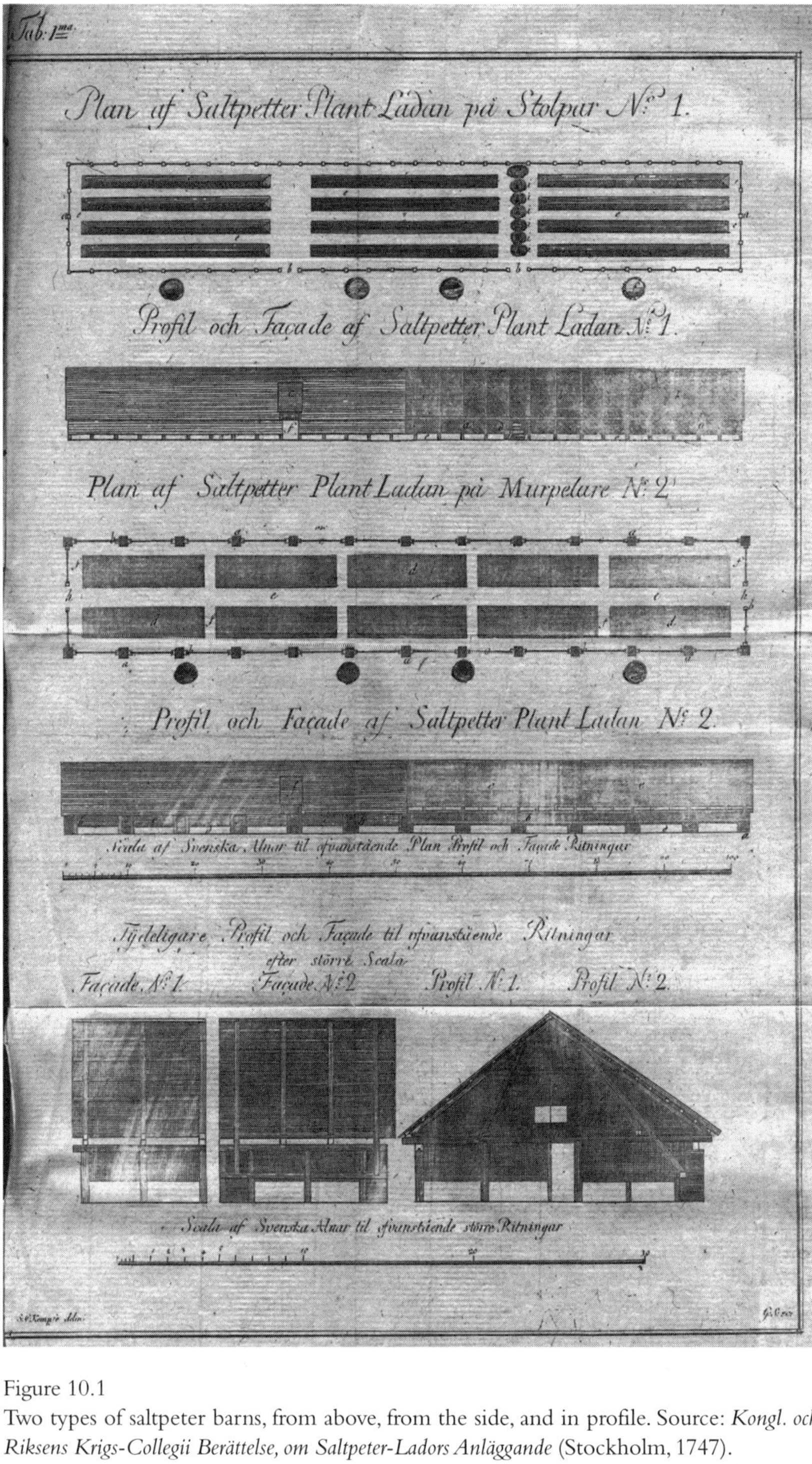

Figure 10.1
Two types of saltpeter barns, from above, from the side, and in profile. Source: *Kongl. och Riksens Krigs-Collegii Berättelse, om Saltpeter-Ladors Anläggande* (Stockholm, 1747).

1739 and operated as a subdivision of the College of Commerce (*Kommers-kollegium*) from 1766.[19] The subsidies, as well as the organizations that administered them, were an established institutional framework for supporting production favored by the government. So the loans and bounties for the building of saltpeter barns was not an institutional innovation, but rather an expansion of the system for supporting the manufacturing of textiles, tobacco, sugar, glass, paper, porcelain, and other desirable goods.

The subsidies seem to have made it financially profitable to produce saltpeter. According to one contemporary statement, a saltpeter barn would give an annual return of 9–12 percent.[20] The bounties were supposed to have been abolished in 1784, when the War College complained that the production of saltpeter prepared in barns did not meet their expectations. Still, some compensation was paid out after that year.[21] According to one source, about one third of the saltpeter produced in Sweden in the 1790s originated from soil that had been prepared in barns. At the same time, it is important to point out that most of the saltpeter was still produced within the older allotment system. In 1815 there were 1,128 saltpeter barns scattered around the country (not including Finland, which had been lost to Russia in 1809).[22]

Along with the expansion of the system with loans and bounties, the number of saltpeter boilers decreased in southern Sweden from the 1760s.[23] Neither reasons nor explanations exist for this decline, which occurred despite a population growth and an increasing proletarization in the countryside. It could, however, be an indication of new production methods and alternative ways to organize work.

In 1805, Parliament abandoned the saltpeter boilers' allotment system, as well as the monetary saltpeter aid, and revoked the king's right to the saltpeter-rich soil. Instead, the county governors were authorized to collect a new tax. As a substitution for the traveling saltpeter men who visited homesteads to extract saltpeter on site, every homestead now had to deliver a specific quantity of saltpeter (*salpetergärden*) annually, which the authorities bought at a set price. Formerly, some homesteads in the richer farming regions had evaded visits from saltpeter boilers because of a lack of wood fuel necessary to evaporate the saltpeter from the lye; this new taxation signified a more equitable system throughout the country.[24]

Swedish Saltpeter Production in the Eighteenth Century

One way to estimate the impact of saltpeter barns on saltpeter production is to examine documented production rates; however, very little is known about the scale of Swedish saltpeter and gunpowder production in the eighteenth century. Nevertheless, according to an unverified statement, the production of raw saltpeter reached a peak at 144.5 tons in 1712, during the Great Northern War (1700–21).[25] But subsequently this fell to about 50 to 60 tons a year.[26] Sixty years later, however, saltpeter production once again rose to a new all-time high. In 1773, 353 tons of saltpeter were produced and delivered to the Swedish gunpowder mills, although the annual deliveries quickly sank to around 250 tons. This still satisfied both military and civil demands.[27]

Even if there is some disagreement about the exact production figures of raw saltpeter, a general trend is clear. A peak in production appears around 1710 followed by a decline. (See figure 10.2.) The need for gunpowder during the Great Northern War may explain this peak, while the ensuing stagnation has been viewed as a result of insufficient deliveries of raw material to the saltpeter works. This, in turn, was a consequence of the conscription of saltpeter men into new regiments. Another rise in production can be discerned during the second half of the eighteenth century, and, despite a decline in the 1770s, production largely satisfied the demand. This is reflected by the decision of the King in 1783 to limit the annual saltpeter production rate to 30,000 lispounds (255 tons).

Gunpowder or Corn: The Public Debate on Manure

Saltpeter production was important for military destruction and agricultural production. The central issue to both communities was how to produce saltpeter most efficiently. Public discourse attacked this problem from many different angles during the eighteenth century, a time when economic doctrines constituted a critical foundation. The general political debate was marked by a cameralism that went hand in hand with an agriculture-friendly form of reform-mercantilism. The essence of this economic doctrine was the use of domestic resources to compensate for the lack of overseas colonies and for the consequent expenditures of foreign trade.[28] Debates were especially influenced by the argument that the fields of a country constituted its only source of wealth, although more specific physiocratic views played a limited role in Swedish eighteenth-century economic policy.[29]

Besides employing economic doctrines, the debaters also underpinned their arguments with chemical matter theories. In the seventeenth century, the predominant view was that saltpeter was a mineral, which—if the circumstances were right—could be grown or generated.[30] The use of saltpeter barns signified one important conclusion arising from this theory. Throughout the eighteenth century, similar ideas were suggested. At the same time, though, other views were also proposed, such as the possibility that different substances

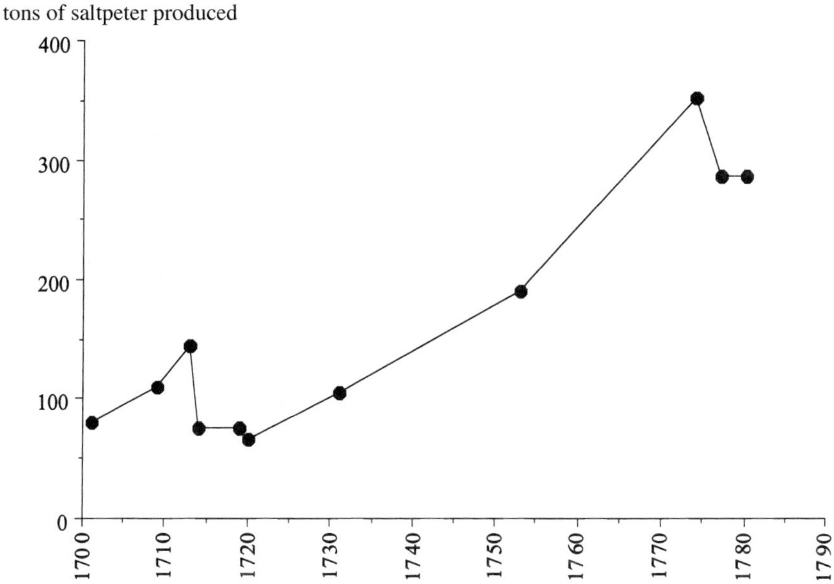

Figure 10.2
Graph showing Swedish production of saltpeter in the eighteenth century, compiled from data in the following secondary sources: Olof Langlet, "Salpetersjuderiet och salpetersjudarna," in *Från Borås och de sju häraderna*, vol. 29, ed. H. Kruse (Borås, 1975); Gunnar Grenander, "Krigsindustrin," in *Kungl. Artilleriet: Frihetstiden och Gustav III:s tid*, ed. S. Claëson (Stockholm: Militärhistoriska förlaget, 1994.)

could be transformed into saltpeter. For example, sulfuric acid could be transformed into nitric acid, which in turn could produce saltpeter.

This mix of economic and scientific arguments mirrors an ideology of utility.[31] This was a thought system connected directly with the Enlightenment and subsequently with ideals associated with Hume, Smith, and the Physiocrats; it was not to be confused with the formal utilitarian ethics of Jeremy Bentham. With roots in the early seventeenth century, the ideology of utility stressed the application of natural philosophy and economics as a means of raising domestic production for the benefit of the whole nation. It not only paved the way for the introduction of economics at Swedish universities but also generated greater demand for the sciences, especially chemistry.[32]

The ideology of utility in Sweden had an institutional foundation in the Royal Swedish Academy of Science. Here the director of the Surveying Office (*Lantmäterikontoret*), Jacob Faggot, formerly an inspector of alun works who worked out chemical methods for determining the quantity of saltpeter in gunpowder, was particularly active.[33] He strongly defended the importance of agriculture from a cameralistic standpoint.[34] A circular from the War College in 1747 also referred to Faggot as the constitutive manager of an experimental barn it proposed for Stockholm. In this role, he investigated "which proportion of the materials in the mix, would be the best in order to promote the production of saltpeter."[35] Likewise, in 1757 he sanctioned another pamphlet in support of saltpeter barns.[36]

Another person interested in these barns was the professor of chemistry at Åbo Academy in western Finland, Pehr Adrian Gadd, the former inspector of saltpeter works in Åbo and the county of Björneborg. After his appointment to this professorship, he continued serving on the side as a director of saltpeter barns. Under Gadd's supervision, one student published and defended a thesis on saltpeter in 1771. Here he argued for the possibility of growing or generating saltpeter in soil under certain circumstances. The use of saltpeter barns was one important conclusion.[37]

During the eighteenth century, alchemy also was a source of ideas for saltpeter production. One of the most important builders of saltpeter barns was the general and alchemist Jacob Johan Anckarström. When he had left the military in 1767, he planted fruit gardens and built saltpeter barns on his estate north of Stockholm. Inspired by alchemy, Anckarström's matter theory stated that ordinary salt, given the proper environment, could be transformed into saltpeter. Although he did not detail how to prepare the soil for this transmutation, he concluded that the barns should be built so that a maximum quantity of the soil's salt could become saltpeter.[38]

After Anckarström had presented his ideas in the middle of the 1770s, the public debate apparently died out. It would take almost two decades, until the middle of the 1790s, for the saltpeter barns to be defended once again.[39] This time it was Lars Cronstrand, the director of saltpeter works in the county of Kalmar, who argued that farmers themselves should be ordered to collect stale urine and manure and to use it to prepare soil in saltpeter barns connected to their cow-houses.[40] Crown boilers with the right knowledge should then evaporate the saltpeter from the soil, compensating the farmer for his work. In this way, Cronstrand calculated, manure would be left over for the farmers to spread in the fields in order to raise crops.

Peter Jacob Hjelm, a Fellow of the Royal Swedish Academy of Science and director of Laboratorium Chymicum of the Board of Mines (*Bergskollegium*), also confronted the issue of saltpeter production. Contrasting other participants in this debate, Hjelm not only advanced his own views but also translated a treatise by Lavoisier on the building of saltpeter barns.[41] This is somewhat remarkable given that Hjelm was a phlogistonist and Lavoisier was one of the founding fathers of modern chemistry. He, along with others, including the converted phlogistonist Claude-Louis Berthollet, formulated an alternative to the theory of phlogiston in trying to explain combustion.[42] The Swedish historian of science, Anders Lundgren, has used this circumstance to show that the dynamics in chemical theory towards the end of the eighteenth century did not imply such a radical change for those interested in practical chemistry as it did for those interested in the theories of matter. This pragmatism among practitioners apparently did not affect chemists at the universities as well as at the Board of Mines, where Hjelm also worked.[43]

Faggot, Gadd, Anckarström, Cronstrand, and Hjelm constituted a group of advocates for saltpeter barns. Despite their mixed backgrounds, they all (possibly including Hjelm) had practical experience with saltpeter production. Faggot had directed an experimental barn, Gadd had been an inspector of saltpeter works, Anckarström had performed experiments at his own establishment, and Cronstrand had directed a saltpeter works. In addition, their experience was combined with an interest in chemical matter theories. Here, however, the similarities ended and the arguments used to promote saltpeter barns were founded on a wide range of theories.

On the opposing side of this debate, Carl Fredrich Lund voiced his discontent with his own saltpeter barns. As the town clerk of Linköping, he had shown an early interest in the building of barns, and in 1748 he received a government loan of 3,000 daler copper coins free of interest. In 1770, as mayor, he found that "it was as impossible to bring out of a salpeter barn, much saltpeter, as it is to suggest the cultivation of 50 acres, where the land holds no more than 10."[44] Lund believed that saltpeter in air could be fixed in soil with the help of minerals and dead plants, and, conversely, that saltpeter in soil could be diffused into the air through the action of light. Lund considered that the dark interior of saltpeter barns was an unsuitable environment for the generation of saltpeter. Instead, he suggested that soil and manure should be treated in the open at the farmhouses so that as much saltpeter as possible could be evaporated by the crown boilers.

Lund's proposal triggered off a press debate in Stockholm in the autumns of 1770 and 1771.[45] Some years later, another circular was disseminated under the title *Swedish Agriculture's Damage and Loss of 360,000 Barrels of Corn Annually, Through the Present Organization of Crown Boilers*.[46] Here, the pseudonym Philopater argued that it would be more productive to spread the manure in the fields than to use it for the production of saltpeter. Saltpeter, the author concluded, should instead be imported from abroad where it could be purchased at a cheaper rate. In the cameralistic climate of eighteenth-century Sweden, the opposition did not prevail; however, its ideas were not discarded altogether.[47] To save the system of crown boilers and at the same time let the farmers keep their manure, an anonymous author suggested some sort of compromise stating that crown boilers should not collect manure at farms but rather at churches, markets, and other public places.[48] According to this suggestion, the saltpeter in Swedish gunpowder would no longer derive from cattle, but from humans.

In general, the opponents to saltpeter barns had as much practical experience as its advocates. What the opponents lacked, however, was academic status. In turn, their arguments were rarely substantiated by chemical matter theories. The only exception was Lund, who cited chemical classifications of saltpeter as a mineral that could not be generated. All other opponents referred to the economic damage any domestic production of saltpeter would supposedly incur. As such, some of the critics of saltpeter barns just as keenly opposed the saltpeter boilers' allotment system. In short, during the second half of the eighteenth century, saltpeter production was subject to the same rationalization as so many other trades and industries of that time. On the one side, supporters of the saltpeter barns relied on chemical theories, although they never reached a consensus about which one was most valid. By contrast, opponents based their arguments on economic doctrines. In this way, both camps stressed the two sides of the ideology of utility that marked much of the public debate in the Age of Enlightenment: natural philosophy and economics. At the same time, virtually all of the debaters had firsthand experience with saltpeter production. This indicates the importance of practice over social position, whether that of a university chemist, a member of the Academy of Science, a disheartened owner of saltpeter barns, or a noble alchemist. Still, theory—both chemical and economic—proved to be a popular rhetorical strategy.

The Complaints of the Peasantry

Despite these public debates among academics and high-ranking officials, the organization of domestic production of saltpeter affected farmers more than any other social group. In Sweden, the channel most commonly used by the wealthier farmers to make their voices heard was the parliament. From the death of Charles XII in 1718 and the new constitution of 1719 to the coup d'état of Gustav III in 1772, Sweden was governed by four estates: the nobility, the clergy, the burghers, and the peasantry. Admittedly, the Estate of the Peasantry was the weakest of the four in the Swedish parliament. Still, by contrast with the peasantry of many other European countries at that time, the constitution gave Swedish peasants a formal means to voice their political opinions.

The Estate of the Peasantry was a homogeneous assembly, representing about half of the population in 1750. Here, peasants who owned their own farms, or who held hereditary leases on Crown farms, were eligible to vote;

but landowners of other estates, as well as peasants who leased land from nobility, were barred from voting.[49] Gradually throughout the eighteenth century, the Estate politically matured and acquired greater influence on domestic policies. Its rise was only temporarily weakened by the overthrow of the government in 1772.

Although representatives of the peasantry were excluded from the most powerful of committees, the Secret Committee, they could bring up parliamentary appeals (*riksdagsbesvär*) on behalf of their voters. These constituted a political channel for the Swedish peasantry in the eighteenth century, especially concerning the most important question for the Estate of the Peasantry at this time: how to keep taxes and other duties within reasonable proportions.[50] More specifically, the parliamentary appeals often addressed local or regional issues rather than ideological ones. In practice, this meant that the most common subject of these appeals was the general allotment system for taxation and other related concerns, such as the stage system for transportation and the redistribution of land holdings. Problems connected to the saltpeter boilers' allotment system also frequently surfaced. Most commonly, farmers complained about the lack of wood fuel, with which they often had to supply the boilers. Others criticized the farmers' duty to deliver saltpeter manufactured in the allotment system to collection sites.[51]

The abolition of this allotment system in 1805 therefore signified a major victory for the peasants. Instead of having saltpeter boilers that exploited soils only at homesteads where there was enough wood fuel, all farms now had to contribute to the saltpeter production of the country.[52] By stipulating that every homestead had to deliver a specific quantity of saltpeter annually, the Royal Ordinance of 1801 transferred the responsibility to the farmers and made it their interest to organize saltpeter production more effectively. The new regulation supplied institutional conditions that ultimately promoted the preparation of soil in saltpeter barns. If we can deduce that the introduction of saltpeter barns during the second half of the eighteenth century relied on a chemical and economic discourse that marked an ideology of utility, then we can conclude that the diffusion of the saltpeter barn resulted from the parliamentary appeals of the peasantry.

Conclusion

Swedish saltpeter production underwent a series of institutional changes during the eighteenth century. In the early decades, saltpeter was a scarce resource, which posed a constant threat to gunpowder production and therefore to national security. From the 1730s, however, the production slowly rose so that by the 1770s its supply met the Swedish demand. The transformation of saltpeter from a scarce resource into an abundant asset occurred despite a decreasing number of saltpeter boilers from the 1760s onward. This paradox is partly explained by the introduction of saltpeter barns and new institutional conditions for saltpeter production. Yet, since it took time to reorganize saltpeter production and to introduce appropriate institutional conditions, the establishment of saltpeter barns was gradual. Also a major factor in the rising saltpeter production was the already established production organization of the saltpeter boilers' allotment system. Couched as it was in an older, well-adjusted institutional framework, it became increasingly efficient in spite of persistent complaints.

If the introduction of saltpeter barns, together with the system of loans and bounties, was a result of arguments eagerly presented in debates involving an ideology of utility, then their institutional confirmation in the nineteenth century was largely an effect of the peasants' opposition to the saltpeter boilers' allotment system. While the saltpeter barns had offered a means to phase out the allotment system, they were eventually given a legal framework that revived the institutional conditions that had been abandoned for the saltpeter boilers' allotment system more than 150 years earlier. Then, the so-called saltpeter aid was collected in goods. Similarly, the Ordinance of 1805 implied a tax that was collected in saltpeter. In effect, this reform constituted a regression to those institutional conditions prevailing before the introduction of the saltpeter boilers' allotment system.

In this way, the saltpeter barns elucidate how a new technology can provide alternatives for dismantling an old and much-criticized organization. The saltpeter boilers' allotment system was eventually replaced—not by a new set of institutional conditions, as suggested by those in favor of an ideology of utility, but by an even older one that transferred responsibility for saltpeter production to the most pronounced opponents of the allotment system: the peasantry. The institutional changes of the Swedish saltpeter production in the eighteenth and early nineteenth centuries demonstrate how interests, which are not scientific, industrial, or militaristic, still play an important role in the militarization and demilitarization of munition production.

Notes

1. Heinz Walter Wild, "Black powder in mining—its introduction, early use, and diffusion over Europe," in *Gunpowder*, ed. B. Buchanan (Bath University Press, 1996).

2. Henning Hamilton, *Afhandling om Krigsmaktens och Krigskonstens Tillstånd i Sverige, under Konung Gustaf II Adolfs Regering*, Kongl. Vitterhets-, Historie- och Antiquitets Academiens Handlingar, no. 17 (Stockholm, 1846), p. 240.

3. Å.G. Ekstrand, "Salpeterindustrien i Sverige," *Svenska Kemisk Tidskrift* 5 (1893): 61.

4. Assar M. Lindberg, "Salpeterframställningen i Sverige fram till 1642," *Ymer: Tidskrift utgiven av Svenska sällskapet för antropologi och geografi* 84 (1964): 277.

5. Lars Hoffmann Barfod and Jens Chr. Balling Jensen, *Bogen om krudt: Fordums krudtvaerker i Norden of folkene bag dem* (Politiken, 1992), 97–98.

6. Olof Langlet, "Salpetersjuderiet och salpetersjudarna," in *Från Borås och de sju häraderna*, volume 29, ed. H. Kruse (Borås, 1975), p. 23; Lindberg, "Salpeterframställningen i Sverige," pp. 278–79.

7. Alf Erlandsson, "Salpetersjudarna i Skåne 1679–1762: Med särskild hänsyn till 1740–talet," in *Vetenskaps-Societeten i Lund Årsbok 1961* (Lund, 1961), p. 36; Bengt Åhslund, "The saltpetre boilers of the Swedish Crown," in *Gunpowder*, ed. Buchanan, p. 172.

8. Erlandsson, "Salpetersjudarna i Skåne," pp. 53–54; Langlet, "Salpetersjuderiet och salpetersjudarna," pp. 24–25.

9. Sven A. Nilsson, "Krig och folkbokföring under svenskt 1600–tal," in *De stora krigens tid*, ed. R. Torstendahl and Torkel Jansson (Almqvist & Wiksell, 1990), pp. 56–80.

10. Erlandsson, "Salpetersjudarna i Skåne 1679–1762," pp. 35–86. Regarding resistance to the introduction of the control system, see Assar M. Lindberg, "När smålänningarna sysslade med olaga kruttillverkning," in *Värendsbygder: Norra Allbo Hembygdsförening 1962* (Alvesta, 1962).

11. Ekstrand, "Salpeterindustrien i Sverige," p. 83.

12. Ottomar Thiele, "Salpeterwirtschaft und Salpeterpolitik: Eine volkswirtschaftliche Studie über das ehemalige europäische Salpeterwesen," *Zeitschrift für die gesamte Staatswissenschaften*, Ergänzungsheft XV (Tübingen, 1905), p. 20.

13. Patrice Bret, "The organization of gunpowder production in France, 1775–1830," in *Gunpowder*, ed. Buchanan. Cf. Charles Coulston Gillispie, *Science and Polity in France at the End of the Old Regime* (Princeton University Press, 1980), pp. 50–73.

14. *Kongl. och Riksens Krigs-Collegii Berättelse, om Saltpeter-Ladors Anläggande* (Stockholm, 1747). On the origin of the report, see Ekstrand, "Salpeterindustrien i Sverige," p. 64.

15. Carl Cronstedt, foreword to *Kongl. och Riksens Krigs-Collegii Berättelse*, p. 7. Quotes in original: "kan tiena för the enfaldigare" and "rikeligen löna mödan."

16. Ibid., pp. 3–4. Quote in original: "at undgå then lasten Riket i förra tider måst widkännas, at låta ansenliga *Capitaler* gå ut för Saltpeterns införskrivande til Landets förswar och nödtorftigt bruk, utan ha man jemwäl warit sorgfällig, at med tilräckeligit krut kunna förse Bergwärken, i synnerhet sedan man märkt, at man med mycken besparing af Skogarne och lindring i arbetet, nästan nu öfweralt begynt at bruka krut wid Bergsprängningar i Grufworne."

17. Assar M. Lindberg, "De småländska salpeterladorna," in *Värendsbygder: Norra Allbo Hembygdsförening 1961* (Alvesta, 1961), pp. 67–72.

18. Ekstrand, "Salpeterindustrien i Sverige," p. 67.

19. Sven Gerentz, *Kommerskollegium och näringslivet: Kungl. Kommmerskollegium 1652–1952* (Stockholm, 1951), pp. 225–271.

20. Lindberg, "De småländska salpeterladorna," p. 70.

21. Ekstrand, "Salpeterindustrien i Sverige," p. 67; Lindberg, "De småländska salpeterladorna," p. 69.

22. Lindberg, "Salpeterframställningen i Sverige," pp. 267–268.

23. In Sweden's southernmost county, Scania, the number of saltpeter boilers decreased 15 percent in the 1760s, and in the southwest county of Älvsborg, the corresponding decrease was 30 percent (Erlandsson, "Salpetersjudarna i Skåne," p. 62; Langlet, "Salpetersjuderiet och salpetersjudarna," pp. 45–48). The number of saltpeter boilers had also been decreasing in the 1710s (Gunnar Grenander, "Krigsindustrin," in *Kungl. Artilleriet: Frihetstiden och Gustav III:s tid*, ed. S. Claëson, Militärhistoriska förlaget, 1994, pp. 310–311).

24. Langlet, "Salpetersjuderiet och salpetersjudarna," p. 27.

25. Grenander, "Krigsindustrin," p. 310.

26. In a contemporary source, the peak of saltpeter production is set at 91 tonnes in 1708 and is thereafter said to have decreased to 50–60 tonnes annually throughout the rest of the century (Anders Gustaf Barchaeus, *Academisk Afhandling, Om saltpeter-sjuderi inrättningen i Sverige*, Uppsala, 1784–1785, volume 2, p. 32). This source was used in Robert P. Multhauf, "The French Crash Program for Saltpeter Production, 1776–94," *Technology and Culture* 12 (1971), p. 166.

27. Grenander, "Krigsindustrin," p. 310.

28. One of the most important advocates of these ideas was Carolus Linnaeus, with his notions of harmony between the natural resources of a nation and the needs of its inhabitants. See Eli F. Heckscher, "Linnés resor—den ekonomiska bakgrunden," Svenska Linné-Sällskapets årsskrift 25 (1942): 1-11; Lisbet Koerner, "Linnaeus' Floral Transplants," *Representations* 47, 1994 (summer): 144–169.

29. Eli F. Heckscher, "Fysiokratismens ekonomiska inflytande i Sverige," *Lychnos* 8 (1943): 1–20; Lars Herlitz, *Fysiokratismen i svensk tappning 1767–1770*, Meddelanden från Ekonomisk-historiska institutionen vid Göteborgs universitet, no. 35 (Gothenburg, 1974).

30. Sigurd Nauckhoff, "De kemiska uppfattningarna om salpetersjuderiet i Äldre tider," in *Kungl. Svenska Vetenskapsakademiens Årsbok 1953* (Stockholm, 1953), pp. 431–442. See also Robert P. Multhauf, "The Constitution of Saltpeter, According to Becher and Stahl," in *Science, Medicine and Society in the Renaissance*, volume 2, ed. A. Debus (Heinemann, 1972); Multhauf, *Neptune's Gift: A History of Common Salt* (Johns Hopkins University Press, 1978), pp. 125–143.

31. This complex ideology with its internal contradictions and inconsistencies is better described as a Zeitgeist than as a well-defined platform for action. See Karl Forsman, "Studier i det svenska 1700–talets ekonomiska litteratur," in *Historiska och litteraturhistoriska studier,*

volume 23, ed. E. Hasselblatt et al., Skrifter utgivna av Svenska Litteratursällskapet i Finland, no. 312 (Helsinki, 1947).

32. Sven-Eric Liedman, *Den synliga handen: Anders Berch och ekonomiämnena vid 1700–talets svenska universitet* (Arbetarkultur, 1986). Chemistry has been seen as an exception to the more general decline of the sciences in Sweden in the 18th century. See Karin Johannisson, "Naturvetenskap på reträtt: En diskussion om naturvetenskapens status under svenskt 1700–tal," *Lychnos* 29 (1979–80): 127.

33. Jacob Faggot, "Sätt Att utröna, huru mycken Salpeter finnes i ett färdigt gjordt Krut: samt Anmärkningar Om Krut-verket i Allmänhet," *Kongl. Svenska Vetenskapsacademiens Handlingar* 16 (1755): 96–117.

34. Börje Hanssen, "Jacob Faggot som ekonomisk författare," *Kungl. Lantbruksakademiens tidskrift* 81 (1942): 51–74.

35. Cronstedt, foreword to *Kongl. och Riksens Krigs-Collegii Berättelse*, p. 5. Quote in original: "låta utröna hwilken *proportion* af *Materialiernes* blandning wore then bästa til Saltpeter-wärkens befordring."

36. *Underrättelse om Saltpeters ymnoga Tilwärkning jemte bifogade Anmärkningar, Til Riketes allmänna Nytta* (Stockholm, 1757).

37. Pehr Adrian Gadd, *Undersökning chemisk och oeconomisk, om medel til saltpetter sjuderiernes förbättring och upkomst i riket* (Åbo, 1771). Regarding the background of the dissertation. See Nauckhoff, "De kemiska uppfattningarna," pp. 436–437.

38. Jacob Johan Anckarström, *Chemiske tankar och rön, i synnerhet rörande kristalleingar och salpeter-sjuderier, samt genväg, til de chemiske operationers lättare verkställande* (Stockholm, 1774), pp. 44–48.

39. An exception was the institutional history of saltpeter production published by the university teacher in economics Anders Gustaf Barchaeus. The cause of the dissertation had been the public debate, but it did not itself contain an explicit position. See Barchaeus, *Academisk Afhandling, Om saltpeter-sjuderi inrättningen i Sverige.*

40. Lars Cronstrand, "Wälmente Tankar om Saltpeter-Sjuderiets förenande med Landtbruket til bådas bättre Trefnad och Förkofran," *Ny Journal Uti Hushållningen* (September-October 1794): 197–236.

41. Peter Jacob Hjelm, *Underrättelser om fördelacktigaste sättet att anlägga salltpeter-lador och att i stort tillvercka salltpeter* (Stockholm, 1799). See also Peter Jacob Hjelm, *Anwisning till bästa sättet att Tilwerka Salltpeter, Lämpad i synnerbet till smärre inrättningar och, till så mycket allmännare nytta* (Stockholm, 1799). The original was possibly Antoine Lavoisier, *Instruction sur l'établissement des nitrières et sur la fabrication du salpêtre* (Paris, 1777).

42. See, e.g., Seymour H. Mauskopf, "Gunpowder and the Chemical Revolution," *Osiris*, second series, 4 (1988): 93–118.

43. Anders Lundgren, "The New Chemistry in Sweden: The Debate That Wasn't," *Osiris*, second series, 4 (1988): 146–168.

44. Carl Fredrich Lund, *Afhandling, Om salpeter-sjuderierne och jord-bruket i Sverige* (Norrköping, 1770), 6–7. Quote in original: "det var så omöjeligit, utur en Salpeter-lada frambringa mycken Salpeter, som det orimmeligit, at göra fösrslag på 50 tunneland jords upodlande, hvaräst sjelva jordkretsen ej innehåller mera än 10 tunneland."

45. Eric Kiellberg, "Upgift, huruledes Riket skal kunna winna millioner Riksdaler, och en ymnoghet af fruktbar jord, för Landets upodling, igenom en f_rslagen Saltpetter-tilwerkning," *Dagligt Allehanda,* no. 246, October 29, 1770; Marcus von Petrus, *Daligt Allehanda,* no. 267, November 24, 1770; I. A. H., *Dagligt Allehanda,* no. 273, 1 December 1770; "Saltpetterhyttan i Jönköpings Län, den 3 Martii," *Inrikes Tidningar,* no. 81–84, October 17–28, 1771. Regarding Kiellberg's proposition, see also Sten Lindroth, *Kungl. Svenska Vetenskapsakademiens Historia 1739–1818,* vol I: 1, *Tiden intill Wargentins död (1783)* (Stockholm: Almqvist & Wiksells, 1967), pp. 252, 348.

46. Philopater (pseudonym), *Swenska Landt-Brukets Skada och F_rlust, af 360. 0000 Tunnor Säd årligen, Igenom Nu warande Saltpetter-Sjuderiets Inrättning* (Stockholm, 1775).

47. Nitrosus Redivivus (pseudonym), *om allmänna Riks-Hushållningens störe båtnad och winst* (n.p., n.d.).

48. "Om Rikets skada af Saltpeter-Sjuderier," *Hushållnings Journal,* March 1780: 3–18.

49. Michael F. Metcalf, "Parliamentary Sovereignty and Royal Reaction, 1719–1809," in *The Riksdag,* ed. M. Metcalf (Riksdag, 1987).

50. Pär Frohnert, "Administration i Sverige under frihetstiden," in *Administrasjon i Norden på 1700–talet,* ed. S. Supphellen, Det nordiska forskningsprojektet Centralmakt och lokalsamhälle—beslutsprocess på 1700–talet, ed. Birgitta Ericsson, no. 4 (Oslo: Universitetsforlaget, 1985), pp. 255–59; Metcalf, "Parliamentary Sovereignty," p. 121.

51. These conclusions are based on peasants' appeals to the parliaments of 1755–56, 1771–72 and 1789, collected and transcribed by Anders Claréus at Stockholm University, who is currently working on a doctoral thesis about the Estate of the Peasantry in Sweden. I am grateful for his kind help and generosity to share his working material with me.

52. This was indeed a rationale mentioned in the regulation itself. See Kongl. Maj:ts Nådiga Förordningen, Rörande Saltpeter-tillwerkningen i Riket, October 26, 1801.

11

Chemistry in the Arsenal: State Regulation and Scientific Methodology of Gunpowder in Eighteenth-Century England and France

Seymour H. Mauskopf

When I mention to fellow academics that I am studying the development of munitions, they invariably ask whether my focus is on Chinese gunpowder or the inception of European gunpowder in the late Middle Ages. When I respond that my focus is on the development of this technology in the eighteenth and nineteenth centuries, and particularly from the perspective of the history of science, my interlocutors are usually puzzled and put off. Surely, they imply, there can be little historical interest in the development of munitions after the initial ascendancy of firearms.

This disparaging view is underscored by much of the literature devoted to (or influenced by) the historical construct of the "military revolution." For instance, there is Michael Roberts' original formulation of a revolutionary change in military tactics around the turn of the seventeenth century in the Netherlands and then in Sweden. Although its original thesis has been contested and altered, there has been agreement that, whatever it was, this military revolution depended on the ascendancy of firearms in military tactics. Major innovations in firearms had run their course by 1700 and were to remain relatively unchanged for the next 150 years.[1] By implication, the propellant used in firearms had to have been a static and uninteresting technology after the seventeenth century.

Bert Hall has made this case in his recent study of munitions and firearms in the early modern period. Locating the critical development and closure of gunpowder weaponry in the sixteenth century, Hall characterizes the next several centuries:

> During the long interval when technological changes were few and far between, technology ceased to act in its most familiar mode—as a stimulant to other changes—and acted instead as a limitation on all actions. Those technologies that did emerge during the sixteenth century and later—edge-roller incorporation of gunpowder's ingredients, the successful casting of

> iron cannon in larger calibers, as well as various innovations in the lock mechanisms for small arms—all served mainly quantitative rather than qualitative ends. They made gunpowder and firearms cheaper, easier to produce, and still more readily available than ever before, but they did little or nothing to alter the basic characteristics of the guns themselves.[2]

In this essay I do not intend to overthrow these prevailing thoughts about a so-called military revolution. In essence, I agree with this hypothesis. Gunpowder technology, if one limits its meaning to the material propellant itself and its relationship to the guns in which it was used, did not change dramatically during the eighteenth and much of the nineteenth centuries.

Although the material aspect of munitions, as well as the guns in which they were used, may not have changed radically, other broader aspects of gunpowder technology did undergo many changes during the centuries after the closure of the military revolution. With a focus on the two most developed nation-states of the era, England and France, this essay will examine two of them: the developing role of the state in the regulation of munitions production and the role of science in the hoped-for amelioration of munitions.

My thesis has several components. The first concerns gunpowder itself. Being a mechanical mixture of three material ingredients—sulfur, charcoal and saltpeter, gunpowder has a complex and often unpredictable behavior. Each of its ingredients moreover had its own set of challenges for procurement and purification. There were also concerns with how to combine, or "incorporate," these ingredients, and in what proportions, to produce the "best" powder. In the eighteenth century, "best" usually meant the fastest burning, and, in the context of conventional smoothbore ordnance, this implied the most ballistically powerful. Beyond these issues lay the challenge of determining how to measure the ballistic force of gunpowder in order to ensure its satisfactory function in the field.

All of these challenges called for experimentation and regulation. The state could certainly provide the latter; science could perhaps provide the former. Up to the last quarter of the eighteenth century, gunpowder production in both England and France was essentially a private enterprise whose output for military consumption was regulated by the state and ballistically tested before being purchased. The eruption of the Seven Years' War in 1756 put an unprecedented strain on the means of producing munitions and led to organizational reforms in both England and France during the last quarter of the century. In both countries, the state took on a more proactive role not only in quality regulation but in actual production and in experimental research efforts to refine production and improve the final product.

The eighteenth century was the era in which "useful knowledge" came to be privileged as the natural outcome of scientific research. Rhetorical hyperboles about the significance of science-based research found their way to munitions, as with this quotation from the Dutch chemist Hermann Boerhaave:

> It were indeed to be wish'd that our art had been less ingenious, in contriving means destructive to mankind; we mean those instruments of war which were unknown to the ancients, and have made such havoc among the moderns. But as men have always been bent on seeking each other's destruction by continual wars; and as force, when brought against us, can only be repelled by force, the chief support of war must, after money, must now be sought in chemistry.[3]

Therefore, it was not unexpected that the methods of science—and scientists themselves—would be utilized by states in their competitive enterprise to improve the production and the quality of military munitions. However, as we will see, the transformation of scientific into useful knowledge was by no means unproblematic in the case of munitions.

Historiography

Before exploring the eighteenth-century challenges of gunpowder and the attempts by the state to meet them, I will address the relevant archival and published sources. The secondary literature has been sparse until recently. Thanks to the work of Jenny West, Wayne Cocroft, and Brenda Buchanan,[4] we are becoming well informed about the commercial, labor, and technological details of eighteenth-century military gunpowder production in England. Through the work of Patrice Bret,[5] the same holds true in France for the reforms of the 1770s and the 1780s and the Revolution. But we have less detailed knowledge of France's gunpowder production at the level of the individual powder mills earlier in the century than we have for England's. Secondary literature concentrates on the upper echelons of the gunpowder bureaucracy, especially the succession of *fermiers* who purchased the monopoly rights to supply the French military with gunpowder. It gives little information about how individual mills were run.[6]

Primary literature on gunpowder production from the middle of the eighteenth century to the early nineteenth century is richer for France than for England. Gunpowder appeared in Diderot's *Encyclopédie*[7] and received extensive treatment in the *Chymie expérimentale et raisonée* by the apothecary and chemist Antoine Baumé.[8] During the Napoleonic period, the massive *Traité de l'art de fabriquer la poudre à canon* by Jean Joseph Auguste Bottée de

Toulmont and Jean René Denis Riffault des Hêtres appeared,[9] as did the slighter but excellent *Instruction sur la fabrication de la poudre ou détails de divers procédés en usage pour la fabrication de la poudre, et la préparation de ses principes constituans.*[10] Much of my account (and that of my predecessors) relies on Bottée and Riffault's text.

Nothing comparable appeared in English until John Braddock's *Memoir on Gunpowder*, which discusses the principles of its manufacture and its proof. This was published in England in 1832.[11] In the late eighteenth century, all that appeared were several essays on the subject of gunpowder production, such as those on saltpeter and gunpowder analysis in Richard Watson's *Chemical Essays*.[12] I will use Watson, Napier, and Coleman as contemporaneous sources, and Braddock as a later control.

Structurally, this essay will be organized around the following topics: the national organization of gunpowder production; the mechanisms for ensuring production and quality control; overall organizational reforms; and the systematic and experimental investigation of gunpowder, its constituents, its explosive reaction, and its ballistic force.

National Organization of Gunpowder Production

France

At the start of the eighteenth century, gunpowder production in France operated under a form assumed in 1665. This was the system of leasing, or "farming" out, the production of gunpowder to a private entrepreneur. After surety had been provided to the government, this enterprise would be responsible for supplying a pre-established annual quantity of gunpowder at a pre-set price paid in advance installments.[13] The tenure of the farm was nine years, after which it was either renewed or passed on to a new entrepreneur. To keep up with rising needs of the Thirty Years' War, the government of Louis XIII experimented during the 1630s with various means of procuring and overseeing saltpeter and gunpowder production.[14] In 1663, Louis XIV, soon after assuming royal power, annulled the previous experiment and established the new arrangement whereby powder and saltpeter production would be under the control of one *commissaire général des poudres et salpêtres*.

The first of these, François Berthelot, began his first nine-year tenure on January 1, 1665.[15] Berthelot was originally commissioned to supply annually 200,000 pounds of powder "casked and barrelled, suitable for muskets and cannon," from thirteen mills or magazines,[16] being paid in advance at nine sous a pound.[17] He apparently carried out his commission with technical success and personal profit. In 1669 he was commended for having "overcome all the abuse that had slipped into the above-mentioned procurement of saltpeter and production of powder."[18] He remained in control of powder manufacture until 1690, except from 1685 to 1688 when a financial consortium managed to secure the farm.[19] The consortium was a failure and Berthelot was called back.[20] Yet one constructive outcome was the *ordonnance* of 1686, which specified procedures for manufacturing and testing.

This arrangement for securing saltpeter and gunpowder lasted through three quarters of the eighteenth century.[21] The consensus has been that it degenerated in the course of the eighteenth century. Charles Gillipsie underscores this view:

> The Powder Farmers of the reign of Louis XV, occupying themselves in finance and speculation, took no interest in technology. They preferred to contract out to local entrepreneurs the fabrication of much of the powder they were required to supply, and allowed the powder magazines and refineries to fall into neglect. They were under no obligation to furnish more than the powder stipulated in their contract.[22]

This tendency was aggravated by the trajectory of geopolitics in the eighteenth century. Powder demands, which had been very high during the wars of the second half of Louis XIV's reign, fell during the peaceful years of the Regency and the first part of the reign of Louis XV.

During the War of the Austrian Succession and the Seven Years' War[23] in the 1750s and the 1760s, *fermiers* failed to satisfy the gunpowder demand.[24] Domestic saltpeter procurement had become an inefficient process. Saltpeter was collected in France, as it had once been throughout Europe, by a corps of scavenging saltpetermen who had the right to search for crystals of saltpeter in basements, outhouses, and barns full of putrefying organic waste. These laborers refined the crude saltpeter once[25] and then sold it to the *fermiers*, who refined it twice more into usable form for making gunpowder. Understandably, saltpetermen were socially marginalized and strongly resented for their rights of domestic intrusion. Additionally, by the middle of the eighteenth century, their living had become very precarious. The gunpowder farm kept the price paid for saltpeter as low as possible and found it more profitable, through government subsidies, to purchase saltpeter imported from India through the Netherlands.[26] As a result of this circuitous operation, the gunpowder *fermiers* continued to make handsome profits, while the government and the saltpetermen were shortchanged. Given England's domination of India after the Seven Years' War, it became very risky for France to depend on this source for one of its most strategic materials.

England

Military powder was produced in eighteenth-century England by privately owned gunpowder mills under contract with the Ordnance Office.[27] A Warrant of 1683 ratified the organization of this office, which persisted until the reforms of the Crimean War. It was headed by an Ordnance Board consisting of five principal officers[28] who controlled the military stock for the armed forces. The entire Ordnance office staff comprised about forty persons, including aides to the board members and storekeepers at the garrisons and outposts. At the Tower of London there were about fifteen staff members, including the storekeeper of the saltpeter; two proof masters of powder were also stationed at Greenwich.

London and Greenwich were the principal centers for the Ordnance Office, whose administrative offices were located at the Tower and the Palace Yard, Westminster. London served as the principal point of entry for saltpeter from India and sulfur from Italy. For most of the century, Greenwich was the principal magazine for proving and storing gunpowder.[29] As a result, the mills supplying gunpowder to the military were all located in the southeast of England, near London. Because of this proximity, mill owners had ready access to the Ordnance Board and to the docks receiving imported raw materials; in turn, they were able to transport the processed gunpowder over short distances with minimal risk of explosion.

Privately owned powder mills were economically precarious operations; they remained vulnerable to the vicissitudes of foreign policy as well as inadvertent explosions in the course of production.[30] During wartime, powder mills had all the government orders they needed (and more); but when peace came, they ceased abruptly. At the start of a new war, neither the Ordnance Office nor the powder entrepreneurs could forecast its duration or its intensity. As a result, powdermakers were usually men "connected with other trades, of prominent social standing, in positions of influence in the City commerce, or a combination of these."[31] Commercial sales of gunpowder on the private market, particularly as barter in the slave trade, often provided lucrative compensation.[32]

Down to the reforms of the early 1780s, the government offered gunpowder contracts to about ten powder mills.[33] With each mill it contracted for a certain quantity of powder of a specified manufacture and quality.[34] Although this arrangement seems to have worked tolerably well during the wars of the first half of the century, it functioned poorly during the Seven Years' War:

> The Ordnance Office, faced with an unprecedented demand for gunpowder, did not receive the quantity expected. Throughout the war it made

> various unsuccessful attempts to obtain more cooperation from the powder makers, who themselves had considerable problems in achieving a product of a satisfactory standard and in the necessary quantities, especially when trying to balance the demands of government and private trade.[35]

Early in the war and continuing until 1759, the Office resorted to purchasing gunpowder from the Netherlands, which was simultaneously selling saltpeter to France.[36] The government also attempted various strategies to ameliorate the problem of domestic gunpowder supply. One was to prevent all contracted mills from selling powder on the private market. Another was to offer financial incentives to powdermakers who met their contracted quota reasonably well.[37]

Finally, in 1759, the government purchased one of the contracted powder mills, the Faversham Mills in Kent, so that powder production at one source would be directly under the control of the Ordnance Office. This tactic for increasing production failed,[38] although it did set a precedent that was to be expanded in the 1780s when the government purchased the Waltham Abbey Mills.

Mechanisms for Ensuring Production and Quality Control[39]

France

France maintained quality control processes on gunpowder production for over a century under regulations laid down by a royal *ordonnance* of 1686. This *ordonnance* seems to have been initiated by the crisis resulting from François Berthelot's temporary expulsion from the gunpowder farm (1685–1688) and by a consequent decline in the quality of powder[40]. The *ordonnance* specified all of the processes of gunpowder making: the refinement and treatment of the ingredients, the means of incorporating them to produce the final product, and the methods to test their ballistic efficacy. Most of the specifications were ratifications of what had been traditional production procedures.

Gunpowder is a mixture of three ingredients: sulfur, charcoal, and saltpeter (or "niter" [KNO_3]). The first, imported from Italy and Sicily, was the simplest and least problematic to refine. The raw sulfur, or rock sulfur (*soufre en roche*), was broken up and melted in iron kettles under gentle heat. After allowing the melt to rest for four to five hours, during which the surface scum was skimmed off, the liquid sulfur was decanted and allowed to crystallize.[41] Sulfur, however, was considered to be the ingredient most deleterious to the guns, and there was considerable interest in determining whether a sulfur-free powder was feasible.

The other two ingredients required more detailed regulations. Charcoal was made from wood by charring it in brick-lined pits or in furnaces. In France, the principal issue concerning this ingredient was the wood source. Craft tradition dictated that only certain types of light, soft wood from young, thin branches were suitable for gunpowder production. The traditional wood in France was black alder (*bois de bourdaine*), which was the only wood source permitted by the 1686 *ordonnance*.[42] In 1701 Louis XIV confirmed the rights of powdermakers to prune young black alder branches from trees in the royal forests.[43] As we will see, during the course of the century, investigators questioned whether black alder was indeed the optimum source of charcoal for gunpowder.

Even when produced naturally under prime conditions, saltpeter is always intermixed with other salts, which must be removed or converted chemically into saltpeter. The *ordonnance* of 1686 specified that crude saltpeter must undergo three refinements[44] to be perfectly free of grease and salt. Each refinement consisted of three stages: (1) dissolving in water the salts contained in the crude saltpeter earth and adding potash (in the form of wood ash) to convert some of these salts into saltpeter; (2) collecting and boiling up the resultant liquid and then skimming off the scum; and (3) cooling the liquid to precipitate out the saltpeter.[45]

The *ordonnance* specified the standard for incorporating these three ingredients into gunpowder. After being wetted, they were pounded together in a stamping mill for a minimum of twenty-four hours.[46] Stamping mills housed batteries of mortars made of hard oakwood and vertical wooden pestles. For proper incorporation, the mortar cup was ovoid in shape to recirculate the powder that spattered on the walls with each stamp. Each pestle was also capped with a pear-shaped bronze bearing. The motive force, originally provided by animal power, evolved into a battery of pestles driven by a waterwheel that stamped at least 54 times per minute.[47]

Although the proportion of the ingredients was not formally specified, it had been (and was to remain into the nineteenth century) 75 percent saltpeter, 12.5 percent sulfur, and 12.5 percent charcoal.[48] Only one grain-size of military powder (small-grained musket powder) was produced instead of two grain-sizes of powder (large-grained for cannon), according to the earlier procedure of making military powder.[49]

Finally, this *ordonnance* and its successor[50] established the procedure for measuring the ballistic force of gunpowder and the threshold standard. The test instrument was a small mortar known as a *mortier eprouvette*, or *mortier d'ordonnance*. One ascertained the force by measuring the range of a standard cannon ball (*boulet de fonte*) of 60 pounds when shot from a *mortier-eprouvette* containing three ounces of powder and elevated at 45°. The standard range was 50 fathoms (*tois*); the threshold for powder made before the new *ordonnance* was 45 fathoms.[51] By 1769, the threshold was increased to 90 fathoms because "the procedure of powder-making had been perfected in the interval so as greatly to exceed this limit."[52] This increase occurred despite the gunpowder farm's severe deterioration. I have no explanation for this disparity. At this time, the use of the *mortier eprouvette* and the procedure of proof were specified in great detail:

> The artillery officer charged with the proof of the powder presented for acceptance by the fermier, chose, at his pleasure, a tenth of the two-hundred pound barrels and a twentieth of those of one hundred pounds. After verifying their weight, he took a sample of powder from each of the barrels for testing. For the powder to be accepted, a charge of three ounces of each of the samples had to give a range of ninety fathoms. The powder was rejected if the range was only seventy-five fathoms. After reworking, powder could only be accepted with a range of eighty fathoms and was rejected if it fell below sixty-six fathoms.

Also detailed was the procedure for barreling the powder (including the nature and thickness of the wood to be used and the quality of the canvas for sacking) and preparing the reports of the proof, which were to be signed by a number of officers, including the presiding officer of the tests, the quartermaster, and the director of artillery.[53] This testing procedure remained in place until the middle of the nineteenth century, although the threshold range was raised once again before the end of the eighteenth century to 100 fathoms.[54]

England

The English directive of the Ordnance Office for gunpowder manufacture corresponded to the French *ordonnance*. In contrast to the French, the English set their proportions of saltpeter, charcoal, and sulfur at 75, 15, and 10, respectively, per 100 parts of gunpowder.[55] West cites an example from 1753 in which the government details powder manufacture for the Ewell Mills:

> The Board reported that each barrel [of 300 contracted for] should be prepared from 80¼ lb double refined saltpetre, 15 lb charcoal, and 12¾ lb refined brimstone [sulfur]; the total composition was to be divided into three separate parcels of 26¾ lb saltpetre, 5 lb charcoal and 4¼ lb. brimstone, each of which was to be worked for five hours to reduce the entire quantity from 108 lb to 100 lb.[56]

By the middle of the century, saltpeter and sulfur were being imported respectively from India and Italy. The chemist Richard Watson, a prominent writer on gunpowder in the late eighteenth century knew of no saltpeter generation or collection in England.[57] Wood for charcoal was obtained from the Wealden areas of Surrey and Kent[58] and posed the main challenge for English powdermakers, since timber was becoming increasingly scarce.

The East India Company exported and sold its saltpeter to the Ordnance Office. Conditions in India, with plentiful village sewage and cow sheds, high temperatures, and a long dry season, were ideal for the formation of earth rich in various nitrate salts. To eliminate the other nitrates and to concentrate the saltpeter, wood ash (containing potash [KCO_3]) was mixed with the earth, and water was slowly trickled through it. The resultant solution was boiled and evaporated to form impure crystals of saltpeter.[59] To calculate the purchase price, this impure saltpeter export (or *grough petre*[60]) underwent inspection, weighing, and "refraction" to approximate the amount of impurities. After its purchase, the saltpeter was issued to the mills according to contractual agreements.[61] The mills were expected to refine the saltpeter at least twice. Like the French procedure, English saltpeter refining involved boiling, skimming, filtering, crystallizing, and drying.[62] Subsequently, crystallized saltpeter was often fused into cakes before use.

Like their French counterparts, the English powdermakers directly purchased the imported sulfur and then refined it.[63] Similarly, there were debates in France and England over the function and necessity of sulfur in gunpowder explosion,[64] especially since sulfur was known to foul gun barrels. John Braddock succinctly promoted the utility of sulfur:

> Gunpowder made without sulfur has, however, several bad qualities; it is not, on the whole, so powerful, nor so regular in its action; it is porous and friable; possesses neither firmness nor solidity; cannot bear the friction of carriage, and in transport crumbles into dust. The use of the sulfur, therefore, appears to be, not only to complete the mechanical combination of the other elements, but being a perfectly combustible substance, it increases the general effect, augments the propellant power, and is thought to render the powder less susceptible of injury from atmospheric influence.[65]

While there was no debate over the necessity of charcoal in gunpowder, two issues concerning charcoal production occupied the attention of both powdermakers and investigators: the optimal wood for gunpowder charcoal and the best manner of charring wood.

Although the British government ordained no specific wood source, English powdermakers subscribed to the traditional preference for light, soft timber consisting of relatively thin branches. This included alder, willow, and black dogwood. However, Watson noted that experiments by Antoine

Baumé in France showed that hard wood could furnish equally effective charcoal for gunpowder. This was promising for English powdermakers, "as it is not always an easy matter for them to procure a sufficient quantity of the coal of soft wood."[66] In the late eighteenth century, both England and France conducted tests to ascertain which wood yielded powder with the greatest ballistic force.

The second issue concerned optimum methods for charring this wood. Traditionally, wood was charred in open pits;[67] but, in 1786, Richard Watson[68] offered an improved way of making charcoal. Adapting the methods of the physician and chemist George Fordyce, he proposed charring by distillation in closed iron cylinders:

> The wood to be charred is first cut into lengths of about nine inches, and then put into the iron cylinder, which is placed horizontally. The front opening of the cylinder is then closely stopped: at the further end are pipes leading into casks. The fire being made under the cylinder, the pyro-ligneous acid, attended with a large portion of carbonated hydrogen gas, comes over. The gas escapes, and the acid liquor is collected in the casks. The fire is kept up till no more gas or liquor comes over, and the carbon remains in the cylinder.[69]

William Congreve tested this new method of charring wood in the autumn of 1786 and discovered that it substantially increased the ballistic force of the resultant gunpowder.[70] In the 1790s, the Royal Gunpowder mills at Faversham and Waltham Abbey adopted "Cylinder powder."[71] It was already noted by 1801 that the more powerful cylinder powder had reduced the charges used in ordnance by one-third.[72]

Because of the perceived danger of explosion from overheating, Britain outlawed stamping mills for incorporating the three ingredients in 1772, except in certain Sussex mills that produced fine sporting powder.[73] By then, most of the powder mills in the London area had already switched over to the alternative method of incorporation: cylinder (or edge-runner) mills.[74] Performed in a "slight wooden building and boarded roof,"[75] this incorporation mill

> consists of two stones vertically placed, and running on a bed-stone. On this bed-stone, the composition [i.e. the preliminary mixed ingredients] is spread, and wetted (not with sal-ammoniac, urine, &c. as some authors state, but) [sic] with as small a quantity of water as will, together with the revolutions and weight of the runners, bring it into a proper body, but not into a *paste*. After the stone runners have made the proper number of revolutions over it, and it is in a fit state, it is taken off. . . . These mills are either worked by water or by horses.

Only a limited amount of material (40–50 pounds) was worked at a time to obviate the danger of explosion from sparks forming between the runners and the bedstone.[76] It was nonetheless difficult to estimate "the proper number of revolutions" for cylinder mills. Napier, who preferred stamping over incorporation, criticized the quality of powder incorporated in these mills, because the powder was processed for only "seven or eight hours instead of twenty-four, which was the usual time when stamping mills were employed."[77] During the second half of the eighteenth century, the incorporated powder was further subjected to a screw press. Here the powder was placed between copper plates and subjected to intense pressure, much like that of a printing press.[78] Until the regime of William Congreve, military powder was of one grain size only. Congreve introduced the production of two grain-sizes: a small-grained powder for small arms and cannon priming, and a larger-grained powder for cannon.[79]

Two proof masters, who were members of the Ordnance Office, conducted the official proving of powder.[80] Unlike their counterparts in France, they apparently did not rely on one standard proof test.[81] During a 1740 public proving at Woolwich of allegedly inferior English powder, the proof masters used three different methods. The first was the "vertical eprouvette,"[82] which involved determining how high a given weight (20 lbs. 7 oz.) could be raised by the firing of a given weight of powder (2 drams) under it. The second and third methods each involved proving powder by determining the range in which a standard quantity of powder could fire a given weight of shot from a particular firing piece: In one case, one-quarter ounce of powder fired a 12-pound shot from a 5⅔-inch mortar; in the other, two drams of powder fired a half-pound shot from a swivel gun.[83]

In response to criticisms of ordnance during the American Revolution, the procedure for proving gunpowder was clarified. The procedure for testing (all carried out at Purfleet) was laid out in detail. Most notably, the supervision of gunpowder proof was assigned to the Comptroller of the King's Laboratory at Woolwich:

> After every proof, a report of the state of the powder proved, signed by the Comptroller, the two Firemasters and the Storekeeper is made to the Master General and the Board; who, in consequence of that report, direct what powder shall be received as serviceable into the King's Magazines.[84]

The report noted major discrepancies in results obtained in recent tests of the vertical eprouvette, formerly "the only established mode of Proof,"[85] and of proof with the mortar. There is no evidence that proving powder by mortar completely supplanted that by vertical eprouvette or other means.[86] In 1811, Congreve asserted that only two methods of proof were being used: the measurement of the mortar's range for the large-grained cannon powder, and the measurement of the musket ball's penetration into a wooden board for the small-grained musket powder. In contrast, Congreve noted the shortcomings of the vertical eprouvette:

> Nothing could be more deceptive than the former mode by the vertical eprouvette, in which a soft and rotten-grained powder would produce a much greater effect than the hard and good serviceable powder; and the results of which would be continually at variance with the effects in the modes of actual service.[87]

Yet in 1830 Braddock claimed that the vertical eprouvette, which he also found to be unreliable, was still in use at the Waltham Abbey Mills,[88] as was the mortar.[89] Braddock also described several other ballistic-testing procedures and instruments; these included flashing and the pendulum eprouvette, which was developed by Charles Hutton[90] and called "carbine proof" (essentially, Congreve's proof for musket powder).[91] This suggests that, despite Congreve's assertion, a variety of methods of proof continued to be employed, even by the government manufacturers of military powder, well into the nineteenth century.

Overall Organizational Reforms

France

As we have seen, French gunpowder production had not been adequate for military needs in mid-eighteenth-century wars. Because France's ready access to saltpeter in India was lost to its perennial enemy, England, French munitions production became highly vulnerable. With the accession of Louis XVI and Controller General A. J. M. Turgot in 1774, reformation of the gunpowder farm became a high priority. In May 1775, Turgot abrogated the lease to the gunpowder farm and, in its place, set up a new gunpowder administration, the *Régie des Poudres*, which Charles Gillispie defined as a "privately financed commission, chartered to serve the public interest and responsible to the controller-general."[92]

Although the monopolistic character of the earlier gunpowder farm was retained, its financial arrangements were changed to suppress the sheer incentive for profit. Four directors were appointed, each of whom had to put up a loan to the government of one million livres upon taking office. Each was paid an annual salary of 2,400 livres and received living quarters in the Paris Arsenal. They also received interest on their loans of one percent above the official rate and bonuses when the production of saltpeter and gunpowder exceeded set minimums.[93] Three of the directors had previous technical or fiscal association with the defunct gunpowder farm. The fourth had none; nevertheless, he was appointed because of his administrative ability as a *fermier general*, or tax farmer, and more particularly because of his distinction in chemistry, so "essential and necessary for this kind of administration."[94] This director was none other than Antoine-Laurent Lavoisier, one of the greatest of all scientists and the founder of the Chemical Revolution.

When Lavoisier joined the *Régie des Poudres* in the summer of 1775, he was well into the research program that would lead to his reformulation of chemistry. In the "crucial year" of 1772, he had carried out what have always been seen as classic experiments on the combustion of phosphorus and sulfur. There he showed that weight gain of the resultant calces of these substances was associated with diminution of the volume of air in which the combustion took place. Combustion seemed to involve the absorption of "something" from the air.[95] By the spring of 1775, thanks in part to a major discovery by the English chemist Joseph Priestley, Lavoisier was in a position to clarify what this "something" was: a gas that seemed "more respirable, more combustible, and consequently . . . even purer than the air in which we live."[96] The wider implications of this clarification for chemical theory were still several years off. Nevertheless, Lavoisier's work had already attracted international notice; by 1775 he had already advanced two of the three steps toward the highest rank in the *Académie des Sciences*.[97]

Whether or not Lavoisier initiated the gunpowder reform,[98] he did assume the lead among the directors and did bring to bear his manifold interests and talents in managing the whole enterprise. Known best as a great scientist, Lavoisier was also a brilliant financier, economist, and administrator, for which his activities as a tax farmer gave plenty of scope. Lavoisier thus represented an early example of a modern government scientist. Of all of his governmental scientific service, that in the *Régie des Poudres* was the longest and probably the most significant. The successful turnaround of the gunpowder industry in France after 1775 probably owed more to Lavoisier's administration than Turgot's structural reforms.[99]

In 1789, Lavoisier wrote triumphantly about the success of the *Régie des Poudres*. France no longer depended on foreign saltpeter, and it had increased its gunpowder production rate to a point at which it managed to supply all of its allies in the American War of Independence. Qualitatively, French gunpowder had become "the best in Europe."[100] Lavoisier attributed these successes primarily to the application of scientific concepts to production techniques. Throughout his tenure in the *Régie des Poudres*, Lavoisier had worked to deploy science in the gunpowder industry. As soon as he was appointed to the *Régie* in August 1775, Lavoisier launched what was to be his most ambitious effort to utilize science in the so-called "crash program" for artificial saltpeter production.[101] He enlisted the French *Académie des Sciences* to support a prize competition for the best process for producing saltpeter. Two publications ensued: a systematic survey of the available literature on the subject, under the auspices of the *Académie*, and a book of instructions for cultivating saltpeter beds, under the names of *Regisseurs des Poudres*.[102] Although prizes were awarded and artificial saltpeter beds established, in 1789 Lavoisier reported "*peu de succès*" in this domain.[103] Through a variety of other means, he was nonetheless able to bring about a dramatic increase in the production rate of saltpeter.[104]

In his 1789 claims to success, Lavoisier wrote specifically about instruction in the sciences, especially chemistry and mathematics, for would-be powdermakers and for members of the gunpowder administration.[105] The research of Patrice Bret details exactly what was instituted.[106] First, in addition to the publications associated with the saltpeter production prize, the *Régie des Poudres* published a number of handbooks for saltpetermen on various aspects of saltpeter refining. Beyond this short-term diffusion of scientific knowledge and methods, Lavoisier envisaged the systematic scientific education of a cadre of gunpowder administrators. Starting in 1783, an "*école des poudres*" was set up in locations in and around Paris.[107] Its program took two to four years to complete and had theoretical, applied, and practical components. By 1789, this course of study in munitions production and administration had produced, in Bret's words, "a veritable corps of powder engineers."[108] In fact, during the Napoleonic regime, this "veritable" corps became a literal one, since all powder mill administrators had to be graduates of the École Polytechnique.

England

In England, too, munitions production had been inadequate during the Seven Years' War. The situation remained serious in the following decades, even though England had access to India's rich source of saltpeter. However, the overall quality of the gunpowder, tested during the American Revolution, proved to be notoriously poor; its ballistic force was deficient and it lacked durability.[109] After a serious explosion at the Faversham Mills in 1781, Pitt considered selling off the state-owned mills, "it having been represented to him that the powder merchants could make better gunpowder, and much cheaper, than the King's servants."[110] Yet Pitt's government was persuaded in fact to do just the reverse;[111] in 1787 it acquired another set of powder mills, those at Waltham Abbey in Essex.

In the reforms of 1783, the supervision of gunpowder manufacture at the Faversham Mills (and the Waltham Abbey Mills, after their acquisition by the government) was placed in the hands of the Comptroller of the Royal Laboratory, Woolwich.[112] In the seventeenth century this laboratory had been established primarily to make and test weapons.[113] By the late eighteenth century, under the Comptroller's supervision, it was also authorized to set the standards for powder contracted with private powdermakers.[114] The agent behind these changes appears to have been Deputy Comptroller, Captain William Congreve (1743–1814)[115], who in effect acted as Comptroller during much of the 1780s and, as a major, was awarded the office in 1789.

Congreve can be appropriately viewed as Lavoisier's counterpart across the Channel. Like Lavoisier, he sought to ameliorate the production of munitions through systematic experimentation as well as closer coordination with and supervision by military agencies. Congreve was a professional artillery officer. Although his father was a civilian, his family did have military connections. Aided by an uncle who was a military officer, Congreve secured a commission in the Royal Artillery in 1757 when he was fourteen. Rising up in the ranks to the office of captain by 1772, he served in both the Seven Years' War and the American War.[116]

Unlike Lavoisier then, Congreve was not principally a scientific investigator, but he did exhibit a strong bent for mechanics and experimentation throughout his life. In 1773, he became a founding member of a "Military Society" devoted to "improving military, mathematical, and philosophical knowledge" at Woolwich, where he was posted.[117] In the early 1770s, Congreve made his first mark by designing an efficient carriage for transporting three-pounder guns up steep ascents and over rough ground. The carriage, first employed in the American War, was the reason his commanding officer commended him to the Board of Ordnance: "His knowledge of the mechanical powers, so very necessary in our service, has always been his delight and study; in which my humble opinion is that he stands foremost among us."[118]

In the late 1770s, the facilities of the Royal Laboratory were expanded, and a "Repository for Military Machines," offering theoretical instruction to artillerymen, was established. This instructional facility appears to have been the brainchild of Congreve himself. By the early 1780s, first under the Comptroller Colonel George Napier and more comprehensively under Congreve, the Royal Laboratory conducted extensive experiments with the intention of ameliorating the manufacture of military powder produced at the newly acquired royal powder mills.[119] Powder making in England, unlike that in France, remained a mix of public and private enterprises; nothing like the French institution of general scientific instruction was made available to powder mill officials.

Systematic and Experimental Investigation

An experimental approach to gunpowder dates back at least to the early days of empirical science in the middle of the seventeenth century. It was recognized that the explosive force of gunpowder arose from the sudden release of a great volume of "air" from the solid powder. Spanning much of the eighteenth century, experimenters measured the volumetric ratio of "air" to gunpowder and determined the exact nature of this "air." They discovered new gases, such as "fixed air" and "inflammable air,"[120] and subsequently developed pneumatic chemistry, which refined their understanding of gunpowder's explosive reaction as well as its by-products.[121] There were also experimental investigations carried out to determine what factors in the nature of the components of gunpowder, their proportions, and processes of combining them produced the "best" powder, or that with the greatest ballistic power.

France

French gunpowder research carried out in the 1750s was virtually a prototype for the blend of scientific and practical research conducted by Lavoisier and some of the English researchers several decades later. I have elsewhere referred to such French efforts as the "d'Arcy-Baumé research program," named after the Irish expatriate military officer and mathematician Patrice d'Arcy and the apothecary and chemist Antoine Baumé. Together they performed some of the first comprehensive and systematic tests correlating the ballistic force of gunpowder with variations in its composition (i.e. the proportions of the three ingredients). They also investigated the nature and function of some of gunpowder's ingredients.[122] Although their research had no obvious institutional connection with contemporary French munitions production, its purpose was clearly to gain practical rather than abstract knowledge about gunpowder. Subsequent testing was carried out at the gunpowder mill at Essonne.

Lavoisier's principal scientific focus on gunpowder was the development of methods, including geological forays, for ensuring the saltpeter supply. Nevertheless, he carried out research from time to time concerning the amelioration of gunpowder in the vein of the d'Arcy-Baumé program. Lavoisier tested the assumption that black alder was indeed the best wood for gunpowder charcoal, along with Baumé's hypothesis that charcoal from other wood, even hard wood, could be substituted without a loss of quality. A fellow *regisseur*, Le Tort, also conducted the same research, carrying out comparative ballistic tests at the powder mill at Essonne in 1785. He determined that charcoal made from poplar wood produced somewhat stronger powder than that made from black alder.[123]

Lavoisier's other concerns included the role of water in generating the ballistic force of gunpowder, as well as the comparative merits of stamping and cylinder (or edge-runner) mills in incorporating gunpowder ingredients. Although Lavoisier asserted that powder incorporated by edge runners produced greater ballistic power than that made in stamping mills, the latter predominated in France well into the nineteenth century.

Beyond these specific issues, Lavoisier's reformulation of chemistry identified the process of gunpowder explosion as one of combustion. Perhaps the most practical of these chemical issues concerned saltpeter, whose nature and refining procedure had come into some controversy in the early and mid 1770s. I have argued that one of Lavoisier's pivotal experiments in the development of the new chemistry—the determination of the nature of nitric acid in 1776—was a by-product of his efforts to comprehend the chemistry of saltpeter. As soon as this was clarified, Lavoisier promoted the substitution of pure potash for the traditional but less pure wood ash in saltpeter refinement. This probably signified the greatest factor in increasing the annual production of saltpeter under the *Régie des Poudres*.[124]

By the time he published *Traité élémentaire de chimie* (1789), Lavoisier was able to combine his own theoretical reformulation of chemistry with experimental discoveries concerning the nature of gunpowder explosion. He thus arrived at the first recognizably modern thermo-chemical explanation for this reaction. The principal discovery, made in France by the chemist Claude-Louis Berthollet and in England by Tiberius Cavallo, was the identification of the "airs," or gases, produced by the explosion.[125] Lavoisier explained the reaction in terms of his oxygen theory of combustion and his caloric theory of heat.[126] It became the standard scientific account of gunpowder explosion until the second half of the nineteenth century. Lavoisier and the *Régie des Poudres* also supported the experimental testing of chemical alternatives to saltpeter, the most important of which was Berthollet's potassium chlorate.

England

English scientists made important contributions to the eighteenth-century research tradition of measuring the volume and determining the nature of the "air" released during gunpowder explosion. The most notable contributors in the first half of the century were Stephan Hales and Benjamin Robins. In the second half, English pneumatic chemists, such as Joseph Priestley, included gunpowder explosion in their studies of gases.

During a brief residency in England, the Dutch-born scientist Jan Ingen-Housz (1730–1799) attempted to understand the chemistry of gunpowder explosion in light of some of the discoveries in pneumatic chemistry; but he did so in terms of the pre-Lavoisian phlogiston theory. He published the results in the *Philosophical Transactions of the Royal Society of London* in 1779. It was well known then that Lavoisier had discovered that nitric acid (and hence saltpeter) contained "dephlogisticated air," or oxygen. English phlogistonists believed that charcoal was a compound substance, one of whose components was "inflammable air," or hydrogen. Ingen-Housz offered a vivid summary of the explosive force of gunpowder explosion:

> Nitre [saltpeter] yields by heat a surprising quantity of pure dephlogisticated air. Charcoal yields by heat a considerable quantity of inflammable air. The fire employed to inflame the gunpowder extricates these two airs, and sets fire to them at the same instant of their extrication.[127]

The explosive reaction of gunpowder was particularly violent because it involved the release and reaction of these two gases from their compressed state in the solid powder. But Ingen-Housz gave no account of exactly what was the resultant gaseous product of this burning of dephlogisticated and inflammable air. Of course, the product of that reaction is water, as Henry Cavendish determined in 1781. However, water (or water vapor) is not the gaseous by-product of gunpowder explosion. Two years later Berthollet and Cavallo separately ascertained that the actual product was carbon dioxide. A foreign resident in London, Cavallo made his determination by a crude yet effective experiment.[128] Like Ingen-Housz, he had no documented connection with the Ordnance Board or the Royal Laboratory—nor, for that matter, with the production of gunpowder in England.

Other investigators had closer military ties. One was the loyalist American expatriate Benjamin Thompson (later Count Rumford), who carried out studies of the interior ballistics of gunpowder explosion with the explicit aim of improving military weaponry. The other was Charles Hutton, a professor of mathematics at the Royal Military Academy, Woolwich. Basing their investigations on Benjamin Robins' research, they were primarily concerned with ballistics rather than with gunpowder *per se*: Thompson with interior ballistics, particularly the determination of the pressure of fired gunpowder, and Hutton with exterior ballistics.[129]

Finally, there were the experiments carried out at the Royal Laboratory in Woolwich, begun by George Napier, Comptroller in 1782, and continued at least through the decade and probably much longer by William Congreve. Prior to West's book and Cocroft's recent preliminary study of Congreve's role at the Royal Laboratory, Congreve was known primarily as the father of the rocket innovator, William Congreve Jr. He has only recently emerged as the major figure in the reform of munitions in England during the late eighteenth century.

Understandably, Congreve was primarily concerned with the practical aspects of gunpowder. There is little evidence that he pursued its more theoretical aspects as did chemists such as Lavoisier and experimental physicists such as Rumford. Ordered by the Master General of Ordnance, Congreve instituted a series of parameter-variation tests of every aspect of powder performance and production. These tests also measured the instruments themselves—namely, those used to determine the ballistic capability of the powder. Here he discovered a discrepancy in the proofs obtained from the vertical eprouvette and from a 13-inch mortar. By 1787 he was recommending the 13-inch mortar along with Hutton's eprouvette.[130] Congreve's experimental results probably lay behind the recommendation of the Parliamentary Commission report noted above. But, as I have also noted, it is not clear to what extent the mortar and the Hutton eprouvette had supplanted the other means of proof by the end of the century.[131]

It is beyond the scope of this essay to enumerate all of Congreve's experiments[132]; therefore, I will provide an overview of some of the most important ones. In 1783, Congreve carried out comparative tests of powders manufactured in the proportions of constituents used in France, Sweden, Italy, Poland, and Russia. In 1785 and 1787, he measured the strength of gunpowder containing charcoal made from different kinds of wood—alder, black and white dogwood, willow, hazel, and lime. From 1783 to 1784 he tested glazed and unglazed as well as angular- and spherical-grained powder.[133]

In the 1788 manuscript titled "A State of Facts Relative to Gun Powder," which recounts many of these experiments, Congreve furnished a list of conclusions. Regarding the comparative tests of various national powders, Congreve concluded that "the English has the advantage of the others."[134] As for the wood sources, when charred in the traditional manner, black dogwood proved the strongest, followed by white willow. With wood charred in Watson's closed cylinders, alder wood of eleven years' growth and small white willow of three, four, and six years' growth yielded powder of nearly equal ballistic strength.[135] In 1791, Congreve listed black dogwood, alder, and white willow as the best types to be used in gunpowder production.[136] Angular powder was found to be stronger than spherical, and powder "moderately glazed and the Angles not broke off their Grains" seemed to be as strong as unglazed powder.[137]

No doubt the most significant innovations in powder making under Congreve were the adoptions of Watson's proposal for closed-cylinder charring of charcoal and the two different grain sizes for musket and artillery powder. In his 1788 manuscript, Congreve laconically suggested that charcoal made in closed cylinders "made much stronger Gun Powder than that which is charred in a common Pit, or after the Method practiced at Battle."[138] Yet one of

Watson's obituary notices provides evidence that Congreve had spoken to Watson about more extensive and quantitative tests. Proof data from these tests using both the vertical eprouvette and the mortar yielded a ratio of ballistic force of 100:60 (in round numbers) in favor of the cylinder powder.[139] Congreve himself noted in 1811 that, even with reduced charges, the more powerful gunpowder resulted in bursting mortars and even loss of life.[140] Less information is available to document how Congreve came to adopt two grain-sizes for military powder. He wrote in 1811 that he had "ascertained by experiment, that although the small-grained powder is stronger in *small* quantities, and therefore fitter for musquetry, the large-grained powder is better for the charges of cannon"; however, the 1788 manuscript offers little insight.[141]

Jenny West points to a source in Ingen-Housz's Royal Society paper, which asserted that "impalpable powder . . . rammed into a squib" did not explode but merely burned rapidly. This quick propagation of fire depended on the interstices between the gunpowder grains. From this, Ingen-Housz concluded that "the size of the grains of gunpowder must be proportionate to the size of the fire arms to which it is destined, the greatest fire arms requiring in general grains of the largest size."[142] More research is needed to determine whether this brief but very suggestive passage did in fact inspire Congreve to adopt different grain-sizes for different guns.

Conclusion

Both Lavoisier as a *Régisseur des Poudres* and Congreve as Comptroller of the Royal Laboratory, Woolwich, realized an Enlightenment ideal of public officials: government administrators who effectively deploy science-based "useful knowledge" to reform and improve industry. Their varied styles reflected larger political and cultural differences. For example, Lavoisier's *Régie des Poudres* was a more centralized, government-controlled industry than the quasi-private gunpowder mills that existed in Congreve's England. Hence, Lavoisier was able to institute comprehensive educational requirements for gunpowder administrators that were not feasible in England. Moreover, Lavoisier had models and precedents for the development of his educational reforms in the science-based practical schools that had been instituted by the state in France during the second half of the century. And, of course, Lavoisier was one of the monumental theoretical and experimental scientists of the century.

Yet, even with these differences, the claims to success of the two reformers were remarkably similar. I have already noted Lavoisier's claim that reform, scientific investigation, and scientific education had enabled France to produce gunpowder that was "the best in Europe." Congreve did not put forth his own claim quite so concisely, but Wayne Cocroft has: "Through his systematic practical research into the manufacture of gunpowder and his ability to enact change Congreve had transformed British powder from one of a notorious quality to a world standard."[143]

These kinds of claims naturally invite historical inspection. Part of the historical problem relates to the scientific/cultural context of these claims; their rhetoric is in the vein of general eighteenth-century claims to the value of scientifically based "useful knowledge"—claims which historians have become wary of accepting at face value. Other contexts in which the claims were advanced may be significant. For instance, Congreve's specific claims for gunpowder improvement were made as part of a polemical defense of the economies achieved under his administration of government gunpowder manufacture. I myself have recently questioned Lavoisier's claim about the improvement of the ballistic force of gunpowder, although I admit that it is difficult at this historical distance either to debunk or to substantiate it absolutely.[144] But I have no reason to doubt his claims for the dramatic improvement of saltpeter and gunpowder production made under the *Régie des Poudres*.

Congreve made three principal claims for his administration: gunpowder became more powerful, better suited to different caliber weapons, and more durable.[145] The first he attributed mainly to the employment of Watson's method of making charcoal, the second to the adoption of two sizes of powder grains, and the third to "a variety of circumstances" involving virtually all of the steps in the manufacturing procedure.[146] I see little reason to doubt the first of these claims, which Congreve's own experiments support. Yet, as noted earlier, we now realize that this increased strength may have owed as much to the increased density of timber sources as to the new style of charring wood. Justification for the second claim about the need for two grain-sizes can be found in Congreve's own experiments and possibly in the theoretical basis of Ingen-Housz's paper. In retrospect, Congreve's work constituted a real advance. On the other hand, his argument for large-grained powder for cannon later contradicted that of later nineteenth-century advocates.[147] His third claim of durability is perhaps the most difficult to substantiate, although he himself cited data. Yet little doubt remains that the comprehensive reforms Congreve instituted, particularly the attention to the purity of ingredients[148] and other aspects of production, did improve the quality and the uniformity of the gunpowder, although in ways of which his contemporaries and he were frequently unaware.

Where do I stand, then, on the role of science in the reform of the gunpowder industry and the improvement of gunpowder? Although I do not see its role in quite the simple, roseate terms of my late-eighteenth-century protagonists, I do think that science—particularly systematic experimentation—fostered both technical and financial improvements. Systematic and quantitative tests are best seen as part of a more general evolution of state supervision of the gunpowder industry. Systemic procedures, uniformity, control, and rational innovation came to permeate the industry far more deeply than they had at the start of the century, both in France and in England. Given the historical conundrum of gunpowder, science, and the Military Revolution, I want to be cautious with my claims. Although I see gunpowder as improving during the last quarter of the eighteenth century, I do not think that its hardware was "revolutionized"—that is, it did not radically change in its basic material composition, its form, or its mode of production. A real transformation only occurred during the second half of the nineteenth century in the wake of the protracted revolution in gunnery.[149] By the turn of the twentieth century, black powder was superseded by high explosives for military purposes.

Although revolutionary change failed to occur in gunpowder during the eighteenth century, the same cannot be said for its producers. The changes instituted for black powder after 1860, and the development of smokeless powders, depended almost entirely on government-supported systematic experimentation in munitions and on men trained as scientists and science-based engineering and artillery officers.[150] In this crucial sense, Lavoisier's and Congreve's deployment of science in munitions production established a revolutionary process.

Notes

1. The principal changes involved the development of the socket bayonet (supplanting the pike) and the flintlock musket (supplanting the matchlock). See Jeremy Black, *European Warfare, 1660–1815* (UCL Press, 1994), p. 236. Significantly, Black does not even list gunpowder in his index, nor does he in *Britain as a Military Power, 1688–1815* (UCL Press, 1999). Geoffrey Parker places the revolution squarely in the sixteenth century: "The sixteenth century still seems of central importance because it witnessed the emergence of three key innovations: the capital ship with its broadside; the development of gunpowder weapons as the arbiter of battles and sieges; and, in direct response to this, the 'artillery fortress.'" (*The Military Revolution: Military Innovation and the Rise of the West, 1500–1800,* second edition, Cambridge University Press, 1996, p. 159) For a general statement of this position, see Martin van Creveld, *Technology and War* (Brassey's, 1991), pp. 96–97.

2. Bert S. Hall, *Weapons and Warfare in Renaissance Europe* (Johns Hopkins University Press, 1997), p. 215. Hall characterizes certain eighteenth-century developments (e.g., the adoption of a single grain size for all artillery gunpowder by the British) as "absolutely regressive."

3. H. Boerhaave, *Elements of Chemistry*, quoted in T. L. Davis, "Chemistry in War: An l8th-Century Viewpoint," *Army Ordnance* 5, no. 30 (1925), p. 783.

4. Jenny West, *Gunpowder, Government, and War in the Mid-Eighteenth Century* (Boydell, 1991); Wayne D. Cocroft, "William Congreve (1743–1814) Experimenter and Manufacturer," ICOHTEC Conference, Budapest, 1996 (unpublished "speaker's text"); Brenda J. Buchanan, "The Technology of Gunpowder Making in the Eighteenth Century: Evidence from the Bristol Region," *Transactions of the Newcomen Society* 67 (1995–96): 125–159. See also Brenda Buchanan, ed., *Gunpowder: The History of an International Technology* (Bath University Press, 1996).

5. E.g., "Une administration non révolutionée? Prosopographie des commissaires des poudres et Salpêtres (1775–1817)" (17e Congrès national des sociétés savantes [Commission d'histoire de la Révolution française], Clermont-Ferrand, 26–30 octobre 1992) and other works cited further on.

6. The principal modern monograph is Régis Payan, *L'Évolution d'un Monopole. L'industrie des Poudres avant la loi du 13 Fructidor An V* (Les Éditions Domat-Monchrestien, 1934), summarized recently in Charles Gillispie, *Science and Polity in France at the End of the Old Regime* (Princeton University Press, 1980), pp. 51–58. An earlier but good treatment in the same vein is found in René Pique, *Le Poudre Noire et le Service des Poudres* (Éditions de la Société des Poublications Colloidales, 1927), chapters X and XI. Very rich material is provided in *Le Service des Poudres* (Numéro spécial de la revue "Croix de Guerre") (Information et Propagande Française, N. D.); see especially First Part: "Évolution des Fabrications dans le Service des Poudres," Section B: "L'évolution des techniques de fabrication," especially pp. 60–68; Second Part: "Histoire des Service des Poudres," Section III: "L'ère des traités" (pp. 129–144) and Section IV: "La Régie Royale des Poudres et Salpêtres" (pp. 145–154).

7. "Poudre à canon," and associated articles, *Encyclopédie, ou Dictionnaire raissoné des sciences, des arts et des métiers, par une société de gens de lettres*, volume 13 (Neufchatel [Paris]: Samuel Fausche, 1765), pp. 190–196.

8. Three volumes (Didot, 1773). volume I, pp. 452–479 ("Poudre à canon"), Vol III, pp. 589–616 ("Sur le Nitre ou Salpêtre" and related topics). Baumé's interest in and knowledge of gunpowder had been developed in his researches of the mid 1750s carried out with the military officer and scientist Patrice d'Arcy.

9. Leblanc 1811. Riffault des Hêtres was himself an official in the French gunpowder administration.

10. By L. Renaud, Chef de Bataillon au Corps impérial de l'Artillerie, Chevalier de la Légion d'honneur. (Imprimée avec l'approbation de S. Ex, Le Ministre de la Guerre.) Magimel, 1811.

11. Full title: *A memoir on gunpowder; in which are discussed, the principles both of its manufacture and proof* (London: Reprinted by Permission of the Hon. Court of Directors, Printed at Madras at the Expense of the Indian Government for Use of the Artillery, 1832. Printed at Madras at the Expense of the Indian Government for Use of the Artillery [1830]). Braddock, son of the Master Refiner of Saltpeter, Royal Powder Mills, Waltham Abbey, of the same name, was trained there but spent his career in India. In the preface, Braddock wrote that "there is not a single work extant in the English language that discusses the

manipulation of Gunpowder, and the best and most accurate methods of ascertaining its strength and quality" (p. iii).

12. Five volumes, printed by J. Archdeacon, Printer to the University, 1781–1787. Watson (1737–1816), Bishop of Llandaff, Regius Professor of Divinity and erstwhile professor of chemistry at Cambridge, developed the method of charring wood by distillation. Essays in volume I (1781): VIII: "Of the Manner of making Saltpeter in Europe, and of its Generation," pp. 283–311; IX: "Of the Manner of making Saltpeter in the East Indies," pp. 313–326; X: "Of the Time when Gunpowder was discovered," pp. 327–349; volume II (1781): I: "Of the Composition and Analysis of Gunpowder," pp. 1–32. Others include George Napier, "Observations on Gunpowder," *Transactions of the Royal Irish Academy* 2 (1788): 97–117, and R. Coleman, "On the Manufacture and constituent Parts of Gunpowder," *Philosophical Magazine* 9 (1801): 355–365. George Napier was Comptroller of the Royal Laboratory, Woolwich, in the early 1780s. Coleman was a member of the Ordnance staff at Waltham Abbey and played a role in the reforms of the gunpowder industry. West, p. 185.

13. The head of the gunpowder farm had control (and obligation to maintain upkeep) of the powder mills during his tenure of office. He also had the monopoly on the sale of hunting powder and powder for export (e.g., for sale in the slave trade).

14. Payan, pp. 26–40, characterized by Payan as "la période 'chaotique' du monopole des poudres et salpêtres" (p. 40).

15. Styled "Berthelot des poudres" by La Bruyère (Bottée and Riffault, p. xxxvi [p. xli for title of "commissaire général de l'artillerie, poudres et salpêtres de France"]). By 1678, the title had become "fermier général de la fabrique et vente des poudre et salpêtres," although he also retained the earlier title (p. xlvii).

16. Paris 6,000; Amiens 10,000; Arras 20,000; La Fère 10,000; Dunkerque ou Graveline 20,000; Sedan ou Mézières 20,000; Metz 20,000; Lyon 30,000; Brouage ou La Rochelle 30,000; Perpignan 10,000; Marseille ou Toulon 24,000 (Bottée and Riffault, p. xxxviii; Payan, p. 42).

17. Bottée and Riffault, pp. xxxviii–xxxix. If deemed necessary, he was obliged to furnish an additional "extraordinary" 200,000 pounds of powder or saltpeter at the same price. The payment price for the ordinary provision was eventually raised to 11 sous per pound (Ibid., p. xlv. Payan, pp. 47–49). He was also expected to convert extra saltpeter into good quality powder at his own expense. He had the right to build and extend powder mills and facilities. The *fermier* had the monoply on the domestic sale of hunting powder.

18. Payan, p. 46.

19. Claude Durié assumed Berthelot's post from 1685–1688. The annual requirement of powder was placed much higher than previously; Durié's contract called for him to supply the state with 800,000 pounds of powder year (Payan, p. 69).

20. He was required to supply annually 600,000 pounds of powder with an extraordinary amount set at 400,000 pounds (Payan, pp. 74–75). Berthelot's family remained financially involved in the gunpowder farm as bondsmen through the first quarter of the eighteenth century (ibid., p. 81).

21. The powder requirement was increased considerably when Berthelot stepped down; his successor initially had to supply 2,200,000 pounds per year (at a price of 5 sous/pound for the first million pounds, 9 sous for the next 500,000 pounds, 10 sous for the next 500,000 pounds, and 11 sous for the remaining 200,000), and this was increased regularly during his term to 3 million or 4 million pounds due to Louis XIV's wars (Payan, pp. 83, 86). In 1703, there was a general streamlining of the gunpowder bureaucracy, with the suppression of various offices that had existed since, in some cases, the late sixteenth century (Bottée and Riffault, pp. lxvii–lxviii).

22. Charles Coulston Gillispie, *Science and Polity in France at the End of the Old Regime* (Princeton University Press, 1980), p. 56. Payan (pp. 130–131) wrote that the gunpowder farm "se désintéressait à peu complètement du côté technique d'une exploitation qu'il soustraitait en presque totalité à des entrepreneurs locaux dont la compétence était généralement très limitée"; and therefore "Aucun progrès notable n'avait été, pendant près de cent ans, apporté aux procédés de fabrication.... Le fermier, n'étant pas assuré de voir son bail renouvelé à expiration, n'avait généralement pas intérêt à étudier et à effecuer des modifications, parfois coûteuses, dont ses successeurs auraient seuls recueilli le bénéfice."

23. Change of alliance: France now allied with Austria against England and Prussia.

24. This was held to be a cause of France's defeat in the Seven Years' War. Patrice Bret, "La Vie des sciences. Lavoisier à la Régie des Poudres: le savant, le financier, l'administrateur et le pédagogue," *Comptes rendus de l'Académie des Sciences*, série générale, 11 (1994): 298 (whole article: pp. 297–317.)

25. By leaching with potash, boiling down, skimming off the common salt, and then recrystallizing.

26. Arthur Donovan, *Antoine Lavoisier, Science, Administration, and Revolution* (first published 1993; reissued, Cambridge University Press, 1996), 193.

27. West, *Gunpowder*, p. 13. See also Major-General A. Forbes, *A History of the Army Ordnance Service*, 3 volumes (The Medici Society, Ltd., 1929), volume 1, chapter V ("The Board of Ordnance") for its founding and early history. Most of the administrative details are taken from this latter work.

28. Master General as chairman (sometimes but by no means always prominent military commanders such as Marlborough and Ligonier), Lieutenant General, Surveyor General (responsible for the proving of powder), Clerk of the Ordnance, Storekeeper, and Clerk of the Deliveries.

29. Later in the century, proving was usually done further down river at the powder magazines constructed in the early 1760s at Purfleet. Wayne D. Cocroft, "William Congreve (1743–1814) Experimenter and Manufacturer," ICOHTEC Conference, Budapest, 1996 (unpublished 'speaker's text'), 2.

30. As a result of which they were also difficult to ensure.

31. West, p. 22. In Appendix 1, West provides brief histories of the mills supplying the Ordnance Office in the eighteenth century and brief accounts of the mill owners.

32. West, p. 36.

33. The following mills had contracts with the government during the Seven Years' War: Bedfont mills, Middlesex; Chilworth mills, Surrey**; Dartford mills, Kent; Ewell mills, Surrey; Faversham mills, Kent [became state owned, 1759]; Hounslow mills, Middlesex**; Molesey mills, Surrey; Oare mills, Kent; Waltham Abbey mills, Essex; Worcester Park mills, Surrey. All these mills except Hounslow and Chilworth had supplied the Ordnance Board during War of Austrian Succession. ** Indicates mills added only during the course of war.

34. During the Seven Years' War, contracts were for either 242½ or 485 100-lb. barrels [per month] (according to choice of the maker). Dates of delivery were not specified (West, p. 34). A good deal of the powder supplied was re-processed old stock supplied to the mills by the Ordnance Board.

35. Ibid., pp. 77–78.

36. West, pp. 50–52. At one point (January 1758) there were 14,470 barrels of Dutch powder in store and only 2,840 barrels of English powder. Diplomatic shifts in favor of France towards the end of the decade, and concern over the safety of transport (in secret) across the sea, led to cessation of gunpowder import.

37. For example, from January 1757 until November 1762 the Ordnance Board offered to pay 2s 6d over the contract price (i.e., £1 per barrel instead of 17s 6d) for every 80 out of 100 barrels of powder that passed proof. The results were disappointing (West, pp. 61–63).

38. See ibid., chapter 9.

39. I will discuss those aspects of powder making that were subject to regulation and/or to discussion and experimentation.

40. Bottée and Riffault, pp. li–lii.

41. Bottée and Riffault (pp. 153–156) noted that the simplicity of refinement enabled the powder commissioners to refine their own sulfur with little expenditure of space. When the sulfur was decanted, care was taken to keep the slag at the bottom of the kettle from coming over. This was subsequently collected and re-refined.

42. Bottée and Riffault, p. lii, speculated that its exclusive use might be as much conditioned by force of habit or by desire of the government to keep powder making an "espèce de secret" as evidenced by the clear superiority of black alder over other similar types of wood (p. 121). The procedure of charring wood in furnaces and pits is described on pp. 125–127.

43. In response to a complaint by the then head of the powder farm that royal officials in charge of waters and forests were obstructing the collection of such young branches (Bottée and Riffault, pp. lxii–lxiv).

44. "Cuites": boilings and recrystallizations.

45. Charles Gillispie describes the process with his customary ironic humor and insight: "When the prospecting party brought a load [of saltpeter] into the master's yard, he would set two husky laborers with sledge hammers onto pulverizing the charge of stone and plaster. In Lavoisier's opinion, no feature more stubbornly enmeshed the industry in toils of

surly routine. By 1785 a stamping mill had been devised; the laborers successfully obstructed it as they did every innovation. Their work done, they would shovel the gritty mass into water barrels for leaching out the salt. Potash (potassium carbonate) was thrown in, sometimes in the form of wood ashes, in order to convert the "earthy based" to "true" saltpetre. Now this mixture went into great copper cauldrons for cooking out the grosser impurities, among them common salt. As the concentration increased, a dose of Flanders paste was thrown in to clarify the liquor. On additional evaporation, the brew precipitated muddy yellow crystals of crude saltpetre (*de première cuite*). No farther did the law allow the saltpetrement to go in purifying their product, lest they be in a position to sell if privately. Their privilege required them to deliver it crude into the Arsenal of Paris or in the provinces to other royal magazines, all under lease to the Gunpowder Farm. There the personnel refined it by means of two further recrystallizations, after which it came out of solution, white as flour, to be dried in loaves and shipped to powder mills for corning, together with sulfur and charcoal, into gunpowder. The saltpetrement of Paris formed the most important single sector of the trade. From 1783 to 1790, years for which full figures exist, they extracted from the powdered stones of the capital an annual average of 750,000 pounds. All the rest of the nation yielded only four to five times that amount to their cohorts of the provinces" (*Science and Polity in France at the End of the Old Regime*, pp. 52–53).

46. Reduced in 1688 to "the time necessary proportionate to the force of the water which operates the mills" (*Le Service des Poudres*, p. 62).

47. Slowed in the first quarter of an hour to 40–45 stamps in order to observe the system (L. Renaud, *Instruction sur la fabrication de la poudre* [Magimel, 1811], 102–109). Normally there were 10 pestles to a battery. The description of the machinery for the incorporation procedure in Bottée and Riffault, pp. 210–220, has also been used.

48. Bottée and Riffault, p. 198 ff. For a time during the Revolutionary and Napoleonic epoch, the proportion was changed. See also, Renaud, p. 98, table.

49. Bottée and Riffault, p. lii. This remained official policy well into the nineteenth century. Cf. Renaud, p. 97: "En France, on ne fait de poudre de guerre que d'une seule espèce, soit pour le grain, soit pour la composition, parceque l'avantage de simplifier les êtres, a paru supërieur à tous ceux que cette diversité peut offrir."

50. The two were dated April 4 and September 18, 1686 (Bottée and Riffault, p. liii).

51. Bottée and Riffault, pp. lii–liv, state that the mortar ordained in 1686 was the same as that described in the body of the book "with very slight differences." That one, made of bronze, weighed about "17 myriagrammes" (170 kilograms), had a caliber of 191 millimeters and a bore length of 239 millimeters (p. 340, p. 509 and planche XXXVI, fig. 1). Each fathom (tois) was 6 royal feet and equal to 2 meters.

52. Bottée and Riffault, p. xciii.

53. Bottée and Riffault, pp. xciv–xcv. Quotation is from p. xciv.

54. In 1791. Bottée and Riffault, p. cxix.

55. Coleman, p. 357. The standard given by the Ordnance Board in 1753 (below, note 25) closely tallies with this proportion. However, there must have been some fluctuation in the

official proportions because Congreve stated that the proportions prior to 1785 were: lbs. oz. drs. saltpeter 74 8 11 7/11 / charcoal 15 7 4 4/11 / sulfur 10 0 0. Since 1785, they were 75 (saltpeter / 15 (charcoal)/ 10 (sulfur). ("A State of Facts relative to the Grounds on which the late and present Master General have had so much Reason to doubt the goodness and durability of the Gun Powder which was delivered into theRoyal Magazine for the King's Service," signed "Wm Congreve, Deputy Comptroller Charlton January 17th 1788," Royal Artillery Institution, Woolwich, pp. 17–18) (This item is subsequently referred to as *A State of Facts 1788.*)

56. West, p. 28. The doubly refined saltpeter came from approximately 83½ lb of grough petre (saltpeter as it arrived from import) (West, p. 172). Brenda J. Buchanan gives the 1753 Ordnance Board's requirement as "Percentage of saltpeter/sulfur/charcoal: 74. 3 11. 8 13. 9" ("The Technology of Gunpowder Making in the Eighteenth Century: Evidence from the Bristol Region," p. 150).

57. Watson, "Of the Manner of making Saltpetre in Europe, and of its Generation," p. 291. He went on to note that there have been schemes to revive this industry, e.g., by the Society for the Encouragement of the Arts and Manufactures during the Seven Years' War. But, although one saltpeter works was set up, it was soon abandoned, "the proprietors having been experimentally convinced, that they could not afford to sell their saltpetre for less than four times the price of that imported from India." Watson attributed the high cost of making saltpeter in England to the costliness of wood ashes (also used in soap manufacture) and of labor to collect and make the saltpeter (pp. 292–293). Watson worried about the lack of domestic production of saltpeter in England in contrast to France: "The French very wisely keep up their establishment for the making of saltpetre; the revolutions which have formerly taken place in India, render it not improbable, that similar ones may take place again; and England would feel the distress which would attend the non-importation of saltpetre from the East Indies, more sensibly than any other state in Europe." ("Of the manner of making Saltpetre in the East Indies," pp. 325–326)

58. West, p. 14.

59. Brown, p. 11. See Watson, "Of the manner of making Saltpetre in the East Indies," pp. 313–321, for contemporary ideas about how saltpeter was produced in India.

60. Coleman, p. 357.

61. West, pp. 172–173. Some saltpeter was also imported from Russia, Germany, and Spain. "Refraction" involved breaking a cake of saltpeter that had been melted and noting the appearance of the crystalline fracture. Saltpeter was supplied to the mills upon a deposit of £1,200 for each contract of 485 barrels of powder.

62. Details are given in a compilation of a manuscript in the Library, Old Royal Military Academy, Royal Artillery Institution, Woolwich, dating from the time of William Congreve. The compilation was done by A. B. Caruana, Captain, Royal Regiment if Artillery: "The Method of Refining Salt Petre in the Years 1784 and 1785: One Ton of Petre, either Grough or that extracted from damaged Powder, is dissolved in 280 Gallons of Water, in a large Boiler, and made to boil as soon as possible, one Man attending to take off the Scum as it rises, and frequently throwing cold Water in small Quantities into it, to make the filth rise

freely, it continues boiling till entirely free of Scum, and judged to be of a proper Thickness, which is known by pulling the Scummer into it, and if the Petre adheres to it, so as to form a drop on the edge of about an Inch in Length it is thought to be of a proper Consistence for Chrystallisation; it is then pumped or ladled into a spout from which it runs into a trough with four Brass Cocks, through which it runs into the Filtering Bags, made of double Canvas, which have sand thrown into them, to prevent the Liquor from passing too freely; when one Bag is full and the Liquor runs through clear the remainder of the Bags are, [sic] filled, and the Liquor that runs from them before it is clear is returned into the Trough, and the clear Liquor put into a large Pan to Chrystallize, in which state it stand during the same time as the Whole Liquor from the Extracting Rooms, and when the Liquor is drawn off it, part or the whole if there is room is carried and mixed with the whole Liquor in the Extracting Rooms. The Petre is rinsed and set to drain four twenty four hours and then refined again, in the same manner as above, when this last Petre is drained it is carried to the Store House to dry, except when ordered to be melted. If there is no Extracting, the Liquor from the refined Petre is reduced, and the Petre produced from it, is considered as a single refined Petre." (*Laboratory Papers 1790* Compiled from a Manuscript in the possession of the Royal Artillery Library by A B Caruana, Captain, Royal Regiment of Artillery, pp. 47–48) George Napier believed in three boilings. Moreover, filtration should be done "through a canvas cloth in the form of a jelly-bag"; the saltpeter should be crystallized "in leaden or copper vessels, exposed to a free circulation of air in a dry situation." (Napier, pp. 100–101)

63. Napier recommended purification and "sublimation" by melting in an iron pot under gentle heat and then straining the melt through a double linen cloth. He did warn that prepared sulfur was often adulterated with wheat flour, leading to fermentation and decomposition, a state of affairs that, in his opinion, was the "principal cause of British gun-powder being less durable now than formerly" (Napier, pp, 102–103).

64. Watson, "Of the Composition and Analysis of Gunpowder," vol. 2, pp. 5–8.

65. John Braddock, *Memoir on Gunpowder; in which are discussed, the principles both of its manufacture and proof* (Reprinted by Permission of the Hon. Court of Directors [Printed at Madras at the Expense of the Indian Government for Use of the Artillery], 1832), 2. Braddock discussed (pp. 2–3) the anomalous eighteenth-century results referred to in the preceding note. That the issue was not really settled is shown by the following comment by artillery and munitions authority Sir Andrew Noble early in the twentieth century: "I have never been able to understand why sulfur was so long retained as a component in gunpowder" ("Fifty Years of Explosives," delivered at the Royal Institution, January 18, 1907 [Spottiswoode & Co., 1907], 7).

66. Watson, "Of the Composition and Analysis of Gunpowder," p. 4.

67. "It consists in the wood [stripped of its bark] being cut into lengths of about three feet, and then piled on the ground in a circular form (three, four, or five cords of wood making what is called a pit), and covered with straw, fern, &c. kept on by earth or sand to keep in the fire, giving it air by vent-holes as may be found necessary." (Coleman, pp. 357–358) Writing after the introduction of closed cylinder charring, he described the method of charring in pits as "uncertain in its operation, and defective in every respect for the purpose of making good charcoal."

68. (1737–1816), one-time Professor of Chemistry and Regius Professor of Divinity at Cambridge University, and Bishop of Llandaff.

69. Coleman, p. 358.

70. Tested between September 1 and December 21, 1786. *A State of Facts 1788*, p. 4. See below, under section of experiments.

71. Cocroft, p. 4. Probably first at Faversham (1792) and then Waltham Abbey (1794).

72. Coleman, p. 358. This led to an annual saving by the government of over £100,000. Cocroft, p. 4. See also p. 9 of *A Statement of Facts, Relative to the Savings which have Arisen from Manufacturing Gunpowder at the Royal Powder Mills; and of the Improvements which have been made in its Strength and Durability since the Year 1783* [William Congreve] (Printed by James Whiting, 1811) (hereafter referred to as *A Statement of Facts (1811)*). But Braddock could still write almost thirty years later "I have met with no writer who offers an opinion why cylinder charcoal is superior to pit charcoal for making gunpowder, nor am I, myself, able to solve the question." (Braddock, p. 18) He adduced two possible reasons: the greater purity of cylinder over pit charcoal (he might have added uniformity as well) and the absence of alkali in it (which he thought was engendered by charring wood in air. This latter explanation had not stood the test of time. It is possible that it was due to the presence of more hydrogenous material in cylinder charcoal than that produced by charring in pits. Although the French eventually adopted cylinder charcoal, early reaction to it was not favorable. Bottée and Riffault gave the following evaluation of comparative tests using charcoal made the English and the French ways: "Les résultats de ces épreuves, sur les méthodes de carbonisation des Anglais, dûrent prouver à l'administration qu'elles n'étaient pas préférables à celles usitées en France, et la confirmer surtout dans l'idée qu'elle avait eue de l'exagération ridicule des prétendus avantages des modes de carbonisation employés en Angleterre." (p. 140)

73. However, Renaud deemed cylinder mills to be "more dangerous" and less efficient than stamping mills. This was one of the reasons he gave for the discontinuation of the use of cylinder mills built at Essonne in 1754 (pp. 109–110).

74. Buchanan, "The Technology of Gunpowder Making in the Eighteenth Century," p. 144.

75. Coleman, p. 359.

76. Coleman, p. 359. Although watermills were very common, horsemills continued to be built, e.g., at the government-owned Waltham Abbey mills because of lower cost and reluctance to put too much demand on the flow of water through the mills. *The Royal Gunpowder Factory, Waltham Abbey, Essex: An RCHME Survey, 1993* (Royal Commission on the Historical Monuments of England, 1994), 34.

77. Napier, p. 106. He also criticized the edge runners for initially compressing the powder too much and therefore not being subsequently capable of proper incorporation. Renaud, p. 111, gave an approximate time of 6 hours. Crocker, p. 12, gave a range of from 2 hours for blasting powder to 8 hours or more for finest sporting powder. Braddock, p. 43, gives the following regulations for early nineteenth-century incorporating mills for Waltham

Abbey (by this time, water driven): 42 lbs. of powder (a "charge") is worked for three hours under runners which revolve at a rate of seven and a half revolutions per minute.

78. The greater density engendered by the press may have been as significant a factor in increasing the ballistic force of gunpowder as the introduction of closed cylinder charcoal. See Robert H. Howard, "The Evolution of the Process of Powder Making from an American Perspective," ed. Brenda Buchanan, *Gunpowder: The History of an International Technology*, pp. 12–13.

79. *A Statement of Facts (1811)*, pp. 22–23. Although grain size was demonstrated to be of great importance ballistically in the middle of the nineteenth century, it was not so apparent early in the century. Braddock, for example, wrote "but neither large nor small sized grain appears to possess any advantage with reference to the regularity of practical effect: in any given number of firings, the ranges of the one vary as much as those of the other" (p. 51).

80. Colonel A. H. Mockridge, "The Proving of Ordnance and Propellants," *Journal of the Royal Artillery* 77 (1950): 82–83.

81. "There are several methods of proving and trying the goodness and strength of gunpowder." (Coleman, p. 360) He goes on to describe the test known as "flashing."

82. Braddock, pp. 78–80.

83. Mockridge, pp. 85–86. In the vertical eprouvette test, English powder raised the weight "from 4 inches to 6 inches and some tenths"; Spanish powder "no higher than 1 inch and some tenths." In the swivel gun test, English powder threw the shot 15$\frac{3}{17}$ feet at an elevation of 61°; the Spanish powder failed to throw the shot out of the gun; "the tryal by the mortar turned out equally in favour of the English powder."

84. Ibid.

85. Ibid., p. 87.

86. Colemen asserted in 1801 that there were a number of methods of proving powder and especially recommended the method known as "flashing": "Lay two or three small heaps (a dram or two) on separate pieces of clean writing-paper; fire one of them by a red-hot iron wire; if the flame ascends *quickly*, with a good report, leaving the paper free from white specks, and does not burn it into holes; and if sparks fly off, setting fire to the adjoining heaps, [!] the goodness of the ingredients and proper manufacture of the powder may be safely inferred; but if otherwise, it is either badly made or the ingredients impure." (Coleman, p. 360).

87. *A Statement of Facts (1811)*, pp. 27–28.

88. Braddock, p. 78. He cited Congreve's criticism of it, p. 80, footnote.

89. "The method of trial best adapted to shew the real inherent strength and goodness of gunpowder, appears to be an eight or ten-inch iron or brass mortar, with a truly spherical solid shot, having not more than one-tenth of an inch windage, and fired with a low charge. The eight-inch mortar, fired with two ounces of powder, is one of the established methods

of proof at His Majesty's works. Gunpowder that range equally in this mode of trial may be depended on as being equally strong." (ibid., p. 84)

90. Ibid., pp. 82–83. He found that "it fails to give the real force of gunpowder of the very same manufacture, in which the ingredients are the same, and every thing alike, except a difference in the size of the grain" (p. 83). Braddock apparently did not realize how critical grain size was to the expression of ballistic force. Versions of the pendulum eprouvette had been developed in the middle of the eighteenth century by Benjamin Robins and Patrick d'Arcy.

91. Ibid., pp. 89–90. He found carbine proof good for rifle and musket powder, and flashing a good test for determining the care in manufacture of the powder.

92. Gillispie, p. 57.

93. Donovan, p. 194.

94. "Aussi [parce qu'il était'connu par ses lumières en chymie, essentiellement nécessaire pour ce genre d'administration." (quoted from my article, "Lavoisier and the Improvement of Gunpowder Production," *Revue d'Histoire des Sciences*, 48 (1995): 96). Original from P.-S. Dupont de Nemours, *Mémoires sur la vie de Turgot* (Philiadelphia, 1782) quoted in Edouard Grimaux, *Lavoisier, 1743–1794, d'après sa correspondance, ses manuscrits, ses papiers de famille et d'autres documents inédits* (F. Alcan, 1888), p. 84.

95. Rather than simply the release of an inflammable principle (phlogiston) from the inflamed substance.

96. Quoted in Frederic L. Holmes' book, *Lavoisier and the Chemistry of Life* (University of Wisconsin Press, 1985), p. 49. This work remains the best-detailed narrative of the development of Lavoisier's chemical work. See also his *Antoine Lavoisier—the Next Crucial Year, or, the Sources of his Quantitiatve Method in Chemistry* (Princeton University Press, 1998). This gas would subsequently be named "oxigène" (oxygen) by Lavoisier. It should be pointed out that the oxygen theory of combustion was only one component (if a central one) in Lavoisier's reformulation of chemical theory. And the wider concept of the "Chemical Revolution" continues to elicit debate.

97. In 1768 he had been elected to the entering rank (adjointe), and in 1772 he had been promoted to the next level (associé) in the division of chemists. In 1778, he would receive the highest rank of "pensionnaire."

98. See Bret, "La Vie des sciences," p. 298.

99. For details of Lavoisier's activities in the Régie des poudres, see Gillispie, chapter 1. 6; Donovan, pp. 188–201; Bret, "Annexe III: La Régie des poudres et salpêtres, 1775–1792," in *Correspondance de Lavoisier*, t.V (Académie des science, 1993), 261–269.

100. "Mémoire sur l'établissement, les produits et la situation de la Régie des poudres et salpêtres," *Oeuvres de Lavoisier*, 6 Vols. (Imprimérie impériale, 1862–1893). He asserted that the test range of military powder had risen to 115 to 130 fathoms instead of 70–80 fathoms before the reform (volume V, pp. 699–700). He did *not* note the rise of the threshold to 90 fathoms in 1769.

101. The term comes from the title of its major study. Robert P. Multhauf, "The French Crash Program for Saltpeter Production, 1776–1794," *Technology and Culture*, 12 (1971): 164–181. "Artificial" production of saltpeter actually meant maximal natural production under carefully controlled conditions. This was used in Sweden and Prussia.

102. Donovan, pp. 195–198.

103. "Mémoire sur l'établissement, les produits et la situation de la Régie des poudres et salpêtres," p. 694.

104. National production grew by 41% between 1777 and 1780. See Patrice Bret, "The Organization of Gunpowder Production in France, 1775–1830," in Buchanan, ed., *Gunpowder: The History of an International Technology*, pp. 261–266.

105. "La fabrication du salpêtre a été éclairée par les travaux, les instructions et les exemples des régisseurs; tous les sujets qui se dévouent à la partie des poudres et salpêtres sont obligés de faire des cours de chimie et de mathématiques, et de subir des examens sur les connaissances nécessaires, non seulement à la composition du salpêtre et des poudres, mais encore à la construction des moulins et raffineries." (Lavoisier, "Mémoire sur l'établissement, les produits et la situation de la Régie des poudres et salpêtres," p. 700)

106. The following is abridged from Bret, "La Vie des sciences," pp. 302–314.

107. The Arsenal, the Paris refinery, the gunpowder mill at Essonnes, and the bureau de la comptabilité.

108. "Un véritable corps d'ingenieurs des poudres." Ibid., p. 313.

109. Cocroft, p. 3.

110. *A Statement of Facts (1811)*, p. 9.

111. "Fortunately, however, for the country, His Grace the late Duke of Richmond, then Master General of the Ordnance, attended to the representations which Gen. Congreve, at that time Comptroller of the [Royal] Laboratory [Woolwich] [in fact, Deputy Comptroller at the time], thought it his duty to make; by which it was proved, that there existed a profit on the powder manufactured at the King's Mills; and that if this profit were for a few years properly expended in improving the works, the Ordnance would be enabled to make both stronger and more durable Gunpowder at the Royal Powder Mills, than had ever been previously made. This statement was confirmed by a course of experiments, proposed and carried on by the Comptroller; and in consequence, the idea of disposing of the Royal Powder Mills was not only abandoned, but the improvements suggested were carried into effect" (Ibid., pp. 9–10). One of the main points of this publication was to demonstrate that, by 1811, a large savings had been realized through government-supervised manufacture of military powder. It is clear from A State of Facts 1788 and from *A Statement of Facts (1811)* that Congreve was fighting a running battle with the private powdermakers for control of production of military gunpowder.

112. Fourteen or fifteen workers were retained, and workmen were brought in from the Faversham mills (now the King's Powder-Mill). Forty-six workers were employed in October 1787 [79, including and attendant clerk, surgeon, carpenter, cooper, and five

apprentices are cited in a Board of Ordnance document of 16 February 1789 as necessary to run the mills in O. F. G. Hogg, *The Royal Arsenal: Its Background, Origin, and Subsequent History*, volume 2 (Oxford University Press, 1963), 1066]. Water power was not substituted entirely for horse power until 1814; in 1791, double horse-mills were still in use. Powder was sent regularly from Waltham Abbey to Purfleet (Thames) for proving. Col. Sir Frederic L. Nathan, "The Royal Gunpowder Factory. Waltham Abbey," in *The Rise and Progress of the British Explosives Industry* (VIIth International Congress of Applied Chemistry) (Whittaker & Co, 1909), 318–320.

113. The only (and brief) account of its history is: Col. Sir Hilaro W. W. Barlow, "The Royal Laboratory, Woolwich," in *The Rise and Progress of the British Explosives Industry*, pp. 307–315. He mentions that the first reference to a "Royall Laboratory" is in 1688 and to it at Woolwich is in 1694, p. 308. In the late eighteenth century it seems to have been concerned with ammunition manufacture.

114. This included such matters as providing powder-makers with fully refined saltpeter. *A Statement of Facts (1811)*, pp. 21–22.

115. 1783–1789. For his role, see West, pp. 177–178; Cocroft; Hogg, volume I, pp. 458–492. See also [Congreve], *A Statement of Facts (1811)*, p. 15 where he says that the Waltham Abbey mills were acquired "in consequence of his repeated remonstrances."

116. He was promoted to major in 1785, eventually reached the rank of lieutenant-general in 1808, and was created a baronet in 1812. His grandfather and uncle (both named William Congreve) were colonels in the Infantry. To add to the confusion, our William Congreve had a son and namesake who came to overshadow his father through the invention of his famous rocket system. The best and only detailed account of a part of his life (down to 1779) is found in a chapter of a study of his son by Major J. P. Kaestlin. This is titled "Firemaster" and exists in typescript at the Royal Artillery Library, Woolwich Arsenal. The chapter devoted to our William Congreve is titled: "Remote Control or The Start of the Royal Military Repository." Royal Artillery Library, MD 213/14, no. 6. There is also a brief article, "Sir William Congreve, 1741 [*sic*]—1814," by W. Y. Carman, *Journal of the Society for Army Historical Research*, 51 (1973): 63–68. Carman (p. 64) gives the year of Congreve's captaincy as 1777.

117. Kaestlin, p. 5.

118. Lieutenant-Colonel James Pattison. Kaestlin, p. 7.

119. The details of the experiments and improvements will be taken up later.

120. Respectively, CO_2, discovered by Joseph Black (1756) and H_2, discovered by Henry Cavendish (1766).

121. S. H. Mauskopf, "Gunpowder and the Chemical Revolution," *The Chemical Revolution: Essays in Reinterpretation,* ed. Arthur Donovan (*Osiris*, volume 4, 1988): 93–118.

122. E.g. whether sulfur was really needed in gunpowder (since it fouled guns badly) and whether black alder was the best (and only) wood source for gunpowder charcoal. For details on this, see my article "Lavoisier and the improvement of gunpowder production," pp. 98–106.

123. Bottée and Riffault, pp. 122–123. The most elaborate tests of this nature were carried out about a decade later by the chemist Joseph-Louis Proust. See my article "Chemistry and Cannon: J.-L. Proust and Gunpowder Analysis," *Technology and Culture* 31 (1990): 398–426.

124. Mauskopf, "Lavoisier and the improvement of gunpowder production," pp. 107–116. For a general overview of the steps Lavoisier successfully instituted to raise saltpeter production, see Bret, "The Organization of Gunpowder Production in France, 1775–1830."

125. N_2 and CO_2 in modern parlance. Neither Berthollet nor Cavallo was an adherent of Lavoisier's new chemistry when he made his discovery. See my "Gunpowder and the Chemical Revolution."

126. Primarily as a union of the oxygen contained in the nitric acid component of saltpeter with the carbon in charcoal. The recognition that carbon was an elementary substance and not a compound of an earth and the inflammable principle, phlogiston, was a major linchpin of Lavoisier's new chemistry.

127. John Ingen-Housz, "Account of a new Kind of inflammable Air or Gas, which can be made in a Moment without Apparatus, and is as fit for Explosion as other inflammable Gasses in use for that Purpose; together with a new Theory of Gunpowder," *Philosophical Transactions of the Royal Society of London* 69, Part 2 (1779): 398.

128. He exploded musket powder, collected the gases in a large bladder, and measured a volumetric ratio of "phlogisticated air" (nitrogen) to "fixed air" (carbon dioxide) of slightly less than 3:1. Being a phlogistonist, he believed that the production of nitrogen (phlogisticated air) was readily explained by the union of the dephlogisticated air (O_2) in the saltpeter with the phlogiston from the highly flammable charcoal. Less explicable was the production of fixed air by this explosion. Tiberius Cavallo, *A Treatise on the Nature and Properties of Air and other Permanently Elastic Fluids* (printed for the author, 1781), 810–815.

129. See Brett Steele, "Muskets and Pendulums: Benjamin Robins, Leonhard Euler, and the Ballistics Revolution," *Technology and Culture* 35 (1994): 348–382; Seymour H. Mauskopf, "From Rumford to Rodman: The Scientific Study of the Physical Characteristics of Gunpowder in the First Part of the Nineteenth Century," in *Gunpowder: The History of an International Technology*, pp. 278–281. The collection of Hutton's research was published as *Tracts on Mathematical and Philosophical Subjects, Comprising, Among Numerous Important Articles, The Theory of Bridges, . . . Also . . . The Force of Gunpowder, With Applications to the Modern Practice of Artillery*. 3 Vols. (printed for F. C. and J. Rivington, etc., 1812). I am unaware of any recent study of Hutton. West, pp. 177–184, gives a brief but very interesting treatment of the work of Robins, Hutton, and above all Thompson.

130. [Cocroft, p. 2] vertical eprouvette mortar / Poorly made "tender-grained" merchants powder 4½″ 1083 yards / Swedish powder 2 7/10″ 1114 yards.

131. Cocroft himself notes that Congreve used the vertical eprouvette at the Waltham Abbey mills in 1789, in addition to carbine proving as well as the mortar and the Hutton eprouvette (Cocroft, p. 2).

132. "Experiment" was the word Congreve himself used to designate these trials (*A State of Facts 1788*).

133. Cocroft notes that Congreve took a very personal interest in this set of experiments and went down to Faversham to supervise the charring of the wood (Cocroft, pp. 3–4). It is not clear from Cocroft whether the experiments established any ordering of suitability. A contemporary chemist, Joseph-Louis Proust, also conducted comparative tests on proportions of ingredients and on different woods but used the methods (and theory) of Lavoisian chemistry as his testing means. See Seymour H. Mauskopf, "Chemistry and Cannon: J.-L. Proust and Gunpowder Analysis," *Technology and Culture* 31 (1990): 398–426.

134. *A State of Facts 1788*, p. 7. Cocroft used this manuscript, among others. Experiments were carried out in 1787 on gunpowder made in the proportion of 76% saltpeter, 15% charcoal, and 9% sulfur, which Congreve asserted to be better than the standard (Cocroft, p. 3, and *A State of Facts 1788*, p. 18).

135. Ibid., pp. 5–7.

136. Cocroft, p. 6.

137. *A State of Facts 1788*, p. 6.

138. Ibid. At the Battle mills in Sussex, wood was charred in iron pots (Cocroft, p. 4).

139. "In particular, a given quantity of gunpowder, made with this kind of charcoal, threw a ball of sixty-eight pound weight, two hundred and seventy-three feet; whilst the same mortar, at an equal elevation, and charged with an equal weight of gunpowder made with charcoal prepared in the best of the ordinary ways, threw an equal ball only one hundred and seventy two feet. In this experiment, the strength of the cylinder, estimated by the horizontal range, is to that of the best sort of other powder, as 100 to 63. By experiments with the Eprouvette, the proportion of the strength of the cylinder to other powder was that of 100:54. In round numbers, it may perhaps be near enough to the truth to say, that the strength of the cylinder powder is to that of other powder, as 100:60, or 5:3" (quoted from "Obituary of Bishop Watson, *Gentlemen's Magazine*, Sept. 1816, p. 274 in Simmons, p. 23).

140. *A Statement of Facts (1811)*, p. 38, and footnote, where it was noted that bursting mortars had occurred in Alexandria, at the Dardanelles, in Basques Roads, and at Cadiz, where twenty men were killed and wounded. The French were to experience the problem of bursting guns when they adopted English gunpowder manufacturing techniques after the Napoleonic Wars.

141. *A Statement of Facts (1811)*, pp. 22–23.

142. West, p. 180; Ingen-Housz, p. 406. However, it is, I believe, going too far to say that "the scientific experiments of Ingen-Housz confirmed the increased effect possible with graded sizes for different purposes" (West, p. 171).

143. Cocroft, p. 6. Congreve did claim that, under his administration, "such has been the increase in strength in the gunpowder, that it was found necessary to reduce the charge for the service of the navy, from one-half the weight of the shot to one-third." He implied that

this was due (in part) to confirmation of his assertions by "a course of experiments" (*A Statement of Facts [1811]*, p. 10).

144. "Lavoisier and the improvement of gunpowder production," pp. 118–119. There is also the puzzling improvement in ballistic force when the industry was supposedly at its nadir shortly before Turgot's establishment of the Régie des Poudres.

145. *A Statement of Facts (1811)*, pp. 19–20, 22–23, 24–25.

146. For Watson's innovation, pp. 26–27; for the issue of durability, p. 24. By "variety of circumstances," he meant: "The great care now taken to render the ingredients most perfectly pure—the increased attention paid to working the powder under the runners—to the pressing of it—and to the improved modes of dusting and glazing it." He also mentioned the production of two grain-sizes, but not in relation to either of these two general improvements.

147. "When large Quantities of Gun Powder are used, large grain'd Powder is stronger than small grain'd" (*A State of Facts 1788*, p. 6). After the work of T. J. Rodman, the grain-size was enlarged for large guns to *control* the rate of burn and the buildup of pressure in the bore of the gun, not to increase the power. See Seymour H. Mauskopf, "From Rumford to Rodman," pp. 285–289.

148. For example, the Board of Ordnance took over the refining of all saltpeter for gunpowder "that they might be certain of having it properly purified" (*A Statement of Facts [1811]*, p. 21).

149. See, among many other things, "Technology and its Impact on Tactics," in Hew Strachan, *European Armies and the Conduct of War* (George Allen & Unwin, 1983), 117ff. A good older treatment of the relationship between change in gunnery and in black powder is found in Major-General Sir Charles Callwell and Major-General Sir John Headlam, *The History of the Royal Artillery*, 3 volumes (Woolwich: Royal Artillery Institution, 1931–1940), vol I (1860–1899), Chapter VII. 1870–1883.

150. The "almost" refers to Alfred Nobel, whose own work nevertheless falls into the general category of systematic experimentation.

IV

Military Engineering and Artillery

12

Eighteenth-Century French Fortification Theory after Vauban: The Case of Montalembert

Janis Langins

Although the concepts of the Military Revolution and the Scientific Revolution have undergone considerable refinement and some revision since they were first proposed, many military historians and historians of science and technology agree that such revolutions in fact occurred over roughly the same period—from about 1500 to 1800.[1] The Military Revolution witnessed a spectacular growth of centralized armies, an increase in the proportion of infantry in battle, a separation of commercial fleets from navies that underwent major technological transformations, and a widespread use of gunpowder in diverse ways on land and sea. Accompanying these were changes in the financing of war and the organization of industrial military production. Profoundly shaping the modern state, such transformations in turn contributed to the consolidation and further development of Western military changes. The Scientific Revolution saw a heliocentric system replace the old Aristotelian worldview and cosmology. Newton and his successors proposed a dynamics that both accounted for the observed planetary motions and integrated them into a general world system that unified the celestial and the terrestrial. At the same time, a mechanistic and materialistic natural philosophy, along with major advances in mathematics, de-animated the universe and overturned the traditional scholastic verbal modes of reasoning.

What were the links between these two phenomena? Facile assertions that modern science arose from the demands of modern war, or that the growth of science caused or even permitted the transformation of war, have been challenged. Historians have become more cautious about such statements, feeling that the experience of the twentieth century has skewed our historical perspective, and have begun to scrutinize more closely the diverse forms of a very complex relationship. Rupert Hall argued that science and the changes in the nature of science had virtually no impact on warfare and, conversely, that the new needs of warfare had only a marginal effect on science.[2]

Galileo's work on the trajectories of projectiles was of little use to practical gunners. His stimulus for work in this field came from a long-standing interest in the properties of motion in general. Only peripherally was it a response to the demands of military men. Certainly Galileo was aware of the military interests of many of his noble students; he offered special courses on fortification theory and produced his "Military Compass," which he sold from his own workshop.[3] Yet his great scientific accomplishments seem to have been neither inspired by nor used for military purposes. Similarly, theoretical analyses of the performance of ships go back to the seventeenth century, but these were never used in practical design.[4]

It was only around the middle of the eighteenth century—towards the end of the Military Revolution, according to Geoffrey Parker—that theoretical analyses could have profound practical implications, as evidenced by the interior and external ballistics of Benjamin Robins.[5] The widespread enthusiasm for useful knowledge during the Scientific Revolution was not translated into the fact of useful knowledge until much later. By the time of Robins we finally see something that had been proclaimed as "fact" for well over a century by enthusiastic prophets and practitioners of science. After Robins, both established and innovative scientific knowledge was routinely applied to practical military problems. His near contemporary Bernard Forest de Bélidor (1697?-1761), without Robins' mathematical brilliance, also distilled the work of engineers and scientists at the French Royal Academy of Sciences. They applied and developed it in the service of practical engineering problems that in many cases, such as the design of arches and revetting walls, proved to be important for military needs. Yet, no matter how significant this work was, it was essentially pioneering. As Navier put it later, the mathematical analysis was often simplistic and erroneous; its value lay more in its attempt than its achievement. Only towards the end of the eighteenth century, and only in some fields such as navigation and artillery, would the impact of science on war become significant.

Many scientists and laymen from the early modern era onwards were eloquent and enthusiastic in their claims that science was useful for the power of the state, human betterment, and military might. They were usually the same people who sought state patronage for science. The classic example is Francis Bacon.[6] But in warfare, as in civilian domains, there was a great gap between claims and performance. Even in the areas where there were unmistakable signs of productive links, such as navigation and artillery, they occurred to a limited degree only in the second half of the eighteenth century. There may, however, have been interactions that were less direct, more subtle, and more difficult to document, and these will be presented briefly in the concluding section of this essay.

The first section of this essay will offer some examples of the interaction of science and practical military needs, particularly in siege warfare and fortification design. Only separate detailed studies on each of these examples can do justice to the issues they raise, but in the present essay they will be sketched to support the point I made earlier. Although ambitious, interactions between war and science in fortification engineering had little tactical effect. My primary focus is on France and the relation between science and military engineering, a central issue for warfare in both the Baroque and the Enlightenment Ages. The second section of this essay will therefore look at the case of the bitter polemics over fortification styles between Marc-René, the Marquis de Montalembert (1714–1800), and the French military engineering establishment during the last quarter of the eighteenth century. This case study will examine to what extent formal science was used by fortifiers in their material as well as rhetorical designs. It will also offer some insight into the manner in which science influenced military thought through the medium of fortresses. Some speculative ideas will be presented on this in the third and concluding parts of this essay.

With the development of strength of materials, theory of structures, and mechanics of soils, there were perceptible links between science and military engineering. The story begins with Galileo: I have already argued that Galileo's interest was primarily academic, even though it was evident to him and many of his readers that his research addressed important military needs.[7] And this story continues with the French Academy of Sciences and the regimental artillery schools, particularly with the publication of *La Science des Ingénieurs* in 1729 by the artillerist-engineer Bélidor. (Bélidor had also published a book of range tables for mortars; perhaps the most important application of Galileo's theoretically derived parabolic trajectories.[8]) In the first section of his book, Bélidor analyzed soil pressure to determine the proper dimensions of retaining walls for embankments. This was obviously a question that was of major concern to military engineers who revetted their earthworks, and the great military engineer Sébastien le Prestre de Vauban (1633–1707) had supplied an empirical table of dimensions for this purpose. Bélidor showed that his theoretical analysis gave values that differed from Vauban's; but, for the more typical situations encountered by military engineers, Vauban's values sufficed and even provided a comfortable degree of safety.[9] Bélidor's research in this area counted as a pioneering work in the field of soil mechanics, which was developed later with the calculus by Charles-Augustin Coulomb (1736–1806) and others.[10]

Bélidor also tackled the problem of arch stability and calculated the optimum thickness of arch columns. Military engineers would use Bélidor to argue that the size of the columns supporting the arches in Montalembert's casemates was inadequate to resist artillery fire. Among other criticisms of Montalembert, those concerning his structural designs played a major role in the military engineers' decision to reject his ideas entirely; but, curiously, they rarely gave detailed mathematical justifications for this rejection.[11] Likewise, Montalembert lacked the competence to counter the engineers' rebuttals with his own calculations; he relied on the anonymous help of the civil engineer Gaspard-Marie Riche de Prony (1755–1839) to do so. It is worth remembering, therefore, that in spite of Bélidor's relative popularity among engineers and artillerists, he encountered resistance in having his methods accepted and was very conscious that he was writing at a level that strained the mathematical and scientific competence of the average engineer.[12] More than forty years later, the work of the mathematically more sophisticated military engineer Coulomb, who tackled with differential and integral calculus most of the classical problems examined by Bélidor, was virtually ignored.[13] It was only towards the end of the century that Prony brought Coulomb's brilliant analysis to the attention of the average engineer.[14] And it was only after the foundation of the Mézières school of military engineering in 1749 that military engineers were trained formally in advanced mathematics by future academicians such as the abbé Charles Bossut (1730–1814) and Gaspard Monge (1746–1818).[15] The calculus-based texts by Étienne Bezout began to appear only in the 1760s.[16] The numerous reprintings of these popular texts after that date indicate that, during the last quarter of the eighteenth century, mathematics (including calculus) finally became commonplace in the instruction of the average technical officer.

Another site of interaction between science and war was topography. Military engineers began to map the frontier areas of France more and more, although the major topographical activity in the eighteenth century was the Cassini family's enormous enterprise of mapping the entire country, which was conducted by a private company.[17] Nevertheless, topography was obviously an important subject for military purposes. Among state engineers, there was a special unit of military topographical engineers that worked for army staffs, while other units worked for the Ministry of Foreign Affairs.[18] The topographical work of the military engineers Pierre Bourcet (1700–1780) and Jean-Claude Eléonor Le Michaud d'Arçon (1733–1800) was much admired and specifically done for military purposes. For a time, Bourcet even set up a special staff school that featured topography as a major subject.[19] Military historians have pointed out that Bourcet and his books on mountain warfare influenced Napoleon.[20] Yet all these activities occurred in the last quarter of the century. The military topographers, who formed a separate corps, suffered from low status, however, and the mathematical education they received was generally inferior to that of military engineers. (British military topography was even less developed than that of the French.[21]) It was only under Napoleon, after the foundation of the École polytechnique, that French military topographers achieved institutional coherence and a first-rate mathematical education.[22]

Connected with the problems of topography and fortification design was the development of descriptive geometry. This mathematical science began with Girard Desargues (1591?–1661) in the seventeenth century but remained relatively ignored by his contemporaries and successors. It was revived and further developed with the work of the military engineer Amédée-François Frézier (1682–1773) and Gaspard Monge, a professor at the Mézières military engineering school who is now considered the founder of the science.[23] Later, at the École Polytechnique, descriptive geometry was acclaimed *the* universal engineering science. At one point, fortification and architecture were classified among its components, along with more traditional applications, such as stone-cutting, linear and aerial perspective, shading, topography, and surveying. Undoubtedly, even when descriptive geometry had abandoned its imperial pretensions, it still contributed to the development of engineering drawing.

The standard story, which needs to be revised in light of recent research, asserted that Monge developed this new science to solve the difficult problem of defilading fortifications. This signified giving fortifications the proper height and arrangement to protect defenders from the artillery fire of attackers located on higher terrain.[24] This was a tricky problem that called for the construction of tangent planes to the terrain and the fortress. While engineers had previously resolved this through tedious trial and error, Monge offered elegant and economical solutions.[25] A student of Monge's at Mézières, the talented military engineer and mathematician Jean-Baptiste Meusnier (1754–1793) developed even finer solutions when he used contour lines to defilade fortifications.[26] Here was a confluence of topography and descriptive geometry that military engineering had inspired. Yet defilading, considered the acme of the military engineer's art and taught at Mézières from the 1760s onwards, was more of a school exercise than a practical technique. It was unnecessary for designing fortresses on the flat plains of Flanders, and it was useless against high trajectory mortar fire. I am aware that absence of evidence is not evidence of absence. However, in the central French military engineering archives I have never come across the use of defilading for practical design work (as opposed to school exercises), and I find this significant. Moreover, Monge's interest in the technique was more motivated by its relevance to general and abstract problems in descriptive geometry than by its usefulness in warfare. Defilading was merely one of many applications of descriptive geometry on which Monge worked. Like the other examples given, it was peripheral both to military engineering practice and to mathematical theory, since it was merely an application of intersecting and tangent surfaces.

Having made this sketch of how science and military engineering intersected, I now need to address how fortification design itself was a distinct science. Claims for that status start in the Renaissance and are implicit in such titles as Jacomo Lanteri's *Due dialoghi . . . del modo di disegnare le piante delle fortezze secondo Euclide. . . .*[27] It is fallacious to believe that fortifications could be designed only through the deductive scientific reasoning demonstrated in Euclidean proofs. Geometrical constructions are used in laying out the plan and profile of the fortress; in this sense, the Euclidean science of geometry is certainly useful. If one accepts axioms such as the desirability of no dead ground (the area protected from the direct fire of defensive weapons and almost universally accepted by fortifiers after the 1530s), then certain geometrical constraints are employed. In this trivial sense, one could speak of a distinct fortification science.[28] Obviously, such axioms lack the epistemological status of Euclidean axioms.

Vauban himself, that paradigmatic figure of fortifiers, never made any elevated scientific claims for fortification design. Clearly concerned about the technical proficiency of French military engineers, he introduced standards for admission into the corps that included basic elements of mathematics and drawing. The apprenticeship that novice engineers served before taking the final examination essentially consisted of working with a senior engineer in the department of fortifications. This sharpened the mathematical tools they would need and furnished the requisite experience. There is no requirement for the study of a formal science of fortification engineering, however. In fact, Vauban mocked such an idea. When requested by his secretary Thomassin to write a fortification treatise, he allegedly responded, "Do you want me to teach that a curtain is between two bastions and that a bastion consists of an angle and two faces, etc.? That is not my way."[29]

Schoolmasters and mathematics instructors were more amenable to treating fortification as a formal scientific discipline. It is unclear, however, whether this reflected self-serving rhetoric, genuine belief, or a loose conception of what a formal science was. The broader meaning of the word "science" before the nineteenth century obviously supported such claims, but it begs the question at hand.[30] As we shall see, a minority of military engineers also argued that fortification was a "science" and explicitly compared it to Newtonian rational mechanics—the quintessential science in Enlightenment France. Specific contingent reasons exist for this comparison, however, which have much to do with Montalembert.

The *Encyclopédie* classifies *architecture militaire* as a subdivision of elementary geometry, itself a department of pure mathematics and deemed a part of natural philosophy. Yet it is listed as a "pure" science rather than a "mixed" science. This indicates to me not a science of design, but merely the manipulation of lines and surfaces—the most obvious activity of geometry. (Certainly this is baffling to good engineers today as it was then.) The schools taught "science" of fortification as a straightforward application of Euclid in order to draw plans of fortresses and to articulate a few definitions and axiomatic design principles.

A thesis on fortification design was authored in 1751 by a student graduating from Voltaire's and Robespierre's alma mater: the prestigious Collège de Louis le Grand. Appearing in the same year as the *Encyclopédie*, it consisted of a long list of formal design propositions on fortification, which were very Euclidean in style but of little practical use.[31] The first of the "General Principles" of this work asserted: "One fortifies a fortress so that without excessive expense one can with few people defend it against a large number of the enemy." "General Principle" No. 5 stated: "The parts which flank [adjacent bastions] are covered as much as possible [from enemy gunfire]." The second part of the thesis, titled "Rules based on the principles," began with the rule that "Angular bastions are better than round towers," and Rule No. 11 continued: "The angle of the bastion must not be obtuse." An ensuing school exercise constructed the plan of a fortress based on these principles and rules.

Vauban—along with most engineers before, during, and after his career—had a faintly veiled contempt for such schoolmasters and "professors of the square and compass." We can detect a lively interest in the science of his time with Blaise-François de Pagan (1604–1665), Vauban's illustrious predecessor and inspirer of much of his military architecture; however, in spite of all his pioneering research in meteorology and social science, Vauban had no ambition to create a science of fortification.[32] On a number of occasions, Vauban said that the supreme art of the fortifier was to fit a fortress to a given topographical and military situation—thereby implying that his science, if there was one, was an engineering discipline that engaged in synthesis, not analysis. Hence, it was hardly a true science based on the canons of eighteenth-century Newtonian science and the rational mechanics of the French artillery and military engineering schools. Vauban rebuilt and repaired far more fortresses than he designed originally. Everywhere, he was forced to adapt himself to an existing architectural heritage.[33] Even with the financial largesse of Louis XIV in the early years of his reign, Vauban hardly had the luxury of building absolutely from scratch in the ideal conditions that were the norm for the school exercises of colleges.

In his classic article on Vauban, Henry Guerlac argued that Louis XIV's great military engineer did indeed apply science to warfare.[34] Yet, once again, it is necessary to question what this means precisely. It is obvious that mathematics is necessary in designing structures, and undoubtedly more mathematics was necessary in the early modern era than in the Middle Ages. But, as Rupert Hall has argued, the mathematics required was not immeasurably more complicated than that of the medieval master masons. Furthermore, mathematical applications were of a straightforward and elementary kind. To say that mathematics was "applied" is like saying that literature was applied when military engineers wrote their reports.[35]

Guerlac contended that the cultural climate of Vauban's time, founded on a classical penchant for order and symmetry, was powerfully enhanced by the development of science. Speaking of Louis XIV's attempt to rationalize his military forces, Guerlac wrote:

> Doubtless this sustained effort to systematize and order the structure of the army reflected what was taking place in other spheres. Throughout French political life traditional rights and confusions sanctified by long usage were being attacked in the interest of strengthening the central power. This cult of reason and order was not merely an authoritarian expedient, nor just an aesthetic ideal imposed by the prevailing classicism. Impatience with senseless disorder, wherever encountered, was one expression, and not the least significant expression, of the mathematical neorationalism of Descartes, of the *esprit géométrique* detected and recorded by Pascal. It was the form in which the scientific revolution, with its attendant mechanical philosophy, first manifested itself in France. And it resulted in the adoption of the machine—where each part fulfilled its prescribed function, with no waste motion and no supernumerary cogs—as the primordial analogy, the model not only of man's rational construction, but of God's universe.... [I]n the age of Richelieu and Louis XIV the reformers were guided by the spirit of the age, by the impact of scientific rationalism, in their efforts to modernize both the army and the civilian bureaucracy, and to give to the state and to the army some of the qualities of a well-designed machine.[36]

Such a causal connection is by no means easy to establish in spite of the attractiveness and plausibility of the notion.[37] One could argue the reverse: the love of order and the strength of classicism in French culture were powerful forces in the increasing popularity of the scientific world view. Vauban's orderly methods of siegecraft could be seen as a formal ballet or an example of classical drama with its three unities and rigorous rules. If Laurence Sterne's Uncle Toby in *Tristram Shandy*, that figure of good-humored ridicule, was present at the second siege of Namur, Racine, in his role as Historiographer to Louis the Great, was present at the first siege of that city. There is a certain symbolic import to both images. Equally plausible and more credible is the view that the rapid development and prestige of science had some effect on the belief that fortification was a formal science. This resonated with the views of the schoolmasters and mathematics teachers just mentioned and with those of many laymen, but it did not with the views shared by Vauban and most working military engineers. However, the final triumph of Newtonianism in France, after the first third of the eighteenth century, and the growing Encyclopédiste movement after the middle of the century, may have strengthened the belief in fortification as an exact science among laymen; it also may have led to greater receptiveness of such claims by military engineers. An example of this may be the classification in the *Encyclopédie* mentioned earlier. But another factor, such as the controversy with Montalembert, also had considerable weight in the departure from the pragmatic and unpretentious view of fortification engineering as a science.

Certainly the love of order, regularity, and a systematic approach to affairs is clearly visible in Vauban, but it is also evident in his political masters, Louvois and Louis XIV. No doubt the rationalist approach of Vauban and his masters to the problems of fortification and siegecraft was a major, if not the primary, cause of their success. It was no mean feat to bring something that could pass for order out of the chaos of French administration (in the domains of transmitting orders and receiving reports), finances (in the domains of budget, tendering, and supply in wartime), and siegecraft operations. It was this social discipline, rather than scientific theory, that made the French Department of Fortifications under Vauban the most efficient and most copied in Europe.

References to Vauban's "three systems of fortification" imply a coherence and formalization that was not at all to Vauban's taste.[38] Vauban's "first system" was borrowed primarily from his predecessor, Pagan. It consisted of a polygonal trace with the exterior angles crowned by the familiar pentagonal bastion developed in sixteenth-century Italy. The dimensions of the "front of fortification"—comprising a section of the trace that began with half of a given bastion, its adjoining curtain, and half of the adjacent bastion—were specified according to geometrical constructions based on the range of the musket.[39] To provide a devastating enfilade fire, the flank of the bastion was angled so its gunfire would defend the face of the adjacent bastion. The bastions were spacious enough to accommodate an adequate density of fire to mutually support the neighboring bastions. Soldiers manning the guns were protected only by the open parapet, which lacked bomb-proof roofs or vaults.

Vauban's "second" and "third" systems were more truly his own: their major feature was a much larger and detached bastion with the external angle of the trace now having only a small guntower of two stories, containing covered cannon. It is noteworthy that Vauban introduced the second system, by his own admission, based on the local conditions he encountered when rebuilding the fortress of Belfort. It is also noteworthy that it was quietly ignored by his successors. Instead, they adhered to Vauban's first system and merely adjusted some of the dimensions. For instance, they expanded the demi-lunes that covered the curtain. In addition, they did not adopt the guntowers of the second system, for they felt covered guns generated too much smoke to ventilate rapidly. Vauban's successors also alleged that soldiers would rather take their chances on an open parapet than huddle in a casemate within a mass of masonry that could collapse on top of them in an artillery duel.

Montalembert was willing to give Vauban credit as a great besieger of fortresses. Indeed, he argued that Vauban's innovations in siegecraft, such as the murderously effective ricochet shot that enfiladed an open parapet, had made Vauban's own military architecture ineffectual. In turn, these innovations required the massive changes that Montalembert was proposing. As a result of his military experiences, Montalembert had begun to think about fortresses and the spectacle of what appeared to be their increasing inability to withstand sieges. The progressive efficiency of artillery and Vauban's systematized techniques of siegecraft made the fall of fortresses inevitable and appeared to diminish how long they withstood assault.[40]

Montalembert had served in an elite cavalry regiment and participated in nine sieges. Later (although this fact is not given exceptional prominence in his work on fortification) he had been the French liaison officer with the Swedish and then the Russian armies during the Seven Years' War.[41] Assigned to inspect the fortifications of the Atlantic coast and to fortify the island of Oléron against an expected English attack in 1761, Montalembert had developed his ideas and submitted a sketchy description as well as a request to publish them to the Minister of War. The Minister had forwarded them for comment to two senior officers in the military engineering corps. One of them was Pierre de Filley (1695–1779), the general who commanded French military engineers in Germany during the Seven Years' War. While requesting more details, Filley responded rather positively, for he himself had developed a system of fortification, called the "mezalectre system," that departed from the traditional bastioned trace. He also asked that Montalembert keep his ideas confidential so that enemy powers could not benefit from them. The other officer, Charles-René Fourcroy de Ramecourt (1715–1791)—who fifteen years later was to become the de facto head of the military engineering corps after a major reorganization by the reforming Minister of War Saint-Germain—dismissed Montalembert's ideas as the lucubrations of an armchair fortifier and thought there was no harm in publishing them. It was only in 1776, after further unsuccessful efforts to convince the army to adopt his system, that Montalembert finally published the first of what was to become an eleven-volume series on fortification. These were eventually considered to be the most original works on the subject in the eighteenth century.

Montalembert was dissatisfied with the bastion, which he wanted to eliminate entirely. (In his first sketch in 1761, he also had wanted to eliminate the ditch and the covered way; but, by the time it went into print, he had diluted his radicalism.) This proscription against the bastion is probably all that remains constant in Montalembert's successive proposals for fortification, because he changed his views during the quarter-century in which debates raged over his system. His original trace was essentially a *tenaille* trace. There were neither curtain walls nor bastions, just triangular salients whose reentrant angles were always to be 90°—hence the name "perpendicular fortification" that he initially used to describe his system. The justification for 90° reentrant angles was a psychological one. Observations confirmed that soldiers behind a parapet had an invincible tendency to shoot straight ahead, no matter from which direction the enemy was approaching. Hence, in Montalembert's system, musketry would sweep the walls of each salient triangle more efficiently because the flanking walls were longer than the relatively short walls of a bastion's flanks. This sweeping would also be more devastating, since the fire would naturally flank the walls.

With his proposal for using other polygonal traces to maximize the effectiveness of heavy artillery, Montalembert would shift his focus away from perpendicular fortifications and small-arms fire. During the Renaissance, artillery of course had been the stimulus for the classic *trace italienne* and had remained salient in both the assault and the defense of fortresses. Vauban's successor Louis de Cormontaigne (1696–1752), however, had reduced the amount of artillery in French bastions. As did virtually all military engineers of the time, he believed that fortress artillery could slow but never overcome the besiegers' artillery. Large stocks of artillery on hand in a fortress would merely provide attractive booty for an attacking army. Montalembert adamantly asserted that the defenders' artillery could dominate that of the attackers if the fortresses were armed with artillery in quantities that greatly exceeded those in classic seventeenth-century fortifications.[42] With great hubris, he went on to claim that his ideas on fortification would make fortresses impregnable.

While his earlier volumes were titled *La fortification perpendiculaire*, Montalembert's later volumes were named *L'art défensif supérieur à l'offensif* to reflect an ever-increasing stress on artillery.[43] By that time he focused considerably less on the *tenaille* trace of the perpendicular fortification than on the polygonal trace, which consisted of long curtain walls bristling with cannon nestled in casemates and receiving flanking support from *caponiers* (solidly built and roofed masonry galleries protruding at right angles from the curtain walls and crammed with cannon and musketeers shooting in opposite directions along the ditch). These were supplemented by masonry gun towers located behind the curtains that swept the countryside with heavy artillery fire. Instead of the fortress huddling in the ground, as in classical artillery fortification, which defended its perimeter by its passive bulk, it was transformed in Montalembert's scheme into an active system of superior fire that would obliterate any siege.

Montalembert also proposed surrounding the main fortress with a ring of detached forts that could mutually support one another yet individually sustain a small siege. The purpose was to keep the attacking army and its artillery out of range of the main *enceinte*, particularly for ports and arsenals that were extremely vulnerable to bombardment and incendiaries. This configuration would become very popular in the nineteenth century, especially in Prussia and the United States, even before rifled artillery would make it indispensable.

In his polemics against the French military engineers, Montalembert betrays an almost inhuman tenacity. No attack on his system went unanswered; sometimes he even rebutted his opponents in the engineering corps with an entire book. In spite of his argumentative nature, Montalembert did have a strong network of supporters in high places. Under the royal regime, he enjoyed the favor of the influential minister Maurepas. Later, under the Revolutionary regime, Mirabeau was willing to name him inspector of fortifications.[44] However, it was more than the deaths of his influential patrons at crucial junctures that vanquished Montalembert's schemes. It was the solid opposition of the official technical establishment in charge of fortifications—the military engineers—who eventually stopped Montalembert both during his lifetime and after his death. Although other countries, such as Prussia, the United States, Austria, and Russia, adopted significant parts of his ideas in the nineteenth century, the French persisted in ignoring their own prophet.

This opposition of the engineers reflects the tension between engineering officers and regular line officers. Montalembert argued that the engineers had arrogated for themselves, on specious grounds of expertise, functions that should belong to regular field officers. A major part of his enterprise was the demystification of fortification engineering. As the polemics grew more bitter and the last shreds of *ancien régime* politeness were stripped away, Montalembert went so far as to propose that the corps of military engineers be abolished and their functions handed over to the civilian *Corps des ponts et chaussées*. He conceded that there are certain parts of a fortification system, especially military architecture and construction, which must be handled by specialists, but that there is nothing distinctly military that justified a separate military corps. The knowledge of the behavior of structures could be easily applied within military contexts by civilian engineers with no claims to the status of officers and, by extension, to the status of nobles.

It appears that Montalembert on the one hand is minimizing fortification as a military science and as an activity needing specifically military applications of science. Walls are walls. Fortress walls merely have to be stronger, and no special branch of construction science is necessary to design them. On the other hand, common sense and experience are all that are necessary to make a good fortifier, who counts on the compelling effect of his drawings and their simple geometry to convince his readers. In what appears to be a presentation copy of Montalembert's first volumes, probably intended for royal eyes, there were 162 high-quality plates that were hand-colored.

Nonetheless, Montalembert was also keen to use science to justify his own credentials and arguments. In 1747, he had become a member of the Royal Academy of Science in the category of *associé libre*. His scientific attainments at that time were limited, and his election most probably came about due to the influence of his patron, the prince de Conti, who was a prince of the blood. Throughout his academic career, his participation in the Academy was minimal. He used it primarily as a forum for presenting his ideas on artillery and fortification. His keen awareness of the value of scientific credentials was evident when he displayed his title as academician of the Royal Academy of Sciences in Paris and the St. Petersburg Academy on all the title pages of his books.[45] In 1779 he was given the only opportunity to demonstrate his fortification theories by building the small island fort at Aix near Rochefort, a major French naval base on the Atlantic. Here, he requested Charles-Augustin Coulomb (1736–1806) to be an observer from the military engineering corps. Coulomb was the only military engineer who was a member of the Academy of Sciences at the time, and it is clear that Montalembert hoped that the cachet of his title would enhance his argument. Unfortunately for him, Coulomb appears to have been just as unimpressed with Montalembert's schemes of fortification, not to mention his personality, as his fellow engineers. Montalembert would later attribute this coolness to the crushing cultural pressure of the *esprit de corps* of the military engineers, who could even intimidate an academician.

After the construction of Montalembert's fort at Aix, the controversy with military engineers burst into the public sphere. Up to that time, it had been confined to the privacy of ministerial offices and memoranda. In 1785 Fourcroy was belatedly elected a member of the Royal Academy of Sciences. Formerly, he had been one of the two military engineers to examine Montalembert's original proposals; he had been attacked by Montalembert in closed committees over a proposal of colonial fortification; and he had become the resident engineer at Versailles, advising the Minister of War the same year Montalembert began to publish. Fourcroy was the model of the plodding bureaucrat. His military career was respectable but not particularly distinguished, and some would claim that he obtained his post because he was adept at intrigue. Others, like Condorcet, would argue that Fourcroy was the man everyone in the corps, even Fourcroy's senior officers, wanted to see in the position as resident engineer at Versailles.[46] He was also a dedicated disciple of Cormontaigne, who had revised some of Vauban's ideas and whose schemes of fortification would become the dominant orthodoxy in the military engineering corps and at its school at Mézières. Although unpopular among younger officers just before the Revolution for his perceived pliancy in the face of cuts to the fortification programme, Fourcroy probably represented the bulk of the senior officers in his corps. His opinions can therefore typify the generation of military engineering officers that had not gone through the scientifically demanding curriculum at Mézières.

Fourcroy began his career in the Academy by reading a paper on the composition of mock-siege journals, which he claimed were first advanced by Vauban.[47] He presented the mock-siege journal as a useful tool in analyzing the performance of fortifications—that is, their ability to withstand assault. With the help of a detailed map of the fortification to be attacked, the military engineer imagines himself the commander of the assault and writes a journal for this imaginary operation. Supplementing this journal was a map of the siege, which included the parallels dug, the batteries positioned, the lines of fire and mines projected, and the eventual walls breached. The result was a figure: the number of days from initiation of the siege to the inevitable capitulation of the garrison. Of course, there were a number of assumptions: there was no treachery, incompetence, and cowardice on both sides; there were adequate supplies and armaments; and the siege was undisturbed by relieving armies or successful sorties. All this, along with the banishment of the goddess Fortune from combat, is rather odd; yet Fourcroy claimed that it was a valuable technique for comparing the performance of fortifications. He is misleading, however, when he attributes this method to Vauban. The engineering marshal had estimated the capability of various fortresses to withstand a siege, but he had offered no detailed calculation and mapping procedures, as Montalembert was quick to point out. After all, it would be out of character for Vauban to use elaborate topographical and quantitative methods to arrive at a precise number that would ultimately prove meaningless for an actual siege.

For Fourcroy, the mock-siege journal was nonetheless fundamental to the science of fortification. He proceeded to use it to analyze Montalembert's proposed fortifications and to compare them to the almost standard techniques of Cormontaigne. Not surprisingly, Montalembert's fortifications were shown to be grossly defective, and Fourcroy concluded that it would be a waste of the King's money to build them. He then went ahead and published a voluminous elaboration of his argument that aggressively attacked the inventor of perpendicular fortification. He now included a concept that Vauban had never mentioned: the moment of fortification. Redolent with the language of eighteenth-century rational mechanics, a science that the Encyclopaedists considered almost on par with Euclidean geometry, the moment of fortification was defined as the quotient of the number of days it would take to capture a fortification, as calculated from a mock-siege journal, and the cost to construct a single front of fortification. This quotient proved even more arbitrary than the mock-siege journal itself. Lazare Carnot (1753–1823) later pointed out that using the logic of moments of fortification to evaluate the performance of fortresses would lead to absurd decisions. For example, imagine a dodecagonal fortress (with twelve fronts of fortification, each costing only 10 units) that holds out for only 80 days and a hexagonal fortress (with six fronts of fortification, each costing 15 units) that holds out for 100 days. According to the theory of fortification moments, the dodecagonal fortress, costing more *in total* than the hexagonal one (120 units vs. 90 units) and capable of a much shorter period of resistance (80 days vs. 100 days), is a better fortress; it has a moment of 8 (80 days ÷ 10 units), which is greater than the hexagonal fortress' moment of 6.67 (100 days ÷ 15 units). Thus, a fortress whose total cost is a third more and is capable of holding out for almost three weeks less than the other appears superior on the basis of this bizarre equation that considers only the cost of a single front of fortification. Yet, proud of this theory, Fourcroy attributed it to Cormontaigne, whom he glorified as the Newton of the science of fortification. He deemed Vauban, by contrast, a mere Kepler, thereby establishing fortification engineering as a *science positive*. The use of fortification moment would furnish unequivocal results for guiding fortress design. Fourcroy went so far as to proclaim that the true principles of fortification had been discovered once and for all. He felt it had been demonstrated quantitatively that Vauban's bastioned traces were the best ones possible and that any innovation in fortification should be proscribed with the same vigor the Academy had shown in rejecting proposals to square the circle or enlist the power of perpetual motion.

True to form, Montalembert did not waste much time in replying. He filled an entire book with a detailed exegesis of Fourcroy's analysis and requested adjudication by the Academy of Sciences.[48] The Academy was in the habit of appointing a committee to examine works by its members. Because of its role as official censor of scientific and technical works for the royal government, its imprimatur was usually a crisp permission to publish. Occasionally, however, this included a few remarks from the report of the commission that had examined the book. Fourcroy's initial paper (but not his expanded book) passed through this procedure; therefore, Montalembert requested that the same academic committee formally evaluate his own book. Because the Academy functioned as the most prestigious oracle of scientific legitimacy, Montalembert used his membership not only to bolster his status as a savant but also to legitimize his theories of fortification.

The Academy was conscious of its invidious position as the arbiter that would receive criticism from the losing party. It had already been embroiled in the debate over the artillery reforms of Gribeauval. Now, at a time when it was coming under attack in the Marat and Mesmer affairs as an institution hostile to novel ideas and budding geniuses, it faced another hornet's nest.[49] The Academy therefore procrastinated in its decision, and, when it did decide, its secretary Condorcet chose a non-committal wording for its committee's report that denied Montalembert unequivocal support.[50] Frustrated by the Academy's attitude, Montalembert launched a veiled attack against it. Through a pamphlet written by a relative, he accused the Academy of incompetence and slavish subordination to the government and its corps of military engineering.[51] The revolutionary undertone of the pamphlet indicates that Montalembert had finally rejected any right of the official guardian of scientific orthodoxy, whose support he had so avidly solicited, to make pronouncements on military disputes.

Montalembert's and Fourcroy's efforts to solicit the support of the Academy of Sciences reveal the extent to which it represented a legitimizing and validating institution in an age of triumphant Newtonianism. However, did they actually believe fortification design was a distinct science founded on mathematical concepts that the Academy could objectively evaluate? Montalembert always claimed to be a demystifier of fortification, posing as an enemy of what he felt was deliberate obfuscation and unwarranted claims of expertise by engineers motivated by a narrow *esprit de corps*. This hardly made him a tenacious partisan of the right of fortification engineering to be a science. Similarly, Fourcroy came from a milieu and a tradition in which, from Vauban onwards, the particular case and terrain overrode any formal systems and recipes that claimed scientific status.

A number of reasons nevertheless support Fourcroy's position that fortification deserved the status of a Newtonian science. First, like Montalembert, he wished to profit from the prestige of science among the cultivated public and the military. He publicly maintained that fortification was a distinct science in his comparisons of Vauban and Cormontaigne to Kepler and Newton, respectively, although his conviction appears questionable. Fourcroy also privileged "experience"—a word italicized throughout his book—in his attacks on Montalembert as an armchair fortifier and "professor of the square and compass." Clearly, experience surpassed deductive reasoning from general principles; Fourcroy's inept handling of abstract concepts such as the moment of fortification indicates his naive grasp of the mathematical sciences. Like the public he was trying to impress, he was essentially an onlooker rather than a competent practitioner of science, including any supposed science of fortification. Fourcroy may have simply been goaded into his claims for the scientific status of fortification by the dangerous onslaught of Montalembert; this again raises questions about the sincerity of his claims. It seems probable, in the final analysis, that if Newtonian science did provide a paradigm for the art of military engineering, it was more to flaunt and influence than to create and analyze.

A younger generation of military engineers graduating from Mézières were better trained in the mathematical and physical sciences than Fourcroy. They easily handled those mathematical techniques that Fourcroy found cumbersome and, in some cases, incomprehensible. Among this generation was the future Revolutionary politician Lazare Carnot, who was the only one to go against his corps publicly and support Montalembert. Carnot mocked the immature mathematization of Fourcroy's moments of fortification, yet he too dreamed of a geometry of fortification that was on par with Euclid's *Elements*.[52] He was not exactly clear about what this science would be, except to say that it would be based as much on intuition as on Reason. At the turn of the century, Simon-François Gayvernon (1761–1822), another military engineer and professor of fortification at the École Polytechnique, also saw fortification as a specific manifestation of descriptive geometry that owed much of its development to military engineering practice.[53]

The status or nature of a "science of fortification" in the late Enlightenment was ambiguous, to say the least. Both Vauban and Montalembert used science mainly in the form of elementary geometry. Vauban apparently had no illusions about the science of fortification; he strenuously worked at systematizing engineering reasoning and military experience. Montalembert, much more dogmatic about the superiority of his system, also never claimed scientific status for it. However, he did seek to legitimize his ideas by publicizing his association with prestigious scientific institutions. The military engineering corps displayed a similar marketing awareness. Somewhat uncharacteristically, in view of Vauban's strong legacy, the senior leadership of the corps argued that fortification was a bona fide science that emulated the Newtonian paradigm. It is questionable to what degree this was sincere and to what degree the men of an older generation, who held senior positions in the corps immediately before the Revolution, were intimately acquainted with the triumphant Newtonian sciences. The younger generation of military engineers were thoroughly conversant with these sciences but had little need to use them. While they may have been more receptive to the idea of a science of fortification than Vauban, as seen in the examples of a Carnot or a Gayvernon, they nonetheless failed to construct a concrete theory of it.

Equally vague was the military art discussed by Louis Lebègue Duportail (1743–1802), a junior officer in the corps of military engineering in the early 1770s. Possibly in response to Guibert's attacks on the utility of military engineering,[54] Duportail argued that fortifiers spent too much time on the technical details of military architecture, thereby validating the accusations that they were not real soldiers. For him, military engineering was an intrinsic part of the art of war because its essence was defense, the inextricable complement of offense. The function of military engineering not only concerned siege warfare but also field operations and topography (i.e. the construction of temporary and field fortifications and the use of natural terrain in planning tactics and strategies). The military engineer's natural place was the general staff, where he planned maneuvers and battles. War was essentially the dominance of space as comprehended by the *coup d'oeil* of the general as well as that of the topographer engineer.[55] Duportail illustrated the increasing importance of topography for the military engineer; he envisioned the true role of the engineer as a manipulator of time and space for all varieties of warfare, rather than exclusively as a builder and destroyer of permanent fortifications.

Vauban's alleged doctrine of the *pré carré* already demonstrated the need to go beyond the technical problems of a particular fortress and consider its entire geographical context.[56] His successors also looked at an entire defensive area, frontier, or country rather than discrete fortified points. In this sense, Montalembert, with his insistence on impregnability, reverted to an almost medieval conception of the absolute defense of a stronghold. One of his great adversaries in the military engineering corps, Jean-Claude-Éléonor Le Michaud d'Arçon (1733–1800), viewed fortification in systemic or operational terms: the engineer not only had to design each individual fortification but also had to examine the relations between a series of fortifications that both supported each other and hindered the enemy.[57] Resources had to be apportioned over an entire defensive system, and the overall strength of the system had to be evaluated. The power of each individual fortification was measured in terms of its usefulness to the overall system and not solely in terms of its individual resistance.

Topography was evidently as important for d'Arçon as it was for Duportail, but Duportail had looked beyond the engineer's art of fortification to the strategy of war in general. Military engineering occupied the central part of a theory of war, and the military engineer was the military theoretician *par excellence*. Although Duportail's arguments were couched in elevated and abstract terms, offering few examples or concrete demonstrations, his ideas reveal a desire for a general military theory. Although he was unique in his defense of military engineering as the core of military theory, he was by no means alone in his conception of a transcendent theory of warfare.

Transcending fortress design, Duportail's conception of military strategy reflects how far-reaching the interaction between science and warfare had become during the Enlightenment. Azar Gat is correct in arguing that early modern military thought demanded more than field experience and a familiarity with weapons, military history, and human nature.[58] Military men who wrote books on the subject of war, such as Maurice de Saxe (1696–1750), often distinguished between the technical aspect of the art of war (e.g., descriptions of weapons; techniques of drilling, combat, and castramentation) and its "sublime" or theoretical aspect. For some, the latter was intuitive and could be grasped only by the native genius of a great military commander; it was not subject to analysis and rules. Others, such as Jacques François de Chastenet, marquis de Puységur (1656–1743), believed that the sublime aspect of warfare was amenable to a scientific approach:

> We have no school where one can instruct oneself in the military art, no teacher who can teach the fundamental rules. Worse, people are almost convinced that such resources are useless, that war is only learned in the tumult of camps and the movement of armies. As if all the arts did not have certain rules and a theory founded on solid principles, without which not only can one not hope to succeed in them but one shouldn't even expose oneself to practicing them![59]

"Military thought" may be considered an oxymoron by anti-militarists, pacifists, and even some generals and soldiers. It is more elusive than artifacts, formal theories and procedures, and methods of creating specific military technologies. The channels of interaction are more difficult to identify, and the ultimate effect on military operations and strategy more challenging to document.

However, Gat demonstrates that science, along with classical canons of order and harmony as well as administrative efforts towards order, discipline, and system, was part of the cultural background that influenced early modern warfare. What Gat calls a "protoscience," or a nascent modern science, also affected influential military thinkers like Raimondo Montecuccoli (1609–1680). T. M. Barker's description of Montecuccoli as a "military intellectual" is apt and could be applied to other military reformers such as Maurice of Nassau (1567–1625) and Gustavus Adolphus (1594–1632).[60] Soldiers and military theorists wrote and read books on war, and the volume of military writings increased considerably during the Enlightenment. Gat mentions a fourfold increase in the publication rate of military theory books following the Seven Years' War.[61]

Did fortification function as a catalyst of this intellectualization of military thought? Here the answers remain speculative. Some military leaders (e.g. Maurice de Saxe) were contemptuous of fortification, seeing it as mere specialized and limited knowledge, others (e.g. the Marquis de Puységur) were not. In his widely read book on the art of war, Puységur explicitly referred to fortification as the model for a more general military science:

> I am undertaking therefore to show that without war, without troops, without armies, and without being obliged to leave one's house, by study alone, with a bit of geometry and geography, one can learn all the theory of field warfare from its most detailed to its most general parts. This [can be done] in the same way that Marshal Vauban, by the theory contained in the books he has left us and the practice he established in conformity [with that theory], teaches us the art of fortifying, attacking, and defending fortresses—something which is taught daily by people who have never been to war nor have supervised work on fortifications.[62]

Puységur reflects the widespread if not dominant view of Vauban as a deductive military tactician, which I have argued contradicts his actual practice. Nevertheless, Puységur's common portrait of Vauban indicates the impact that science had on military thought—even if putative. Even Guibert, prophet of the new war of movement and nemesis of the old war of position, used the metaphor of fortification to describe his new theory of warfare. Thus, fortresses should be "marshalling points, points of supply and support, bastions whose curtain is a good and maneuverable army."[63]

Those who groped for a new kind of military science privileged the status of fortifications. This is understandable given its central role in early modern warfare, and the deep relations between fortification design and mathematics that commenced with the Archimedean revival of the Italian Renaissance. Although the connection between advanced science and the military arts had been evident to superior military minds, run-of-the-mill mathematical practitioners, and mathematics teachers, it appears that only after the final triumph of Newtonianism in France in the 1730s did this idea begin to percolate down to the average line officer. This coincided with the transformations of French military education from the middle of the eighteenth century onwards and the progressive dominance of a new ethic of service and efficiency over an older one of charisma and dash.[64]

I have argued that the links between science and warfare, including military technology, were tenuous, although the eighteenth century witnessed an intensification of these links in areas such as ballistics and navigation. As opposed to rational organization and efficient bureaucracy, scientific theory neither increased the killing rate nor enhanced the military success of armies, at least not significantly in fortress warfare. The links between science and war in military thought, however elusive, are more suggestive if not profound. Military engineering in general had a prominent germinal role in this interaction in spite of misconceptions about the alleged scientific nature of fortification design.

Notes

1. On the "military revolution" see Geoffrey Parker, *The Military Revolution: Military Innovation and the Rise of the West, 1500–1800* (Cambridge University Press, 1988) and Bert S. Hall and Kelley deVries, "The 'Military Revolution' Revisited," *Technology and Culture, 31* (July 1990): 500–507. Also valuable on this subject is John A. Lynn, ed., *Tools of War: Instruments, Ideas, and Institutions of Warfare, 1445–1971* (University of Illinois Press, 1990). On the "scientific revolution" see A. Rupert Hall, *The Revolution in Science, 1500–1750* (Longman, 1983); I. B. Cohen, *Revolution in Science* (Belknap, 1985).

2. See A. Rupert Hall, "Science, Technology, and Warfare, 1400–1700," in M. Wright and L. Paszek, eds., *Science, Technology, and Warfare* (Government Printing Office, 1983); Rupert Hall, *Ballistics in the 17th Century: A Study in the Relations of Science and War with Reference principally to England* (Cambridge University Press, 1952).

3. Galileo Galilei, *Operations of the Geometric and Military Compass*, 1606, ed. S. Drake (Smithsonian Institution Press, 1978).

4. A. Rupert Hall, "Architectura Navalis," *Transactions of the Newcomen Society*, 51 (1979–80): 157–174.

5. Brett Steele, "Muskets and Pendulums: Benjamin Robins, Leonhard Euler, and the Ballistics Revolution," *Technology and Culture*, 35 (April 1994): 348–382.

6. One could also mention Samuel Hartlib and the group connected with him, the early Royal Society, the Académie royale des sciences, and numerous authors of mathematical texts in the sixteenth and seventeenth centuries. For an eighteenth century example, see Larry Stewart, "The Selling of Newton: Science and Technology in Early Eighteenth-Century England," *Journal of British Studies* 25 (April 1986): 178–192.

7. Interest in mechanics in seventeenth-century Italy was at a high pitch and there were at least two foci of activity and half-a-dozen intellectual traditions dating from Antiquity and the medieval period. Possible practical applications of mechanics to warfare as well as other fields appear to have been only one of several factors attracting workers in the field. See S. Drake and I. E. Drabkin, eds., *Mechanics in Sixteenth-Century Italy: Selections from Tartaglia, Benedetti, Guido Ubaldo, & Galileo* (University of Wisconsin Press, 1969). See also W. Roy Laird, "Patronage of Mechanics and Theories of Impact in Sixteenth-Century Italy," in B. Moran, ed., *Patronage and Institutions: Science, Technology, and Medicine at the European Court 1500–1750* (Boydell, 1991), 51–66.

8. Bernard Forest de Bélidor, *Le Bombardier françois, ou nouvelle methode de jetter les bombes avec precision* (Aux depens de la compagnie, 1734).

9. Bernard Forest de Bélidor, *La Science des ingénieurs dans la conduite des Travaux de Fortification et d'Architecture civile*, 6 volumes in one (Claude Jombert, 1729), 79–91, especially p. 89. This book had numerous reprintings, including one with corrections and comments by Navier (Firmin Didot, 1830).

10. Jacques Heyman, *Coulomb's Memoir on Statics: An Essay in the History of Civil Engineering* (Cambridge University Press, 1972).

11. One exception was Pierre-Dominique de Grenier (1730–1782?), whose paper on the subject was included in Fourcroy de Ramecourt's attack on Montalembert.

12. One example among several: after deriving an algebraic expression to find the thickness of the piers of an arch sustaining a given force, Bélidor indicates a simple arithmetic procedure to do the same thing and provides a numerical example because "les calculs, tout aisés qu'ils sont, pourraient embarrasser ceux qui n'ont point l'habitude" (Bélidor, 1830 reprint, p. 114).

13. Coulomb's classic memoir of 1773, "Essai sur une application des règles de maximis et minimis à quelques problèmes de statique, relatifs à l'architecture," *Mémoires de Mathématique et de Physique présentés à l'Académie Royale des Sciences, par Divers Savans* (Paris, 1776), 343–382, one of the pioneering works of engineering science, did not pass much beyond the rather narrow audience of academicians to which it was delivered.

14. On Prony's popularization of Coulomb's work, see C. Stewart Gillmor, *Coulomb and the Evolution of Physics and Engineering in 18th Century France* (Princeton University Press, 1971), p. 106ff. According to Navier, in his edited reprint of Bélidor's *La Science des Ingénieurs* in 1830, Prony's presentation of Coulomb significantly occurred in his lectures at the École Polytechnique in the 1790s.

15. On Mézières, see René Taton, "L'École royale du génie de Mézières" in René Taton, ed., *Enseignement et diffusion des sciences en France au XVIIIe siècle* (1964; reprint, Hermann, 1986), 559–615.

16. Étienne Bezout, *Cours de mathématiques à l'usage des gardes du pavillon et de la marine, par M. Bezout . . .*, 5 parties en 6 vols. (J. B. G. Musier, 1764–1769). This was followed by his *Cours de mathématiques à l'usage du corps royale de l'artillerie* (Imprimerie royale, 1770–72) and numerous reprintings, translations, and modifications well into the nineteenth century.

17. Josef N. Konvitz, *Cartography in France, 1660–1848: Science, Engineering and Statecraft* (University of Chicago Press, 1987).

18. On the ingénieurs géographes, see Col. Berthaud, *Les ingénieurs géographes militaires 1624–1831: Etude historique*, 2 vols. (Paris, 1898–1902). For engineers of the Ministry of External Affairs occupied in demarcation of frontiers, see also Konvitz, pp. 92ff.

19. Anne Blanchard, *Dictionnaire des Ingénieurs Militaires, 1691–1791* (Montpellier: Centre d'histoire militaire et d'études de défense, 1981).

20. Jean-Lambert-Alphonse Colin, *L'Education militaire de Napoléon* (Chapelot, 1901); and James Donald Hittle, *The Military Staff: Its History and Development* (1944; reprint, Westport, Conn.: Greenwood, 1975). Most of Bourcet's work was published posthumously: Pierre de Bourcet, *Mémoires militaires sur les frontières de la France, du Piémont et de la Savoie depuis l'embouchure du Var jusqu'au Lac de Génève* (Berlin: Decker, 1801).

21. Douglas W. Marshall, The British Military Engineers 1741–1783: A Study of Organization, Social Origins, and Cartography, Ph.D. thesis, University of Michigan 1976.

22. Patrice Bret, "Le Dépôt général de la Guerre et la formation scientifique des ingénieurs-géographes militaires en France (1789–1830)," *Annals of Science,* 48 (1991): 113–157.

23. On Monge, see René Taton, *L'Oeuvre scientifique de Monge* (Presses Universitaires de France, 1951).

24. The standard story is told by François Arago, "Notices Biographiques: Monge" in *Oeuvres Complètes de François Arago*, volume 2 (Gide et Baudry, 1854), 426–459.

25. Bruno Belhoste, "Les Problèmes de défilement" in Jean Dhombres, ed., *L'École Normale de l'An III: Leçons de Mathématiques* (Dunod, 1992), Annexe 16, 541–546.

26. On Meusnier, see Charles Coulston Gillispie, *Science and Polity in France at the End of the Old Regime* (Princeton University Press, 1980).

27. Published in Venice in 1557. See Horst de la Croix, "The Literature on Fortification in Renaissance Italy," *Technology and Culture,* 4 (1963): 30–50. On page 41, de la Croix calls this "the first work in which the planning and design of fortifications is treated as a purely abstract and geometrical problem."

28. A rather rare example of an exception to universal public support of this "axiom" was Jacomo Fusto Castriotto (ca. 1510–1563), an experienced soldier who felt that the dead ground argument against round bulwarks was not decisive in practice, no matter how attractive it was in theory. Christopher Mallagh, Science, Warfare and Society in the Renaissance with Particular Reference to Fortification Theory, Ph.D. Thesis, Leeds University 1981, volume 1, pp. 24–26; volume 2, pp. 59–64.

29. "Voulez-vous que j'enseigne qu'une courtine est entre deux bastions, qu'un bastion est composé d'un angle et deux faces, etc. ? Cela n'est pas mon fait." A. Rochas d'Aiglun, ed., *Vauban: Sa Famille et Ses Ecrits, Ses Oisivetés et sa Correspondance - Analyse et Extraits*, 2 volume (1910; reprint, Geneva: Slatkine, 1972).

30. Still useful for a sketch of the plethora of intellectual traditions, some of which were later classified as pseudo-science, which fed into the mainstream of the Scientific Revolution is Hugh F. Kearney, *Origins of the scientific revolution* (Longmans, 1964).

31. *Thèses de Mathématiques, de Géométrie, de Trigonométrie rectiligne, et de Fortifications . . .* (Thibout, 1751).

32. Blaise-François Pagan, *Les fortifications du comte de Pagan* (Nicolas Langlois, 1689).

33. For overviews of the history of artillery fortification, see Christopher Duffy, *Siege Warfare* (Routledge and Kegan Paul, 1979–85); James Quentin Hughes, *Military Architecture* (Hugh Evelyn, 1974).

34. Henry Guerlac, "Vauban: The Impact of science on war" in Peter Paret, ed., *Makers of Modern Strategy: Military Thought from Machiavelli to the Nuclear Age* (Princeton University Press, 1986), 64–90.

35. Hall, pp. 13–15.

36. Guerlac, pp. 67–68.

37. Unlike Rupert Hall, Guerlac believes that developments in science and mathematics did have more direct effects on war, one of which, besides the improvement of artillery and cartography, was the "revolution" in the "art or science of military architecture." For him, it was "a new field of applied of science." Guerlac, p. 69.

38. On Vauban see P. Lazard, *Vauban 1633–1707* (Alcan, 1934); as well as F. J. Hebbert and G.A. Rothrock, *Soldier of France: Sebastien Le Prestre de Vauban, 1633–1707* (New York: Peter Lang, 1989). For a taste of Vauban's style, see Sébastien le Prestre de Vauban, *A manual of siegecraft and fortification*, ed. G. Rothrock (University of Michigan Press, 1968). The literature on Vauban is a rich one and shows no signs of drying up. For an example of more recent books in French see Bernard Pujo, *Vauban* (Albin Michel, 1991) and a much better one by Anne Blanchard, *Vauban* (Fayard, 1996).

39. For a geometrical construction of a front of fortification according to the rules laid down by Vauban and others, see Ian Hogg, *Fortress: A History of Military Defence* (MacDonald and James, 1975), 54ff.

40. This was the opinion of people such as the chief of the Swedish military engineering corps General Jean-Bernard Virgin in his *La défense des places mise en equilibre avec les attaques savantes et furieuses d'aujourd'hui . . .* (Stockholm, 1781). The opinion of Virgin, who witnessed many of the French sieges in Flanders during the War of the Austrian Succession (1740–1748), and felt that the war had demonstrated the increasing weakness of the defence in fortress warfare, seems to be supported by the actual lengths of successful sieges. During the War of the Spanish Succession (1701–1713), the average length of sieges for both the French and the Allies was about 40–45 days (although the median length of sieges seems to have been about a month for the French and about a month and a half for the Allies). In the War of the Austrian Succession the average length of French sieges was about two weeks with a median that was a few days lower. Admittedly such numbers must be viewed cautiously: the state of maintenance of fortifications and the quality of the troops defending them at different times is not easily quantifiable and is but one of the factors that makes a

"science" of fortification so elusive. But Mons was taken by the French in the War of the Austrian Succession in a mere 16 days, while it had taken the Allies 25 days to take it in 1709. It had taken the great Vauban himself 27 days to take Namur in 1692 but it took the French half that time to take it during the War of the Austrian Succession. Figures compiled from Antoine-Marie Augoyat, *Aperçu historique sur les Fortifications, les Ingénieurs, et sur le Corps du Génie en France*, 3 vols. (Ch. Tanera; J. Dumaine, 1860–64) and Colonel Pierre Rocolle, *2000 ans de fortification française*, 2 vols. (Charles-Lavauzelle, 1973). There are usually references to the increasing power of artillery, but there is no mention of the causes of this increase. By the time of the Peninsular War, Wellington's chief of engineers was claiming that breaches were possible from formerly impossible distances and he attributed this to better boring of cannon and improvements of gunpowder manufacture. See Sir John Thomas Jones, *Journals of the Sieges carried on by the Army under the Duke of Wellington in Spain*, volume 1, 3rd. ed. (John Weale, 1846), 379.

41. He published three volumes of correspondence with the court during his duty as a liaison officer (Marc-René de Montalembert, ed., *Correspondance de Monsieur le Marquis de Montalembert, Étant Employé par le Roi à l'Armée Suédoise, avec Mr. le Marquis d'Havrincour, Ambassadeur en Suéde, Monseigneur le Maréchal de Richelieu, Les Ministres du Roi à Versailles, MM. les Généraux Suédois et autres, etc. Pendant les campagnes de 1757, 58, 60, et 61*, 2 vols. (Londres, 1777). He claimed that he was responsible for urging the Russians to attack and capture Berlin—the low point in Frederick the Great's fortunes during the Seven Years' War. Although Montalembert always asserted the absolute novelty of his ideas and never mentioned any Swedish inspiration, it seems more than a mere coincidence that his guntowers are strikingly similar to those of Swedish coastal fortifications. Duffy (p. 157) feels, rightly I think, that the evidence is overwhelming for Swedish input into Montalembert's ideas.

42. For one of his fortresses (Fort Royal in volume 3 of the *Fortification perpendiculaire*), Montalembert had proposed an armament of 605 guns (serviced by a garrison of 1500 men). A typical Vauban fortress with six bastions would have had about a maximum of ten pieces per bastion for a total of 60 pieces in all. There would have been even fewer guns in a Cormontaigne fortress.

43. The full title of Montalembert's first published volume was *La Fortification perpendiculaire, ou Essai sur plusieurs manières de fortifier la ligne droite, le triangle, le quarré et tous les polygônes, de quelqu'étendue qu'en soient les côtés, en donnant à la défense une direction perpendiculaire. Où l'on trouve des méthodes d'améliorer les places déjà construites, et de les rendre beaucoup plus fortes. On y trouve aussi des Redoutes, des Forts et des Retranchemens de campagne, d'une construction nouvelle* (Philippe-Denys Pierres, 1776). After the fourth volume the title varies but the title of eighth volume in 1790 begins with the proud claim *L'Art défensif supérieur à l'offensif. . . .*

44. Jean Frédéric Phélypeaux, comte de Maurepas (1701–1781) was Minister of the Navy and the Colonies under Louis XV. Disgraced because of a feud with Madame de Pompadour, he returned to power under Louis XVI and was the most influential minister of the King. Until his death, André Boniface Louis Riquetti, vicomte de Mirabeau (1754–1792) was the most influential Revolutionary politician in France.

45. It is significant that Montalembert took the time to get elected to the Petersburg academy during his stay in Russia and that he preceded figures like d'Alembert and Buffon in this honor.

46. M.-J.-A. Condorcet, "Éloge of Fourcroy," *Oeuvres de Condorcet*, 12 vols. (Firmin Didot, 1847), t. 3, 400–453.

47. Charles-René Fourcroy de Ramecourt, *Mémoires sur la fortification perpendiculaire, par plusieurs officiers du génie* (Nyon, 1786).

48. Marc-René de Montalembert, *Réponse au Mémoire sur la Fortification Perpendiculaire par plusieurs Officiers du Corps Royal du Génie, Présenté à l'Académie Royale des Sciences* (Philippe-Denys Pierres, 1787). This is usually catalogued as volume 7 of his multi-volume *Fortification perpendiculaire.*

49. On Marat and Mesmer, see Gillispie, *Science and Polity*; Robert Darnton, *Mesmerism and the End of the Enlightenment in France* (Harvard University Press, 1968).

50. Dossier Montalembert. Archives de l'Académie des Sciences, Paris.

51. Jean de Montalembert, *Lettre de M. le Baron de Montalembert à M. de Kéralio, en Réponse au Compte qu'il a rendu dans le Journal des Savans, du Mémoire sur la Fortification perpendiculaire, publié sous le nom de Plusieurs Officiers du Corps Royal du Génie en 1786* (Spilsbury, n.d. [1787]).

52. Carnot, Lazare, "Éloge de M. le maréchal de Vauban, Discours qui a remporté le prix de l'Académie des sciences, arts et belles-lettres de Dijon, en 1784" reprinted in Charnay, Jean-Paul, ed., *Lazare Carnot: Révolution et mathématique*, 2 vols. (L'Herne, 1984–85).

53. Janis Langins, "L'Enseignement d'une science éphémère: la fortification" in Bruno Belhoste, Amy Dahan Dalmedico, et Antoine Picon, eds., *La Formation polytechnicienne 1794–1994* (Dunod, 1994).

54. Jacques-Antoine-Hippolyte de Guibert, *Essai général de tactique*, 2 vols., 2e éd. (Londres: Libraires associés, 1773).

55. On coup d'oeil, see article under that heading in Denis Diderot and J. d'Alembert, ed., *Encyclopédie* (Paris, 1751); and Edwin L. Dooley Jr., "Gayvernon and Military Education at the École polytechnique" (unpublished talk at the Western Society for French History Meeting, New Orleans, Oct. 1989).

56. Vauban was by no means the only person with ideas on natural frontiers and their defensibility. On the authorship of this doctrine and its coherence see Gaston Zeller, "La monarchie de l'ancien régime et les frontières naturelles," *Revue d'histoire moderne,* 8 (1933): 305–333; and *L'organisation défensive des frontières du Nord et de l'Est au 17e siècle* (Berger-Levrault, 1928).

57. Jean-Claude-Éléonor Le Michaud d'Arçon, *Considérations Militaires et Politiques sur les Fortifications* (Imprimerie de la République, an III [1795]).

58. Azar Gat, *The Origins of Military Thought from the Enlightenment to Clausewitz* (Clarendon, 1989).

59. "Il n'y a parmi nous aucune école où l'on puisse s'instruire dans l'art militaire, aucun maître qui en enseigne les règles fondamentales: il y a plus, on est presque persuadé que de pareilles ressources sont inutiles, que la guerre ne s'apprend que dans le tumulte des camps et dans le mouvement des armées, comme si tous les arts n'avoient pas de règles sûres et une

théorie fondée sur des principes solides, sans lesquels non-seulement on ne peut espérer d'y réussir, mais on ne doit pas même s'exposer à les pratiquer." Jacques-François Chastenet de Puységur, "Avertissement de l'Auteur," *Art de la Guerre, par Principes et par Règles*, 2 vols. (Charles-Antoine Jombert, 1748).

60. T. M. Barker, *The Military Intellectual and Battle: Raimondo Montecuccoli and the Thirty Years' War* (State University of New York Press, 1975).

61. Gat, p. 25.

62. "J'entreprends donc de faire voir que sans guerre, sans troupes, sans armées, et sans être obligé de sortir chez soi, par l'étude seule, avec un peu de géométrie et de géographie, on peut apprendre toute la théorie de la guerre de campagne depuis les plus petites parties jusqu'aux plus grandes, et cela en la même manière que le maréchal de Vauban par la théorie enfermée dans les livres qu'il nous a laissés, et par la pratique qu'il a établie en conformité, nous apprend l'art de fortifier, d'attaquer et deffendre les places, ce qui même journellement est enseigneé par des personnes qui n'ont jamais été à la guerre, ni fait travailler à fortifier les places." Puységur, p. 2.

63. "[P]laces d'armes, des points d'entrepôt et d'appuis, des bastions dont une armée bonne et manoeuvrière est la courtine. . . ." (Guibert, v. 2, p. 90)

64. See David D. Bien, "The Army in the French Enlightenment: Reform, Reaction and Revolution," *Past and Present* 85 (1979): 65–98; and André Corvisier, "Hierarchie militaire et hierarchie sociale à la veille de la Révolution," *Révue internationale d'histoire militaire* 30 (1970): 77–92.

13

MILITARY "PROGRESS" AND NEWTONIAN SCIENCE IN THE AGE OF ENLIGHTENMENT

Brett D. Steele

The military technology of eighteenth-century Europe does not enjoy a high reputation among military historians and historians of technology. Some scholars, including Robert O'Connell and the Brodies, dismiss the entire era as being utterly regressive by respectively pointing to the lack of fundamental changes in weapons hardware and the unwillingness to employ the new scientific knowledge.[1] Martin van Creveld and Russell Weigley likewise concur that the military technology was evolutionary at best. Von Creveld concluded that "in principle the guns with which Napoleon invaded Italy in 1796 did not differ greatly from those with which Charles VIII had done the same almost exactly 300 years before,"[2] and Wiegley noted that "the early modern era brought few technological advances worth destabilizing existing tactics to accept."[3] Geoffrey Parker and Jeremy Black are more impressed with the innovative qualities of eighteenth-century military institutions. What appears to some as only minor evolutionary changes emerges for them as almost revolutionary breakthroughs. These changes include the employment of the flintlock musket and bayonet, skirmishers, rationalized logistical systems, standardized field artillery, and new projectile weapons including optimized canister shot and spherical case shot (shrapnel).[4] Yet Parker and Black argued for the significance of these innovations primarily by comparing Europe with Asia. Black, for instance, observed the consistently poorer performance of the weapon systems employed by Ottoman, Mameluke, and Maratha forces as they confronted Western armies.[5] Parker, likewise, described China's unwillingness to develop firearms, and Japan's formal suppression of them.[6] More recently, Ken Alder, in *Engineering the Revolution*, argued for the technical shortcomings of the Gribeauval artillery regime. His focus was on the French Artillery Corps' abortive efforts to manufacture muskets with interchangeable parts during the French Revolutionary and Napoleonic Wars. The Corps enjoyed much success in transferring the Austrian technique of manufacturing artillery carriages with interchangeable parts and, through its prodigious tooling efforts, solved the major technical problems of manufacturing muskets with interchangeable parts. Nevertheless, the resistance of the traditional musket production centers was high enough to convince Napoleon to scuttle the Corps' hierarchically controlled manufacturing program. Alder showcased how the Corps used a rhetoric of scientific objectivity and sophistication to achieve professional power in the realm of bureaucratic politics. Yet Alder also maintained that the Corps' success occurred despite its ineffectual application of the advanced ballistics theories of the Enlightenment era.[7] Not only did the new gun designs of the Gribeauvalists fail to enhance ballistic accuracy, but the advanced mathematics education promoted at their professional military schools furnished only marginal tactical utility.[8]

Such historical interpretations of Enlightenment military technology may be viewed as having teleological or presentist orientations. To a significant degree, they invoke the standards of modern technology, especially its relatively rapid rate of hardware obsolescence. This is most apparent in O'Connell's and the Brodies' accounts. Because of the relative speed of changes in late-nineteenth and twentieth-century weaponry and scientific applications, they dismiss the eighteenth century as an era of stagnation. Parker and Black are less condemning: when compared to the military systems of the non-European world, those of the Enlightenment West prove to be remarkably progressive if not modern. In their interpretation, however, the Asian forces emerge as the unprogressive "other" who failed to contribute to modernity.

There is also Ken Alder's example in *Engineering the Revolution.* Here he depicted how the reforms of Gribeauval created many cultural features of a modern military system, specifically technocratic bureaucracy; hence, significant progress did occur. Their success had decided technical limitations, especially with respect to the development of Newtonian mechanics for gunnery practice. Alder, however, did not criticize such scientific efforts within the context of eighteenth-century rational mechanics, experimental physics, and military practice. Instead, he implicitly used the ideal technical standard of modern gunnery practice. Alder specifically betrayed his teleological orientation when he described the ballistics tables of the Gribeauval system as "uncertain" and "inadequate".[9]

I propose in this essay to turn away from the methodology of comparing eighteenth-century military technology to the "platonic" ideal that twentieth-century technical performance, scientific accuracy, and innovative enthusiasm invariably represent. More illuminating is to historicize eighteenth-century artillery by asking how close its organizers conformed to the notion of "progress" according to the standards of their own era. After all, progress in the twentieth century was largely measured by technical and economic parameters with little regard for deeper social principles.[10] As Peter Gay observed in his classic *The Enlightenment: The Science of Freedom*, and as Margaret Jacob articulated more recently in *Scientific Culture and the Making of the Industrial West*, Enlightenment notions of progress had explicit ethical and epistemological components. Because the notion of progress in the Enlightenment is complex, I have focused on four relevant developments: a research methodology that seeks to build upon the experimental and theoretical framework of Newtonian science from celestial mechanics to other areas of human knowledge; a union of Newtonian scientific theory with technical practice; the promotion of mathematical-scientific education to ensure the spread of rational thought throughout society; and the employment of such reason for the promotion of social liberty and political autonomy in general, and for the reform of inefficient, corrupt, or egoist institutions of power in particular. The first three developments embody Kant's famous argument that the notion of Enlightenment centers on one's ability to reason for oneself, while the fourth reflects the political orientation of the French philosophes.[11] By evaluating the activities of individual ballistics researchers as well as artillery institutions through such a framework, I will argue that artillery institutions throughout Christian Europe represented a success story for the execution of Enlightenment ideals associated with progress. I will further maintain that such a development occurred in spite of the philosophes' widespread condemnation of warfare.

Building upon Newtonian Science

The success of mathematicians in developing Newtonian science for the solution of basic military problems is well known. Tobias Mayer is the most famous example. With his employment of Euler's non-linear celestial mechanics analysis of Jupiter, and his extensive experience as a cartographer, he constructed the first lunar tables suitable for determining longitude at sea in the 1750s. Coulomb followed Mayer's example in employing the calculus to analyze fundamental problems of military technology not at sea, but on land. After the Seven Years' War, he served as a military engineer at Martinique. There he conducted the first calculus-based analysis of basic fortification design elements, including retaining walls, earthen slopes, arches, columns, and beams.[12] Naval historians have downplayed the significance of Meyer's achievement, however, due to the simultaneous innovation of Harrison's highly accurate marine chronometers. Early modern military historians likewise have ignored Coulomb, since engineers only began employing his theories during the nineteenth century.

Even less acknowledged is the ballistics research of Daniel Bernoulli, Benjamin Robins, and Leonhard Euler during the late 1730s and early 1740s.[13] They expanded Newtonian mechanics, above all its kinetic aspects, to make it applicable to military projectiles, especially high-speed cannon shot.[14] Daniel Bernoulli initiated this process in his famous treatise *Hydrodynamica* (1738). He showed how to calculate the initial velocity of a projectile that is propelled inside a gun barrel by an elastic fluid that was initially compressed in the breech. Bernoulli's work represented an early example of the conservation of *vis viva* being employed in a theoretical engineering analysis. Benjamin Robins, a mathematical disciple of Newton, engaged in a full-fledged theoretical and experimental analysis to supply empirical coefficients to Bernoulli's work, as described in the first chapter of *New Principles of Gunnery* (1742).[15] By employing the hydraulic trough technique of the British pneumatic chemists, such as Stephen Hales, he obtained the data needed to calculate approximately the initial pressure of the elastic fluid or gunpowder gas following its ignition. Robins' subsequent analysis of the muzzle velocity relied explicitly on proposition 39 from book I of Newton's *Principia*. Likewise, Robins invented the ballistic pendulum to measure a musket ball's velocity, and used it to demonstrate implicitly the validity of Bernoulli's original approach.[16] He revealed a remarkable correlation between the calculated and measured values of muzzle velocity, in spite of his numerous simplifying assumptions. A demanding instrument, the ballistic pendulum required such fundamental mechanical principles as the conservation of linear momentum and Huygen's theory of rigid-body motion before a user could deduce the bullet's impact velocity from the amplitude of the pendulum's swing. Few instruments of that era demanded such abstract knowledge for their employment.

Robins also used the ballistic pendulum to correct Newton's theory of air resistance. In this case he employed Newton's Second Law of Motion to deduce the air resistance from the change in velocity with respect to distance. Measuring the decline in the bullet's impact velocity at different ranges accomplished this. Although a polemical defender of Newton's mathematics in his early career, Robins observed that the master's air-resistance theory was limited to low velocities. When a projectile approaches the speed of sound, he discovered that the relationship between air resistance and velocity becomes grossly non-linear. The coefficient relating velocity squared with air resistance increases by three times. With a pioneering coupling of his interior and exterior ballistics results, Robins then showed how to calculate the ballistic trajectories of cannon-shot approximately for elevation angles below 10°. This demonstrated how woefully inadequate Galileo's vacuum theory of projectile motion was for cannons and muskets. Such kinematic theory served as the theoretical framework for the leading artillery treatises since the late seventeenth century, including those of Blondel, Bélidor, and Dulacq.[17] Robins thus demonstrated in *New Principles of Gunnery* that Newtonian mechanics was essential for a rational comprehension of the gunnery of high-speed projectiles such as musket balls, cannon shot, and howitzer shells.

Subject to considerable scrutiny and criticism, Robins' research program in interior and exterior ballistics was nevertheless confirmed analytically by Leonhard Euler, and experimentally by Patrice D'Arcy, an Irish cavalry officer and mathematician who became a French subject in his youth. In response to a request by Frederick the Great to provide a first-rate artillery treatise for Prussian artillery units, following the First Silesian War, Euler engaged in an extensive critique and expansion of *New Principles of Gunnery* using his powerful command of rational mechanics.[18] Dissatisfied with the many simplifying assumptions Robins employed, but relying exclusively on his ballistic-pendulum data, Euler developed rigorous differential equations of both interior and exterior ballistics. With respect to interior ballistics, he considered such complicating factors as a finite explosion rate, windage, the touch-hole, bullet-on-barrel resistance, the inadequacy of Boyle-Marriote's law for gunpowder gas expansion, the mass of non-combustable matter in gunpowder, and other perturbing factors that Robins ignored. In spite of such complexities, Euler solved his non-linear differential equations of interior ballistic motion with awe-inspiring approximate solutions. With such results, he then computed some interior ballistics tables that optimized the muzzle velocity of a wide range of artillery pieces as an aid to artillery design methodologies.

Exterior ballistics was tougher. Euler obtained a solution for the differential equation of high-speed projectile motion subject to Robins' highly non-linear function of air resistance; however, this solution was restricted to very low speeds, which made it virtually useless for military projectiles. Not until 1753 did Euler publish a general solution for projectile motion in a resisting medium.[19] To solve the non-linear differential equations, he had to modify Robins' empirical equation of air resistance to make it proportional to velocity squared only. Euler's general solution also required an extensive set of numerical tables that could only be obtained through numerical integration. Lieutenant Paul Jacobi of the Prussian artillery initially provided this labor. Yet those tables were lost in 1758 during the Siege of Olmütz, where Jacobi was killed in action. Following the Seven Years' War, the exhaustive exercise was repeated by Henning Friedrichs, Graf von Graeveniz. He published the results in 1764.[20]

Bernoulli, Robins, and Euler transformed ballistics from a Galilean science applicable only to mortar gunnery at low ranges into a Newtonian science applicable to guns in general. In the process, they resolved demanding theoretical and experimental dilemmas that stymied the great natural philosophers of the seventeenth century, and united experimental and mathematical science as never before. The scientific success of such Enlightenment ballistics was demonstrated by the subsequent mathematical studies of exterior ballistics by Charles de Borda and Georg Friedrich von Tempelhoff, and in the experimental studies of Papacino D'Antoni, Count Rumford, and Charles Hutton.[21]

Given the celebration Newton's mechanics received from Voltaire, D'Alembert, Hume, and others, this revolution in Newtonian ballistics represents a progressive achievement for the Age of Enlightenment.[22] Robins, in particular, followed closely Newton's example of synthesizing empirical observation with high-level mathematical analysis, and he ultimately superseded it. Whereas Newton relied on the observational technologies and activities of Galileo, Flamsteed, and other astronomers before deducing mathematically the trajectories of planets, Robins invented his own "telescope"—the ballistic pendulum—before deducing the trajectory of military projectiles in the atmosphere. While little mathematical theory is required to operate a telescope, fundamental mechanical principles and analyses are absolutely essential for the ballistic pendulum. As the philosophe Dumarsais suggested, a philosopher embodies "a spirit of observation and exactness, which relates everything to true principles." The ballistics revolution is an unmistakable example of the reification of the Enlightenment's epistemological ideals associated with progress.[23]

Unifying Theory and Practice

The Enlightenment program involved much more than generating new scientific or philosophical endeavors along Newtonian lines. As Diderot demonstrated with the *Encyclopédie*, it sought to influence technical practice through rational, if not Baconian, methodologies. He repeatedly urged that theoretical and practical thinkers unite to overcome the "resistance of nature." Other philosophes hailed James Watt for his spectacular union of scientific reasoning and technical practice with respect to his innovation of the steam engine with the separate condenser.[24] As Margaret Jacob concluded, "The Watts lived the Enlightenment just as much as any French philosophe lived its rather more abstract vision."[25] Benjamin Franklin's celebrated status as the American icon of the Enlightenment likewise rested in part on his success in applying his experimental observations to practical inventions. Count Rumford's scientific and practical obsession with heat operated along similar lines as well. It was Condorcet who most clearly linked social progress with scientifically modeled technical practice. In *Sketch for a Historical Picture of the Progress of the Human Mind* (1793), he wrote:

> If we turn now to the arts, whose theory depends on these same sciences, we shall find that their progress depending as it does on that of theory have no other limits; that the procedures of the different arts can be perfected and simplified in the same way as the methods of the sciences; new instruments, machines and looms can add to man's strength and can improve at once the quality and accuracy of his productions, and can diminish the time and labour that has to be expended on them.[26]

Social progress for Condorcet depended on more than this Baconian unity of theory and practice with respect to science and technology, of course. It also involved politics and administration. He asked: "Will not the general welfare that results from the progress of the useful arts once they are grounded on solid theory, or from the progress of legislation once it is rooted in the truths of political science, incline mankind to humanity, benevolence and justice?"[27] That, of course, is a question that continues to challenge, if not mock, policy analysts and social scientists.

A central feature of Enlightenment ballistics was the rapid influence it had on the practice of military institutions throughout Western and Central Europe. Such success represents a prime example of the union of scientific theory and technological practice that the philosophes valued so highly. This union contrasts with most attempts to apply formally Newtonian or rational mechanics to change technological practice in the eighteenth century.[28] Euler's studies of naval architecture, Charles de Borda's water wheel analysis, Coulomb's solid mechanics studies of fortification elements, and Lazare Carnot's analysis of hydraulic machines are prime examples of theory failing to influence practice during the eighteenth century. They appear to confirm Thomas Hankins' assertion that "only in astronomy did the new analysis show immediate practical results in the increased precision of astronomical tables and in the creation of new theories concerning the shape and motions of the earth and other heavenly bodies."[29] Ballistics research, as early as the War of the Austrian Succession (1740–1748), is a powerful exception to this rule, as demonstrated by Robins' professional relationships with military institutions.

Robins developed a close relationship with Lord Anson, following his victorious return from his circumnavigation of the world. This resulted in increased funding for Robins' experimental studies of heavy ordnance, which demonstrated the validity of the prior ballistic-pendulum data. It also reinforced

Anson's own mission to increase the rationalization of British naval gunnery practice, including the minimization of range and the maximization of firing rates. Robins' relationship with the Royal Navy motivated his proposal for a new naval gun based on interior and terminal ballistic considerations: a short-barreled, thin-walled gun that fired heavy rounds at low velocities to maximize the splintering of enemy hulls.[30] The recommendation helped inspire the innovation of the carronade during the 1770s, a gun the Royal Navy used through the War of 1812.[31] Also significant was Robins' critique of conventional artillery practice that sought to optimize power charges and barrel lengths to maximize projectile ranges from the scientific perspective of Galilean vacuum ballistics. Based on his consideration of air resistance due to the smooth-bore projectile's linear and random angular velocity (which generated a significant lateral aerodynamic pressure that was directly proportional to the muzzle velocity), Robins argued instead that a rational artillery practice should seek to employ the minimum muzzle velocity to obtain the desired level of destruction with maximum certainty. He made such assertions in a letter that responded to queries by Charles Emmanuel III, the monarch of Piedmont-Savoy, through his ambassador to London, Chevalier D'Ossorio.[32]

Piedmontese interest in Robins' opinions during the concluding year of the War of the Austrian Succession is hardly surprising. They had a tradition of unifying scientific theory with military engineering practice that dated from the middle of the sixteenth century. The Dukes of Savoy, including Charles Emmanuel III, demonstrated this personally with their scientific engineering studies. As far as Robins is concerned, it seems not entirely coincidental that Papacino D'Antoni initiated his Newtonian ballistics research agenda as a junior officer in the Piedmontese artillery corps in 1743, just a year after *New Principles of Gunnery* appeared. As described by Walter Barberis, Robins' work rapidly contributed to debates in the scientific military circles of Piedmont on the optimization of artillery operations and designs.[33] From the late 1740s through the 1780s, this evolved into a full-fledged research program. It included interior-ballistics experiments using the ballistic pendulum and other velocity-measuring inventions, formal applications of Robins' and Euler's ballistics principles to optimize barrel lengths and gunpowder charges, and extensive range tests to confirm new gun designs.[34] Even more remarkable was Papacino D'Antoni's practical techniques for determining artillery trajectories in the field, where the muzzle velocity depends on environmental conditions. By firing a cannon ball into compacted earth, measuring the resulting penetration depth, and comparing the result with the penetration depth of a musket ball of known muzzle velocity, he showed how to scale up the result to determine the initial velocity of a large cannon shot.[35] It requires the solution of a simple square root. Such elementary mathematics refutes Alder's assertions that the military limitations of Robins' experimental research were due to his exclusive use of muskets. The scaling effects of spheres is a relatively straightforward problem in experimental fluid mechanics, as D'Antoni demonstrated.[36] Because D'Antoni was the director of Turin's Royal Academy of Artillery and Military Engineering—as well as a senior commander in the Piedmontese army and an artillery analyst of international reknown, in spite of his common social origins—one may presume his detailed recommendations for approximate yet practical calculations of ballistic trajectories in the field were actually executed in practice.

The War of the Austrian Succession witnessed another artillery research and development program: the Liechtenstein artillery reforms of Austria. These reforms established the first heavy field-artillery system that could effectively maneuver in combat. In spite of all the glories heaped on Gribeauval's artillery reform efforts in France, it was Liechtenstein and his technical assistant, Anton Feuerstein, who first designed a heavy field-artillery system along consciously rational considerations.[37] Their innovation was shaped by theoretical principles, empirical observations, and tactical demands. Liechtenstein's reforms thus ranged from the continuous training and scientific education of officers and troops, and the manufacturing of artillery carriages with interchangeable parts and gun barrels with machined bores, to establishing its new tactical role as a dominant weapon of field warfare. Gribeauval's subsequent reform, in contrast, was highly imitative—the experiment with manufacturing muskets with interchangeable parts and Lombard's gunnery tables being its only unique features. Gribeauval, after all, was not only a product of French Enlightenment culture. He was a disciple of Liechtenstein, having served as a senior artillery officer in the Austrian service throughout the Seven Years' War.

Gribeauval openly embraced Robins' argument that the rational design and operation of artillery must consider simultaneously interior, exterior, and terminal ballistics considerations.[38] A more crucial issue is the relevance of *New Principles of Gunnery* on the Liechtenstein reform. After all, Liechtenstein's drive to reconstruct the Austrian artillery system commenced in 1742, in direct response to its disastrous performance at the Battle of Chotusitz. Nevertheless, the maneuverable yet sufficiently powerful field artillery that emerged from Feuerstein's designs clearly conforms to Robins' dictates. The apparent absence from Liechtenstein's collected papers of material related to his artillery reforms, as well as the loss of Feuerstein's papers, makes this connection difficult to confirm. Likewise, of the hundreds of artillery experiments documented in Vienna's Kreigsarchiv from the eighteenth century, none involve a ballistic pendulum. The first documented reference to Robins and Euler in that archive, in fact, occurs in 1759, during the Seven Years' War. The report, "A Memoir on an erroneous principle in the otherwise astute theory of Robins," authored by the mathematician Grummert, was presented to Liechtenstein by the Graf von Pergen.[39] It focuses on the limitations of Robins' interior-ballistics theory, including its failure to formally consider the effect of changing environmental conditions. Needless to say, this critique of Robins' and Euler's ballistics assumes considerable familiarity with the subject.

The assertions of Anton Semek make a direct connection between *New Principles of Gunnery* and Liechtenstein even harder to dispute. An archivist in the Kriegsarchiv before World War I, he maintained that Liechtenstein based his artillery reforms on theoretical ballistics calculations as well as extensive field tests.[40] Such theory involved more than the Galilean vacuum ballistics, as presented in the gunnery tables of Bélidor's *Le bombardier français* (1731). According to Semek, Liechtenstein conducted exterior ballistics experiments to determine the relationship between elevation angle and range, and how weather changes may affect it. This reveals an unmistakable awareness of the significance of air resistance. From an interior ballistics perspective, Semek noted how the elastic expansion of gas served as a basic assumption in Liechtenstein's work to improve the quality of gunpowder—yet another hint of his familiarity with Robins. Finally, Semek observed how Liechtenstein's stated goal was to employ a gun with only enough muzzle velocity for its particular combat purpose, while minimizing its weight. Again, this design philosophy conforms with Robins' "Practical Maxims relating to the Effects and Management of Artillery, and the Flight of Shells and Shot."[41] Liechtenstein himself received a first-rate education in both the practical and the theoretical aspects of the art of war under the auspice of Eugene of Savoy (whom even Leibnitz applauded for his progressive support of the sciences).[42] Liechtenstein's intellectual sophistication is revealed by his extensive correspondence with his close personal friend and great military adversary, Frederick the Great. Furthermore, Liechtenstein was already the chief of artillery when he commanded the Austrian forces in Northern Italy that crushed the French and Spaniards at Piancenza (1746). This brought him in close contact with the allied Piedmontese army, among whom Robins' work was already well known. He also recruited highly talented officers throughout Europe for his artillery development staff during the 1740s and early 1750s. Liechtenstein likewise wrote in his artillery regulations that artillery practice must be based on the science of mechanics:

> The science of gunnery is grounded in the mathematical and physical sciences, and its promotion and advancement is dependent on the solid comprehension of these two subjects. For this reason the mathematics school

> of the Corps has been established: to make these sciences better and more generally known to officers, and others who possess the necessary talents to study them.[43]

Therefore, it is virtually inconceivable that he pushed through his reforms at enormous personal expense without seriously considering the profound implications of *New Principles of Gunnery*.

The close union of ballistics theory and artillery practice was forcefully demonstrated by Robins' own service as a military engineer at the Siege of Bergen-Op-Zoom, and with the East India Company. The same could be said about Euler, through his professional interaction with Paul Jacobi—the leading scientific officer of the Prussian artillery corps between the War of the Austrian Succession and the Seven Years' War. He conducted numerous ballistics studies, including experiments to determine the muzzle velocity of cannons and the air resistance of projectiles, as well as to optimize mortar, breach, and ricochet fire.[44] Jacobi appears to have been a principal participant in the mortar experiments Euler observed in 1748 that measured the impact angle of shells for numerous ranges and elevations.

Following Jacobi's death at Ölmutz, Georg Friedrich Tempelhoff emerged as Prussia's leading military intellectual. In addition to a number of mathematical textbooks, and a translation of Papacino d'Antoni's *Essame delle polvera* (1768), he published *Le bombardier prussien* (1781), an exhaustive theoretical analysis of projectile motion in the atmosphere. Yet Tempelhoff also succeeded in translating such theory into practice. He designed a light-weight mortar that was introduced into Prussian service in 1789, as well as another mortar designed for mountain service. A special battery of these pieces, which could be transported on pack horses, was also established in Prussia before the War of the First Coalition.[45] Georg Vega followed Tempelhoff's example. A professor in Austria's Bombardier Corps who taught the theoretical artillery course, Vega fought with distinction in the Turkish War of 1788–91 and in the War of the First Coalition. While serving under the command of Duke Albert von Sachsen-Teschen in the 1794 campaign of West-Central Germany, he designed a new siege mortar. A 30-pounder with a conical gunpowder chamber, this launched a projectile with almost twice the range as the existing 60-pounder.[46] Vega's mortar proved to be the only standard Austrian gun of the Napoleonic Wars that did not originate in the Liechtenstein reforms.

The success of theoretically informed artillery officers in innovating new weapons was especially pronounced in Great Britain. William Congreve Sr., the commander of the Woolwich Arsenal, was responsible for substantially improving the quality of British gunpowder, as well as inventing the standard field artillery carriage that the British used in the Napoleonic Wars. Its tail was made from a single stock of wood, and was significantly lighter than those of the Gribeauval and Liechtenstein reforms. His son, William Congreve Jr., is famous for his rocket innovation. Yet the younger Congreve's theoretical sophistication, as well as those of Britain's senior artillery generals, was disclosed during the controversy at the end of the Napoleonic Wars. Congreve proposed to replace the heavy guns of the Royal Navy with guns that he designed in part by employing the interior ballistics theories of Euler.[47] Full-scale experiments with a giant ballistic pendulum revealed the marginal inferiority of Congreve's prototype, however.

Then there is the example of Henry Shrapnel and his innovation of spherical case shot. A student of Charles Hutton, who used *New Principles of Gunnery* as a textbook at the Royal Military Academy at Woolwich, Shrapnel confronted a fundamental dilemma that the high degree of air resistance presented. Canister fire was devastating against infantry targets due to the sawed-off-shotgun effect it gave to cannon. Yet air resistance limited its lethal range to less than 300 yards. As Robins demonstrated, the retarding effect of such resistance significantly declined for larger shots of greater mass. This is a consequence of Newton's Second Law of Motion and that air resistance is proportional to the projectile's diameter. To use his particular argument, "Small balls cannot have so great velocity or be projected to very considerable distances, unless enclosed in a heavy spherical case, which from its form and weight is not much influenced by the resistance of the Air or directed from its Direction."[48] Shrapnel thus sought a means to hold the canister shot together in order to reach the longer ranges of round shot and still deliver the same lethality of canister fired at short ranges.

A thin shell filled with shot and powder, and then ignited with a timed fuse, proved to be the answer. Its successful employment in the Napoleonic Wars, however, occurred only after Shrapnel's exhaustive theoretical and experimental ballistics studies on optimizing its lethality. As Timothy Dubé has argued, "Through his study of the factors governing the action of his shell in the gun, in flight and at the target, Shrapnel applied a scientific methodology, which was then just beginning to emerge. The survival of Shrapnel's calculations and the detailed range tables documents two essentials of any scientific theory—validity and reliability." ("Henry Shrapnel: His Artillery, Small Arms and Spherical Case Shot," *Canadian Journal of Arms Collecting* 29, 1991, no. 4: 123–131) Such work led Shrapnel to make some powerful discoveries:

> . . . the velocity of these Musquet Balls then will exceed that of the same kind of shot when impelled from a Musquet with full 1/2 as much again

> velocity, as the resistance of the Air cannot so much impede a large ball in passing through it as a small one, or can so great a velocity be given with an ounce of powder, as 3 or 4 pounds, any sized leaden ball may be fired within a Spherical Case in this manner.[49]

Although the most successful, Shrapnel was hardly the only artillery officer concerned with extending the range of ordnance through ballistic innovations. Colonel Louis Villentroy, the prominent interior ballistics expert who translated Charles Hutton's experimental ballistics studies into French in 1801, was authorized by Napoleon to develop long-range mortar for the defense along the English Channel in 1804. Based on an advanced theoretical ballistics study using the most up-to-date data and theories, Villentroy concluded that the range could be significantly increased over standard French siege mortars if the barrel's diameter was enlarged in equal proportion to its expanded length and coupled with an enhanced thickness in the region of the bore.[50] Although a dozen or more such mortars were eventually employed at the Siege of Cadiz, and launched almost 500 shells at ranges more than 4,000 yards, the destructive impact was limited due to problems with the long-range fuses.

Promoting Scientific Education

One can hardly over-estimate the significance of educational reform for Enlightenment thought. D'Alembert, Rousseau, Condillac, Condorcet, and numerous Physiocrats all devoted serious attention to this issue, having received significant inspiration, especially from Locke's *Some Thoughts Concerning Education*. They recognized that their central objective for replacing traditional philosophical and religious authority with empirical and deductive reasoning depended heavily on a reconstructed educational system. It would help transform children into reasoning adults capable of thinking autonomously on both social and technical issues.[51] With his famous conclusion that *l'éducation peut tout*, Helvetius even went so far as to argue that education alone may be used to achieve such perfectability of man. The philosophes never rejected classical learning entirely, but they were sharply critical of the deadening process that the study of Latin usually achieved in traditional ecclesiastic schools. As Voltaire concluded almost seriously, he himself learned only Latin and nonsense from his Jesuit educators. The philosophes had high praise for those institutions that emphasized mathematical and experimental science, such as the universities of Scotland. They admired their ability to link scientific learning with industrial practice, as well as with moral thought. Such education was crucial for the widespread distribution of scientific ideas that Progress requires. "Until the present epoch," Condorcet observed, "the sciences have been the patrimony only of a few; but they have already become common, and the moment approaches in which their elements, their principles, and their most simple practice will become really popular. Then it will be seen how truly universal their utility will be in their application to the arts, and their influence on the general rectitude of the mind."[52] The practical linkage of informal scientific learning and industrial innovation was especially pronounced in England. In addition to widespread lectures and experimental demonstrations, many prominent innovators, including James Watt, John Smeaton, and Josiah Wedgwood, pursued intense personal studies of scientific principles. Such learning, in fact, was an explicit objective for many Newtonians in the early eighteenth century, especially Thomas Desaguliers. Margaret Jacob argued that, with respect to the hydraulic machinery in particular, "relatively sophisticated mechanical knowledge had to be a part of one's mental world before such mechanical devices could be invented and, more to the point, effectively exploited."[53]

Artillery institutions went far in executing these educational ideals during the Enlightenment. Given the direct utility that Newtonian mechanics began acquiring in field and siege warfare, following the ballistics research of Robins and Euler, artillery regiments throughout Europe responded by increasing the mathematical training their officers received. By the end of the War of the Austrian Succession, numerous states had established military academies for their artillery and engineering officers. This was hardly a French innovation, however, as many historians of science and engineering have maintained.[54] Numerous scientific military schools were institutionalized by the late sixteenth century, as John Hale has described.[55] They furnished the basic mathematical knowledge centered around Euclidean geometry that a successful military engineering practice demanded. Such schools existed in formal academic settings, including the Universities of Leiden, Tübingen and Kassel, and numerous Jesuit colleges, especially La Fleche. As Descartes' biographical record discloses, the Dutch army of Maurice of Nassau also provided instruction to young noblemen seeking to master the art and science of siege warfare. There Simon Stevin served as the senior lecturer. The French regimental artillery school of La Fère, in turn, developed a pioneering scientific curriculum under the guidance of Bélidor. By employing the statics and hydraulics of Stevin, and the kinematics of Galileo, Bélidor introduced the cadets to the statics of machines, fortification elements, and hydraulic operations, as well the theory of mortar gunnery or ballistics in a non-resisting medium. The cadets also received significant practical training in maneuvering and operating guns, in addition to fortification engineering. Nevertheless, they had no formal exposure to experimental physics, not to mention calculus or rational mechanics.

France persisted with such a non-Newtonian curriculum until the 1770s. Gribeauval then initiated the teaching of Robins' and Euler's ballistics, along with differential and integral calculus.

In comparison to the academies in France, the military academy of Piedmont-Savoy was even more progressive from an Enlightenment perspective. Charles Emmanuel III founded the *Regie Scuole Militari teoriche e practiche di Artiglieria* in 1739, initially along the lines of the Regimental Artillery Academy at La Fére.[56] The Royal Military Academy at Woolwich and Liechtenstein's artillery academy in Budweiss soon followed during the War of the Austrian Succession. They too followed the French example of maintaining a non-Newtonian curriculum through the 1770s. The Turin academy was the first to change, however. The influence of the ballistics revolution on its curriculum was unmistakable as early as 1755 when Charles Emmanuel III appointed Luigi de Lagrange as a professor to teach differential and integral calculus. It was a necessary prerequisite for the study of gunpowder and its explosive force inside a gun barrel, as specified by academic regulations.[57] Unfortunately, Lagrange's abstruse research in analytical mechanics rendered him a confusing teacher. After Lagrange left for the Berlin Academy of Science in 1766, Papacino d'Antoni wrote the following:

> The sublime talent of the substitute professor La Grangia, which placed him rightfully among the outstanding Academicians in Europe, did not make it possible for him to reduce the fundamentals necessary to us to their elements. Thus the individuals, who in the earlier course criticized his treatises as too advanced, metaphysical, diffused throughout with extraneous material and lacking applications to the professions of artillery and fortification engineering, had judged correctly. However, if someone made this identical pronouncement on the contents of the most recent course, he would display his ignorance of the subject matter.[58]

Pioneering endeavors always come with risks and invariably require corrections, as D'Antoni ultimately achieved.

By 1764, the Turin Academy's regulations were prescribing an experimental study of interior ballistics by employing instruments to measure the gunpowder explosion force and the muzzle velocity of the projectile.[59] Papacino d'Antoni's *l'Esame della polvere* (1765), his calculus-based treatise on interior ballistics that builds on Robins' work, was the textbook that helped fulfill such a requirement. In addition to the ballistic pendulum, it described the different instruments and techniques developed in Turin for measuring the muzzle velocity of both muskets and artillery. Likewise, in d'Antoni's introductory scientific treatise, *Institutions physico-mécaniques* (1773), the concluding section addresses the dynamics of the ballistic pendulum in terms of differential calculus.[60] Turin's absolutist government did more than maintain the most theoretically advanced mid-eighteenth-century artillery school; it also established a formal school of artillery practice and a metallurgical research laboratory. As Seymour Mauskopf has described, Turin also boasted an advanced gunpowder research program in chemistry that was centered around Saluzzo.[61]

The Piedmontese effort to achieve a union of Newtonian science and artillery practice through formal education was soon imitated by the main European powers, including France, England, Austria, and finally Prussia. With Charles Hutton's appointment as mathematics professor at Woolwich (1775), and the permanent installment of the Gribeauval reforms in France (1776), Newtonian ballistics as developed by Robins and Euler was established in their military curricula. The founding of Austria's Bombardier Corps in 1786 marked a similar turn, while Prussia followed by founding its Artillery Academy in 1791 under the directorship of Tempelhoff. The École polytechnique, founded in 1794, was therefore hardly unique in offering advanced courses in rational mechanics and calculus for prospective engineering and artillery officers. Those offered in the Bombardier Corps were especially comparable, as demonstrated by Joseph Hohnal's *Versuch einen theoretischen Artillerie Unterrichts* (1793) and the fourth volume of Georg Vega's *Vorlesungen über die Mathematik* (1800). One may therefore conclude that as far as teaching students to reason with advanced theoretical and experimental tools, in addition to applying these tools to practical situations, the artillery schools of both the first- and the second-rate military states went far in actualizing the educational ideals of the Enlightenment philosophes.

While the conservative Vallierists artillerists in France blocked the military instruction of calculus-based ballistics as developed by Bernoulli, Robins, and Euler, the same cannot be said about the Gribeauvalists. They arranged for Bézout to write his *Cours de mathématique* in the late 1760s to furnish an up-to-date mathematics text that culminated with the advanced ballistics analysis of Charles de Borda's.[62] Taking Robins' as well as his own ballistics observations into consideration, Borda furnished the first closed-form algebraic and trigonometric expression of a ballistic trajectory subject to high-speed air resistance.[63] Likewise, the mathematics professor at the École Royales d'Artillerie de Grenoble translated *The Mathematical Tracts of the Late Benjamin Robins* into French in 1771. By 1772, instructors at the École Royale d'Artillerie d'Auxonne were employing it as a textbook.[64] In 1783, Auxonne's senior mathematics professor, Jean-Louis Lombard, translated Euler's *Neue Gründsatze der Artillerie* for the benefit of his artillery students.[65]

Jean-Louis Lombard

Known primarily in military history as the mathematics professor who instructed the young Napoleon Bonaparte in Robins' and Euler's ballistics, Lombard computed the standard gunnery tables of the Gribeauval artillery system, *Tables du tir des canons et des obusiers* (1787).[66] Perhaps the most innovative aspect of the Gribeauval artillery reforms, they furnished French gunners throughout the French Revolutionary and Napoleonic Wars with detailed data concerning the relationship between gunpowder-charge weight and muzzle velocity, as well as the relationship among elevation angle, range, and impact or terminal velocity. Furthermore, the tables were based on experimental observations and Euler's mathematical analysis. To ensure that young artillery officers would be fully enlightened to its scientific and mathematical foundations, Lombard's teaching treatise, *Traité du mouvement des projectiles*, was posthumously published in 1796–97. There he set up and solved the differential equations of ballistic motion subject to air resistance on which the tables relied.[67]

Napoleon Bonaparte himself hinted at the practical utility of Lombard's tables. In an order from 1801 on the education of technical officers, he wrote: "The exercises that were taught in the engineering school before the revolution are available and leave nothing to be desired. This material, when conveniently organized, and some theoretical principles which are provided by Lombard and Robins, would be appropriate for the artillery."[68] The power of Lombard's tables may have also inspired the politically active artillery officer, Jean-Pierre Lacombe Saint-Michel, to question whether aspiring wartime artillery officers really needed the highly theoretical mathematical education offered at the École polytechnique. He argued, as a member of the Council of Ancients in the Directory, that since the available ballistics tables are sufficient for most combat situations, junior artillery officers should focus less on mathematics and more on military discipline and leadership.[69]

In spite of the pioneering conjunction of rational mechanics, experimental physics, and direct combat utility that Lombard's work represented, he is almost entirely forgotten in the history of science and technology.[70] The few comments Alder makes about Lombard are almost disparaging:

> In the 1780s, Professor Lombard drew up new range tables based on Robins' equations, which he normalized for the dimensions of the Gribeauval cannon and its *hausse*. The Revolutionary and Napoleonic artillerists went to war with these tables. Shooting was not, however, a matter of simply plugging in numbers. Lombard admitted that his tables remained inadequate. Unlike astronomical tables, they were based on "hypotheses too often belied

> by experience." To account for the idiosyncrasies of their particular weapon, gunners still needed to fire a test shot before looking up the right adjustment. Moreover, this adjustment would be uncertain in that gunners were still plugging in the range of the shot. Unfortunately, as Lombard pointed out, the muzzle velocity was unavailable to gunners in the field.[71]

Furthermore, Alder added that Lombard confessed, in the introduction to *Tables du tir des canons et des obusiers*, that his gunnery tables were inaccurate and that "subsequent experiments through the Revolutionary period did little to alleviate the limits of Lombard's approach."[72]

No doubt, Lombard responsibly disclosed the simplifying assumptions used to compute his tables. Taking a cue from Euler's *Neue Grundsätze der Artillerie*, he ignored the vertical component of air resistance; taking a cue from Borda's exterior ballistics analysis, he also assumed that air resistance is proportional to the velocity squared. This is described in *Traité du mouvement des projectiles*. Euler conducted his analysis using an air resistance function that was proportional to both the velocity squared and the velocity to the fourth power to ensure maximum conformity to Robins' ballistic pendulum data. What Alder neglected to mention was Lombard's remark that, while his results could not boast the accuracy of astronomical tables (even with the high manufacturing precision of the Gribeauval guns, they still have their mechanical inconsistencies), the tables sufficed for practical gunnery purposes. Professionally trained artillery officers armed with them in combat, Lombard asserted, will always outperform enemy counterparts who lack such analytical tools.[73] It is important to note that the simplifying assumptions Lombard held did not go unnoticed during the Napoleonic Wars. Clément, a veteran artillery officer, confirmed the utility of Lombard's tables in 1800 for cannon which fired solid shot only at low elevation angles. He complained, however, that the tables were insufficient for howitzers, those short-barreled guns that were particularly useful for launching explosive shells at high-elevation angles in field engagements.[74]

Alder's argument that French artillery officers had no means of determining the muzzle velocity of their guns in the field also does little justice to Professor Lombard. Lombard specifically calibrated his tables so that the muzzle velocity could be determined by measuring the range of a standardized test mortar and the gunpowder charge. More serious, however, is Alder's critique that the Gribeauvalists' entire ballistics research program was irrational, because they lacked the means to measure directly the muzzle velocity of cannon for themselves. He observed that the French state refused to fund a large-scale ballistic pendulum that could handle cannon shot until the Restoration, while Robins' experimental data was supposedly valid only for musket shots, not heavy artillery. Such limitations led Alder to conclude that the Gribeauvalists "ignored the best scientific knowledge of their day."[75]

Actually, Lombard had a very good reason for not investing in full-scale ballistic-pendulum experiments: he possessed a far simpler and less-expensive methodology. By firing a cannon shot horizontally into a tissue-paper target and measuring how much its projectile fell vertically, he was able to calculate back to determine the muzzle velocity. He thus employed the same air-resistance relationship he used in his exterior ballistics tables.[76] This was the means with which Lombard determined the muzzle velocity for one datum point in each table—the remaining points were determined using a simple proportionality relationship confirmed by Robins' and Euler's analyses. Given the vast improvements in gunpowder quality, thanks to Lavoisier's rigorous administration, Euler's and Robins' interior ballistics equations were obsolete for Lombard.

Using Scientific Reasoning for Liberty and Autonomy

The Enlightenment philosophes were concerned with scientific means, but also with humanitarian ends. Yes, they advocated the reasoning associated with Newtonian mechanics, and its utilization for technological affairs, but not as an end in itself. They viewed these epistemological means as necessary for the reform of corrupt or egoistic social institutions, as well as the promotion of liberty.[77] Yet they were keenly aware of how it could be misused for suppression. How then could an institution devoted to rationally optimizing the destruction of human life, such as an artillery corps, possibly be considered progressive? After all, a persistent theme for the philosophes, especially Voltaire and Diderot, was the irrational and reprehensible nature of warfare. Artillery reformers sought to change their organizations by increasing the tactical performance of their weapons, as well as minimizing the costs in terms of friendly human deaths, wasted ammunition, and squandered time learning lessons already available through scientific theory. In other words, they sought to increase the rationality of their organizations in terms of eliminating traditional yet inefficient and even self-serving practices, thus conforming to the basic reform criteria of many Enlightenment philosophes, above all Voltaire. But how does such internal tactical reform promote greater social autonomy and freedom in general? On the surface it would simply promote the ability of soldiers to kill each other with greater ease, as Voltaire mocked in *Candide* with respect to the Bulgarian and Abarian armies. Military commanders still remained "master-butchers," to use Diderot's phrase, even when they engaged in scientifically informed cutting. Such destruction could hardly represent Enlightenment progress.

The philosophes were not strict pacifists, however. Voltaire, in fact, strongly supported the military conflict of the American Revolutionary War, and died a happy man knowing the war was shifting to the side of the rebellious Americans—the "forces of philosophy," to use his words. The French philosophes, as Peter Gay argued, considered the victorious conclusion of the American Revolutionary War for the colonialists to be the greatest practical achievement of the Enlightenment.[78] Less well known, however, is Voltaire's enthusiasm for the Russian wars under Catherine the Great against the Ottoman Turks. He apparently could never forgive the Ottomans' conquest of Greece in the name of God. Not only did he demand their outright military defeat from Catherine, but also their extermination as a state, if not as a culture.[79] Voltaire even lobbied hard for the Russians to employ his classically inspired war-chariot design against the Ottomans, having failed to get the French authorities to use it in the Seven Years' War. Voltaire apparently could not grasp the idea that field artillery and simple palisades would have rendered such a vulnerable combat system a suicidal endeavor.[80] What Voltaire's behavior ultimately suggests is that it is hardly an oxymoron for a military institution to be enlightened. Even though it is devoted to perfecting the art of war, its destructive capability may serve the cause of liberty, as the American Revolution represented. The question is whether there was a special feature of the scientific changes embraced by eighteenth-century artillery institutions that promoted such ideals, including the resistance to tyrannical rule, corrupt institutions, and foreign oppression.

I believe there was. The high degree to which artillery technology was intellectualized following the ballistics revolution lowered the practical experience needed to command batteries efficiently. But this benefit required that the commanders be already knowledgeable of Newtonian science and be willing to invest that knowledge in the study of scientific ballistics and gunnery treatises. If such conditions existed, then the advantage a powerful military state could have over weaker neighbors or rebellious colonists would decline. The resistors to such hegemonic power, in other words, could quickly acquire a great deal of military "experience" by properly leveraging their abstract scientific knowledge.

This is precisely what happened in the Continental Army. Henry Knox assumed its command during the Siege of Boston and exhibited highly competent military leadership throughout the war. Yet he had never commanded an artillery battery before in his life and assumed command with minimal militia experience. As a bookseller specializing in military literature before the war, however, he acquired a solid knowledge of the theory and practice of gunnery, as well as fortification engineering.[81] This impressed Washington sufficiently to give him the artillery command. On a smaller scale, Alexander Hamilton received his commission as an artillery officer from New York State primarily due to the mathematical competence he acquired from Princeton University.[82] Even John Adams, as a member of the Continental Congress during the war, studied various artillery and military engineering treatises to make more competent military policies. The success of the colonialists in wearing down British resolve depended on their ability to adapt European military techniques, as Van Steuben's troops demonstrated with the mechanized close-order drill, and Knox displayed with Bélidor's if not Robins' gunnery theory. Yet the adaptation occurred rapidly enough to win the war because the knowledge was highly amenable to the "Western" scientific mindset that the colonists already possessed.

This transferable feature of the Western art of war to cultures that already embraced the mechanical philosophy of the Scientific Revolution was not unique to the eighteenth century, of course. It was a critical feature of the broad-based Military Revolution of the early modern era. It would be absurd to suggest that a Christian knight, a Mongolian horse-archer, or an Ottoman janissary could train to become an effective combat commander by studying high-level mathematics and natural philosophy at the expense of combat training.[83] Each belonged to a decidedly "craft-based" warrior culture. Yet by the late sixteenth century we have the example of the Marquis de Spinola, who began his victorious leadership of the Army of Flanders with only mathematical and historical studies on his military resume. This intellectual prerequisite for early modern military command applied to a similar extent to the Duke of Parma, Maurice of Nassau, Gustavus Adolphus, William of Orange, the Dukes of Savoy, and Charles XII. They devoted considerable effort to studying mathematics, mechanics, and optics to become masters if not creators of the Western art of war. Let's also not forget about the hundreds of Jesuit priests in Spain whom Phillip IV recruited as military engineers because of their mathematical competence. The ballistics revolution and its rapid embrace by Western military organizations only accelerated this trend of leveraging mathematical and scientific mechanical knowledge into tactical, operational, and acquisitional competence.

The equalization of military power that resulted from the intellectualization of Western warfare had decided limitations, however. Those cultures that initially rejected the mechanical philosophy in general and Newtonian science in particular found themselves at an enormous disadvantage when forced to compete with Western European armies during the Enlightenment and Revolutionary eras. These included the feudal cavalry of Poland, the Janissary Corps of the Ottoman Empire, the Mamlukes of Egypt, and the Marathas of

India. Could their inevitable conquest and colonial domination possibly represent progress from an Enlightened philosophe's perspective?

As David Ralston described in *Importing the European Army*, successful resistance to Western military domination demanded the adaptation of the Western art of war. Such a transformation, however, ultimately generated enormous cultural upheavals.[84] This was due in part to the Newtonian world view that had become deeply imbedded in the mechanized military culture of the West by the late eighteenth century. As Selim III and Muhammed Ali grasped in the early nineteenth century, and the leaders of China's Self-Strengthening movement (especially Li Hung-chang) and Japan's Meiji's Restoration soberly comprehended during the latter half, Westernized military officers had to be indoctrinated into Newtonian science before they could compete successfully against Western forces.[85] Without such a transformation in their metaphysical, epistemological, and ultimately cultural outlook, they could never match the ruthless discipline and efficiency of Western armies. It took the bitter defeats represented by the Russian conquest of the Crimea, the French invasion of Egypt, and the Opium Wars to clarify this point. Once the Western military reforms were initiated, however, it was through the newly Westernized officers that much of the secular and rational ideals of the Enlightenment philosophes and their nineteenth-century heirs first penetrated into the ruling classes of non-Western cultures.[86] As Ralston articulated with respect to Turkish military cadets,

> By being introduced to the European academic milieu, the students were swept up in the "self-sustaining effervescence" of the European world of ideas. They could also thereby develop a new frame of reference in accordance with which they might seek, if they were so inclined, to elucidate and understand current Ottoman realities. By their schooling, then not a few of the students found themselves increasingly critical of the ruling establishment and even disaffected from it. They were examples of a phenomenon to be seen again and again not only in the Ottoman Empire but in many societies subjected to the powerful influence of European ways and institution.[87]

The teaching of Newtonian mechanics and calculus, for the sake of accurate gunnery practice, helped construct the mentalities needed to comprehend Locke, Smith, Voltaire, Rousseau, and ultimately Marx. As Ralston noted, the Young Turks of the Ottoman Empire demonstrated this dramatically during the nineteenth century. There is also Ali Pasha Mubarak of Egypt, famous for his numerous scientific textbooks and treatises on engineering, fortifications, and topography, as well as his Enlightenment-inspired administrative reforms. He received his initial scientific education as an aspiring engineering officer in

Mohammed Ali's army during the 1840s and went on to study military engineering and artillery theory at the École du Metz in France.[88] Yan Fu is an even more extreme example. He gained fame for formally introducing the concept of Western progress to China, and for being the first Chinese translator of and commentator on classic British liberal thought, including that of Spencer, Mill, Smith, Huxley, and Darwin. Yan Fu received his initial education in Newtonian science as a young naval cadet at the Fuzhou Shipyard School and subsequently studied at the Royal Naval College of Greenwich in 1877.[89] The close connection between Newtonian scientific theory and military practice achieved in Europe during the eighteenth century thus helped spearhead the successful introduction of Enlightenment thought throughout the non-Western world. The irony was that the traditional power structures of the leading Islamic and Asian cultures were ultimately destroyed by permitting Western thought to penetrate their realms as a means to preserve themselves. The overthrows of the Tokogawa and Ch'ing dynasties, as well as the constitutional revolution of the Ottoman Empire in 1908, are just some of the examples. Given the contempt that philosophes had for "despotic" governments dominated by traditional religious mentalities—especially those of the Ottoman Sultans—such defeats would represent progress. One may also add that the astonishing success of the anti-colonial movements during the twentieth century rested not only on the embrace of Enlightenment political frameworks. The Vietminh gunnery commanders under Giap's leadership at Dien Bien Phu may have been motivated by Marx and Engels, but they executed their decisive assault against French Imperialism using the ballistics of Robins and Euler—thanks to their first-rate French mathematical education. The liberating effect of scientific reason, as far as national autonomy is concerned, proved greater than even Condorcet could have imagined.

Acknowledgments

I wish to express my appreciation to Janis Langins of the University of Toronto and Margaret Jacob of UCLA for their valuable insights. I am also indebted to Sharon Traweek, Sandra Harding, and Joyce Applebee of UCLA for encouraging me, as a new postdoctoral lecturer, to articulate the cultural relevance of eighteenth-century ballistics and gunnery.

Notes

1. Robert L. O'Connell, *Of Arms and Men: A History of War, Weapons, and Aggression* (Oxford University Press, 1989), pp. 148–166; Bernard Brodie and Fawn M. Brodie, *From Cross-Bow*

to H-Bomb:The Evolution of the Weapons and Tactics of Warfare (Indiana University Press, 1973), pp. 100–123.

2. Martin van Creveld, *Technology and War from 2000 B.C. to the Present* (Free Press, 1989), p. 87.

3. Russell F. Weigley, *The Age of Battles: The Quest for Decisive Warfare from Breitenfeld to Waterloo* (Indiana University Press, 1991), p. xvi.

4. Geoffrey Parker, *The Military Revolution: Military Innovation and the Rise of the West, 1500–1800* (Cambridge University Press, 1988), pp. 146–154; Jeremy Black, *European Warfare, 1660–1815* (Yale University Press, 1994), 41–60.

5. Black, *European Warfare*, pp. 199–209

6. Parker, *Military Revolution*, pp. 136–145.

7. Ken Alder, *Engineering the Revolution, Arms and Enlightenment in France, 1763–1815* (Princeton University Press, 1997), pp. 103–108.

8. Alder, *Engineering the Revolution*, pp. 310–311, 344.

9. Alder (ibid., p. 107) writes that "Lombard admitted that his tables remained *inadequate*" and that the gunner's adjustment for the idiosyncrasies of their particular gun "would be *uncertain* in that the gunners were still plugging in the range of the test shot." What Alder doesn't say is the degree to which it is uncertain relative to the existing gunnery techniques it proposes to replace. Instead, he only offers that they are simply more uncertain than astronomy tables.

10. For a description of the 20th-century trend towards viewing progress largely in terms of technical power for its own sake, see David H. Hopper, *Technology, Theology and the Idea of Progress* (Westminster John Knox Press, 1991), pp. 55–76. For some of the classic accounts of the idea of progress in the Enlightenment era, see J. B. Bury, *The Idea of Progress: An Inquiry into its Growth and Origin* (Dover, 1955); R.V. Sampson, *Progress in the Age of Reason: The Seventeenth Century to the Present Day* (Heinemann, 1956); Charles Frankel, *The Faith of Reason: The Idea of Progress in the French Enlightenment* (King's Crown Press, 1948). Any discussion of military institutions as agents of progress are completely absent in these monographs, however.

11. Immanuel Kant, "An Answer to the Question: What Is Enlightenment?" in J. Schmidt, ed., *What Is Enlightenment?* (University of California Press, 1996).

12. C. Stewart Gillmor, *Coulomb and the Evolution of Physics and Engineering in Eighteenth-Century France* (Princeton University Press, 1971).

13. This neglect may be related to Prof. Hall's classic argument that denied the mechanical possibility of having a useful science of smoothbore ballistics due to supposed irregularities and lack of uniformity in early modern casting and powder-making techniques. See A. Rupert Hall, *Ballistics in the Seventeenth Century* (Cambridge University Press, 1952).

14. Although Robins and Euler's work in ballistics has been analyzed by W. Johnson primarily in engineering research periodicals, only recently has it attracted the attention of historians. See Brett Steele, "Musket and Pendulums: Benjamin Robins, Leonhard Euler, and

the Ballistics Revolution," *Technology and Culture* 35, no. 2 (1994): 348–382; Steele, The Ballistics Revolution: Military and Scientific Change from Robins to Napoléon, Ph.D. dissertation, University of Minnesota, 1994. Much of the following discussion of Robins' and Euler's work is based on this dissertation.

15. Benjamin Robins, *New Principles of Gunnery: Containing the Determination of the Force of Gunpowder, and An Investigation of the Difference in the Resisting Power of the Air to Swift and Slow Motions* (London, 1742).

16. For a study of efforts to use the ballistic pendulum before Robins, see William Johnson, "The Origin of the Ballistic Pendulum: The Claims of Jacques Cassini (1677–1756) and Benjamin Robins (1707–1751)," *International Journal of Mechanical Science* 32, no. 4 (1990): 345–374. For additional studies by Johnson on Robins' ballistics research, see "Benjamin Robins' New Principles of Gunnery," *International Journal of Impact Engineering* 4, no. 4 (1986): 205–219; "Benjamin Robins (18th Century Founder of Scientific Ballistics): Some European Dimensions and Past and Future Perceptions," *International Journal of Impact Engineering*, 12, no. 2 (1992): 293–323.

17. Bernard Forest de Bélidor, *Le Bombardier francais, ou nouvelle methode de jetter les bombes avec precision* (Paris, 1731); Dulacq, *Théorie nouvelle sur le mecanisme de l'artillerie* (Paris, 1741).

18. Leonhard Euler's translation and analysis of *New Principles of Gunnery* is published in F. Scherrer, ed., *Leonhard Euler, Opera Omnia*, volume XIV, second series (Berlin, 1922). It was translated into English by Hugh Brown as *The True Principles of Gunnery Investigated and Explained* (London, 1777).

19. Leonhard Euler, "Recherches sur la véritable courbe que décrivent les corps jetés dans l'air, ou dans un autre fluide quelconque," *Mémoires de l'Académie de Berlin*, t. 9 (1753).

20. Henning Friedrichs, Grafen von Graeveniz, *Akademische Abhandlung von der Bahn der Geschuzkugeln, nebst praktischen Tabellen und Regeln* (Rosens Schriften, 1764).

21. Charles de Borda, "Sur la courbe décrite par les boutlets et les bombes, en ayant éngard à la résistance le l'air," *Histoire de l'Académie Royale des sciences* (1769): 247–271; Georg Friedrich von Tempelhoff, *Le bombardier prussien* (Berlin, 1781); Benjamin Thompson, "New Experiments upon gun-powder, with occasional Observations and practical Inferences; to which are added, an Account of a new Method of determining the Velocities of all Kinds of Military Projectiles, and the Description of a very accurate eprouvette for Gun-powder," *Philosophical Transactions of the Royal Societ* (read March 29, 1781), LXXI: 229–328; Charles Hutton, *Tracts on Mathematical and Philosophical Subjects* (Rivington, 1812).

22. For the significance of Newton on Enlightenment thought, see Peter Gay, *The Enlightenment: An Interpretation* (Norton, 1969), pp. 128–150. For a specific account of the penetration of Newtonian thought into the French Enlightenment, see Geoffrey Sutton, *Science for a Polite Society: Gender, Culture, and the Demonstration of Enlightenment* (Westview, 1995), pp. 241–272. On the ambiguity of "Newtonianism," in eighteenth-century philosophical circles, see Thomas Hankins, *Science and the Enlightenment* (Cambridge University Press, 1985), pp. 9–10.

23. Cesar Chesneau Dumarsais, "Definition of a Philosophe," in I. Kramnick, ed., *The Portable Enlightenment Reader* (Penguin, 1995), p. 22.

24. Gay, *The Enlightenment*, p. 10. Gay argues that medicine as best fulfilling this Baconian ideal of unifying theory and practice during the 18th century, and completely ignores military developments, as most Post War-era interpreters of the Enlightenment continue to demonstrate with the obvious exception of Michel Foucault.

25. Margaret Jacob, *Scientific Culture and the Making of the Industrial West* (Oxford University Press, 1997), p. 130.

26. Antonine-Nicholas de Condorcet, "Learning How to Reason for Human Progress," in J. Chambliss, ed., *Enlightenment and Social Progress* (Burgess , 1971), p. 14.

27. Ibid., p. 18.

28. Margaret Jacob and Larry Stewart have published outstanding studies of the relevance of Newtonian mechanical philosophy on the cultural foundations of the British Industrial Revolution (Jacob, *The Cultural Meaning of the Scientific Revolution*, Knopf, 1988; Stewart, *The Rise of Public Science: Rhetoric, Technology, and Natural Philosophy in Newtonian Britain, 1660–1750*, Cambridge University Press, 1992). Nevertheless, it remains uncertain whether the actual mechanics directly involved Newtonian kinetics, as opposed to Galileo's kinematics and Stevin's statics as developed in Bélidor's engineering treatises.

29. Hankins, *Science and the Enlightenment*, p. 23.

30. For a detailed discussion of Robins' gun design proposal, see William Johnson, "Robins on Boring out Cannon to Accommodate Heavier Shot," *International Journal of Impact Mechanics* 8, no. 3 (1989): 281–285.

31. Benjamin Robins, *Mathematical Tracts of the Late Benjamin Robins*, I (London, 1762), pp. 283–294.

32. See Benjamin Robins, "A Letter to Martin Folks, Esquire, President of the Royal Society, in answer to one of his, inclosing a written Message from the Chevalier D'Ossorio, Envoy from the King of Sardinia," in *The Mathematical Tracts of the Late Benjamin Robins*, volume I (London, 1762).

33. Walter Barberis, *Le armi del Principe: La tradizione militare sabauda* (Guilio Einaudi, 1988), pp. 215–217.

34. The three artillery treatises of General Papacino D'Antoni contain many references to the Piedmontese artillery research program in the 18th century. They were translated into English by Captain Thomson of the Royal Regiment of Artillery as *A Treatise on Gun-Powder; a Treatise on Fire-Arms; and a Treatise on the Service of Artillery in Time of War* (Egerton, 1789).

35. For a discussion of his theoretical and experimental analysis of the muzzle velocity of both muskets and cannons, see D'Antoni, pp. 74–98.

36. Ibid., pp. 85–86.

37. For a basic outline of the Liechtenstein reforms, see Anton Dolleczk, *Geschichte der Österichischen artillerie von den frühsten Zeiten bis zur Gegenwart* (Graz, 1973), p. 292. For a concise account of the Liechtenstein reform's historical significance, see Christopher Duffy, *The Army of Maria Theresa* (Hippocrene, 1997). The best source, however, is an obscure book by

an archivist in the Kriegsarchiv: Anton Semek, *Die Geschichte der Artillerie, ihr Werdegang, hire Entwicklung bis heute* (C.W. Stern, 1908).

38. I made this argument in The Ballistics Revolution (pp. 207–212). Ken Alder seconded it in *Engineering the Revolution* (pp. 104–108).

39. Grummert, "Erinnerung eines unrechtigen Hauptunstandes in der sonst überaus scharfsinnigen Theorie des Robins," Kriegsarchiv Wien, Altes Artillerie Archiv, XIX C2, No. 6.

40. Semek, *Die Geschichte der Artillerie*, pp. 51–52.

41. Robins, *Mathematical Tracts*, p. 245.

42. Manfred Rudersdorf, "Josef Wenzel von Liechtenstein (1692–1772): Diplomat, Feldmarschall und Heeresreformer im kaiserlichen Dienst," in D. Willoweit, ed., *Liechtenstein—Fürstliches Haus und staatliche Ordnung* (Volker, 1988), p. 357.

43. Liechtenstein, *Reglement für das Kaiserlich Königliche gesammte Feld-artillerie corps* (Wien, 1757), p. 188.

44. Max Jähns, *Geschichte der Kriegswissenschaften vornehmlich in Deutschland*, volume II (München and Leipzig, 1890), pp. 2351–2354.

45. Oskar Albrecht, "Georg Friedrich von Tempelhoff," *Soldat und Technik*, September 1966, pp. 493–494.

46. Friedrich Gatti, *Geschichte des k. k. Bombardier-Corps* (Wien, 1905), p. 58.

47. For a detailed description of this gunnery controversy, see Steele, The Ballistics Revolution, pp. 217–223.

48. "Explanation of the Effects & advantages which may be derived by firing case shot, in the mode proposed by Major Shrapnel—1803," MD 1047/1, Royal Artillery Library, Woolwich.

49. "Enumeration of the advantages of Spherical Case Shot," MD 1047/1, Royal Artillery Library, Woolwich.

50. James P. Herson Jr., "French Innovation versus British Technology at Cadiz: Missile Warfare in the Peninsula War," presented at 25th Consortium on Revolutionary Europe, New Orleans, 1995.

51. For an overview of the issue of education for Enlightenment thought, see James Brown, *A History of Western Education: The Modern West, Europe and the New World*, volume 3 (St. Martin's Press, 1981), especially the chapter "Education and the Enlightenment: The Conceptual Revolution." See also "The Politics of Education," in Gay, *The Enlightenment*.

52. Condorcet, "The Utility of Science," in I. Kramnick, ed., *The Portable Enlightenment Reader* (Penguin, 1995), pp. 68–9.

53. Jacob, *Scientific Culture and the Making of the Industrial West*, p. 132.

54. For example, Ken Alder maintains in *Engineering the Revolution* (p. 57) that "the artillery schools of *ancien régime* France were the first institutions in Europe where students received a scientific education."

55. John Hale, *Renaissance War Studies* (Hambledon, 1983), pp. 225–246.

56. On the founding of this academy, see *Sotria dell'Artiglieria Italiana* Parte Prima (Dalle Origini al 1815), volume II (Dalla Fine del Secolo XVII al 1815) Edita a Cura della Revista D'Artiglieria e Genio (Roma -XIII), 1070–1080.

57. Archivio di Stato di Torino—Artiglierie—Carte antiche, volume XVI, 1731–1750.

58. J. L. Lagrange, *Analytical Mechanics*, ed. A. Boisssonnade and V. Vagliente (Kluwer, 1997), p. xvii.

59. Archivio di Stato di Torino—Sezione 4a—Artiglieria—Carte antiche, volume III, pp. 334–343.

60. Papacino D'Antoni, *Institutions physico-mechanqiues à l'usage des écoles royales d'artillerie et du genie de Turin*, volume II (Strasbourg, 1777), pp. 289–314.

61. Seymour Mauskopf, "Gunpowder and the Chemical Revolution," *Osiris*, second series, 4 (1988), pp. 100–103.

62. Etieene Bézout, *Cours de Mathématiques, a l'usage du Corps de l'Artillerie. Tome quatrieme, Concernant l'application des Principes généraux de la Méchanique, à différens cas de mouvement & d'équilibre. Nouvelle edition, revue et corrigée* (Paris, 1797).

63. Charles de Borda, "Sur la courbe décrite par les boutlets et les bombes, en ayant éngard à la résistance de l'air," *Histoire de l'Académie royale des Sciences, avec les Mémoires de Mathématique & de Physique, pour la méme Année* (1769): 247–271.

64. Service Historique des Armees de Terre (SHAT), XD 253.

65. Jean-Louis Lombard, ed., *Nouveaux principes d'artillerie, commentés par Léonard Euler* (Frantin, 1783).

66. Jean-Louis Lombard, *Tables du tir des canons et des obusiers* (Auxonne, 1787).

67. Jean-Louis Lombard, *Traité du mouvement des projectiles* (Frantin, year V [1796–97]).

68. N. Bonaparte, *Correspondance de Napoléon Ier publiée par ordre de l'Empereur Napoléon III*, volume VII (Paris, 1861), p. 233, document #5621.

69. Corps legislative. Counseil des Anciens. "Opinion de Lacombe Saint-Michel sur l'École polytechnique. Séance du 7 floréal an VI" [Texte imprimé] (Impr. Nationale, an VI).

70. For example, Gillespie ignores him in his otherwise exhaustive *Science and Polity* (Princeton University Press, 1980).

71. Alder, *Engineering the Revolution*, pp. 107–108.

72. Ibid., p. 377.

73. Lombard, *Tables du tir*, pp. 9–10, 31.

74. Clément, "Remarquer sur le tir des bouches à feu en campagne," SHAT, 2. d-1 (b), box #5.

75. Alder, *Engineering the Revolution*, pp. 106–108. Interestingly enough, Papacino D'Antoni maintained the following with respect to determining the muzzle velocity of bullets by measuring their penetration depth into compacted soil (*A Treatise on Gun-Powder; a Treatise on Fire-Arms; and a Treatise on the Service of Artillery in time of War*, p. 85): "The initial velocity of cannon balls may be easily found by this method of determining the velocities of bullets projected from fire-arms of small caliber: not only the charge that gives the longest range, and the law of pressure of the fluid on the shot in passing along the bore may be ascertained; but the greatest elasticity of the fluid, and the point in the length of the bore where it is produced, may be determined: in a word, all the solutions of the problems relative to small pieces from the 167 paragraph to the present, are equally applicable to the largest cannon."

76. For a derivation of this relationship, see *Traite du mouvement des projectiles*, pp. 109–113.

77. For a discussion of the basic incompatibility that occurred when executing these goals, see Gay, *The Enlightenment*, pp. 497–499.

78. Ibid., pp. 555–558.

79. Henry Meyer, *Voltaire on War and Peace* (Cheney, 1976), pp. 48–52.

80. Ibid., pp. 116–122.

81. Fairfax Downey, *Sound of the Guns: The Story of American Artillery from the Ancient and Honorable Company to the Atom Cannon and Guided Missile* (D. McCKay, 1956), p. 21. Also see North Callahan, *Henry Knox: General Washington's General* (Rinehart, 1958), pp. 35–36.

82. Robert Hendrickson, *Hamilton I (1757–1789)* (Mason/Charter 1976), p. 92.

83. The notion that a relationship exists between mathematical knowledge and military power was still difficult to accept in late nineteenth-century China as evidenced by Grand Secretary Wo-jen. He declared that he head never heard of anyone using "mathematics to raise a nation from a state of decline or to strengthen it in a time of weakness." See David Ralston, *Importing the European Army: The Introduction of European Military Techniques and Institutions into the Extra-European World, 1600–1914* (University of Chicago Press, 1990), 126.

84. David Ralston, *Importing the European Army*.

85. For evidence in of the teaching of calculus the Ottoman Empire's technical military schools during the late eighteenth century, see Stanford J. Shaw, *Between Old and New: The Ottoman Empire under Sultan Selim III, 1789–1807* (Harvard University Press, 1971), p. 148. Also see Ekmeleddin Ihsanoglu, "Some Critical Notes on the Introduction of Modern Sciences to the Ottoman State and the Relation Between Science and Religion up to the End of the Nineteenth Century," in *Comité international d'etudes pré-ottomanes et ottomans, VI Symposium, Cambridge, 1rst-4th July 1984* (Divit, 1987), p. 236; Niyazi Berkes, *The Development of Secularism in Turkey* (C. Hurst, 1998), pp. 75–77. For a discussion of Mohammed Ali's

creation of the first Western-style education institutes in Egypt, designed specifically to furnish scientifically grounded military officers, see P. J. Vatikiotis, *The History of Egypt* (Johns Hopkins University Press, 1986), pp. 56–58, 491. The T'ung-wen Kuan school for the study of foreign languages aside, the earliest state-sponsored efforts to furnish a Western scientific education in China occurred through the efforts of Li Hongzhang and Tseng Kuo-fan in developing the Kiangman Arsenal. Its educational institute was founded in 1866 to train both naval officers and engineers. See K. Biggerstaff, *The Earliest Modern Government Schools in China* (Cornell University Press, 1961), pp. 154–165.

86. Mohammad Farhad Atai's study of the first Persian students sent to study in Europe during the nineteenth century (The Sending of Iranian Students to Europe, 1811–1906 (The Student Missions), Ph.D. dissertation, University of California, Berkeley, 1992) reinforces this observation. They went primarily through individual effort and not as a part of a military Westernization campaign, given the relative conservatism of the Persian military establishment. As a consequence, these students had little success in transferring Western scientific and intellectual ideas back to Persia.

87. Ralston, *Importing the European Army*, pp. 68–69.

88. Vatikiotis, *The History of Egypt*, pp. 109–110.

89. David Wright, "Yan Fu and the Tasks of the Translator," in M. Lackner et al., eds., *New Terms for New Ideas* (Koninkilijke Brill, 2001), pp. 235–236. Also see Michael Gasster, *China's Struggle to Modernize* (Knopf, 1983), pp. 25–27.

About the Authors

GÁBOR ÁGOSTON teaches in Georgetown University's Department of History. His publications include *Gül Baba és bektasi rend Magyarországon* [Gül Baba and the Bektashi Order of Dervishes in Hungary] (with Balázs Sudár) (2002), *Magyarország a 17. században* [Hungary in the Seventeenth Century] (with Teréz Oborni) (2000), *Magyarország története 100+1 tételben* [History of Hungary] (with Róbert Hermann) (2000), and *A hódolt Magyarország* [Hungary under Ottoman rule] (1992).

AMIR ALEXANDER received his Ph.D. from Stanford University in 1996 and has taught history, philosophy, and the history of science at Stanford and UCLA. He is the author of *Geometrical Landscapes: The Voyages of Discovery and the Transformation of Mathematical Practice* (2002).

FREDERIC J. BAUMGARTNER, Professor of History at the Virginia Polytechnic Institute and State University, received his Ph.D. from the University of Wisconsin in 1972. He has authored seven books: *Radical Reactionaries: The Political Thought of the French Catholic League* (1976), *Change and Continuity in the French Episcopate: The Bishops and the Wars of Religion* (1986), *Henry II, King of France* (1988), *From Spear to Flintlock* (1992), *Louis XII* (1994), *France in the Sixteenth Century* (1995), and *Longing for the End: A History of Millennialism in Western Civilization* (1999).

BRENDA J. BUCHANAN, a visiting research fellow at the University of Bath's Centre for the History of Technology since 1987, has been Convenor of the Gunpowder Section of the International Committee for the History of Technology since 1994 and Chair of the Gunpowder and Explosives History Group (UK) since 2000. She edited and contributed to *Gunpower: The History of an International Technology* (1996).

Lesley B. Cormack teaches in the Department of History and Classics at the University of Alberta and is the author of *Charting an Empire: Geography at the English Universities* 1580–1620 (1997).

Kelly DeVries, Professor of History at Loyola College in Maryland, holds a Ph.D. in Medieval Studies from the Centre for Medieval Studies at the University of Toronto. He has authored six books, including *Guns and Men in Medieval Europe, 1200–1500: Studies in Military History and Technology*, *Joan of Arc: A Military History*, *The Norwegian Invasion of England in 1066*, *Infantry Warfare in the Early Fourteenth Century: Discipline, Tactics and Technology*, and *Medieval Military Technology*, and several articles on medieval historical and technological subjects. He serves as the Secretary-General of the United States Commission of Military History, as editor of the *Journal of Medieval Military History*, and as Series Editor of The History of Warfare for Brill Publications.

Tamera Dorland, a communications analyst, has taught literature and rhetoric at UCLA and the University of Utah. She has published on narrative and queer theory.

Barton C. Hacker is Curator of Military History at the National Museum of American History. His publications include *West Point and the Making of America* (with Margaret Vining) (2003), *World Military History Bibliography* (2003), *Elements of Controversy: The Atomic Energy Commission and Radiation Safety in Nuclear Weapons Testing, 1947–1974* (1994), and *The Dragon's Tail: Radiation Safety in the Manhattan Project, 1942–1946* (1987).

Alexzandra Hildred, Ordnance Director of the Mary Rose Trust, is the editor of *Weapons of Warre: The Armaments of the Mary Rose Trust* (forthcoming). In addition to her continuing work on the *Mary Rose* site, she has worked on numerous underwater archaeological projects in the United Kingdom and elsewhere. She has been Chair of the Maritime Affairs Group of the Institute of Field Archaeologists. She teaches Nautical Archaeology at the University of Plymouth. She is archaeological advisor on three of the fifty wrecks designated under the 1973 Protection of Wrecks Act as being of archaeological, historical, or artistic importance. She also holds the position of advisor on ordnance on the Advisory Committee for Historic Wreck Sites, the government committee that lobbies for underwater archaeology.

Thomas Kaiserfeld received his Ph.D. in the History of Technology from the Royal Institute of Technology in Stockholm in 1997. He has since worked on a variety of projects ranging from the establishment of district heating in Sweden in the 1940s to saltpeter production in the eighteenth century. He is currently researching the use of expert knowledge in political decision making from the eighteenth century to the present.

JANIS LANGINS teaches the history of technology at the University of Toronto's Institute for the History and Philosophy of Science and Technology. He is the author of *Conserving the Enlightenment: French Military Engineering from Vauban to the Revolution* (2003).

MICHAEL S. MAHONEY teaches history of science and technology in the Program in History of Science at Princeton University. He is the author of *The Mathematical Career of Pierre de Fermat, 1601–1665* (second revised edition, 1994) and of a series of monographs on the mathematics of Rene Descartes, Isaac Barrow, Christiaan Huygens, and Isaac Newton. His articles include "The Mathematical Realm of Nature" (1998), "The History of Computing in the History of Technology" (1988), "Boys' Toys and Women's Work: Feminism Engages Software" (2001), and "Software as Science—Science as Software" (2002).

SEYMOUR H. MAUSKOPF received his Ph.D. from Princeton University in the history of science. His publications include *Crystals and Compounds* (1976), *The Elusive Science* (with Michael R. McVaugh) (1980), and *Chemical Sciences in the Modern World* (1993). In 1998 he received the American Chemical Society's Exter Award for Outstanding Contributions to the History of Chemistry. He has taught history of science at Duke University since 1964.

BRETT D. STEELE, a research analyst, has taught courses in engineering strategy and the history of science and technology at UCLA and Stanford. His publications include *Military Reengineering between the World Wars* (2004), *Vertical Envelopment, Rotorcraft, and Operational Considerations for the Objective Force* (with Jon Grossman et al.) (2003), "An Economic Theory of Technological Products" (1995), and "Muskets and Pendulums: Benjamin Robins, Leonhard Euler and the Ballistics Revolution" (1994). Steele is currently a researcher at ANSER's Homeland Security Institute.

Index